PRESERVATIVES IN THE FOOD, PHARMACEUTICAL AND ENVIRONMENTAL INDUSTRIES

A complete list of titles in the Society for Applied Bacteriology Technical Series appears at the end of this volume.

THE SOCIETY FOR APPLIED BACTERIOLOGY
TECHNICAL SERIES No. 22

Preservatives in the Food, Pharmaceutical and Environmental Industries

Edited by

R.G. BOARD

School of Biological Sciences,
Bath University, Bath, UK

M.C. ALLWOOD

Central Pharmacy, Addenbrooke's Hospital,
Hills Road, Cambridge, UK

J.G. BANKS

Chipping Campden Food
Preservation Research Association,
Chipping Campden, UK

BLACKWELL SCIENTIFIC PUBLICATIONS

OXFORD LONDON EDINBURGH

BOSTON PALO ALTO MELBOURNE

Editorial offices:
Osney Mead, Oxford, OX2 0EL
8 John Street, London, WC1N 2ES
23 Ainslie Place, Edinburgh, EH3 6AJ
52 Beacon Street, Boston
Massachusetts 02108, USA
667 Lytton Avenue, Palo Alto
California 94301, USA
107 Barry Street, Carlton
Victoria 3053, Australia

First published 1987

Typeset by Clowes Computer Composition, printed and bound in Great Britain by William Clowes Limited, Beccles and London

DISTRIBUTORS

USA and Canada
Blackwell Scientific Publications Inc
PO Box 50009, Palo Alto
California 94303

Australia
Blackwell Scientific Publications (Australia) Pty Ltd
107 Barry Street,
Carlton, Victoria 3053

British Library
Cataloguing in Publication Data

Preservatives in the food, pharmaceutical and environmental industries.—(The Society for Applied Bacteriology technical series; no. 22)
1. Food preservatives 2. Materials—Deterioration
I. Board, R.G. II. Allwood, M.C.
III. Banks, J.G. IV. Series
664′.06 TP371.2

ISBN 0–632–01727–9

Contents

Contributors

M.C. ALLWOOD, *Research and Development Laboratory, East Anglia Regional Hospital Board, Central Pharmacy, Addenbrooke's Hospital, Cambridge CB2 2QQ*

J.G. BANKS, *Campden Food Preservation Research Association, Chipping Campden GL55 6LD*

N.S. BATTERSBURY, *Water Research Centre, Henley Road, Medmenham, PO Box 16, Bucks SL7 2HD*

R.G. BOARD, *School of Biological Sciences, University of Bath, Bath BA2 7AY*

A.F. BRAVERY, *Building Research Establishment, Princes Risborough, Aylesbury, Buckinghamshire HP17 9PX*

FIONA S. CARLILE,[1] *Agriculture Development and Advisory Service, Ministry of Agriculture, Fisheries and Food, Kenton Bar, Newcastle upon Tyne NE1 2YA*

D.G. CHAPMAN, *School of Pharmacy, Robert Gordon's Institute of Technology, Schoolhill, Aberdeen AB9 1FR*

S.P. DENYER, *Department of Pharmacy, The University, Nottingham NG7 2RD*

CAROLE DURANT, *Johnson & Johnson Ltd, Southampton Road, Cosham, Portsmouth, Hampshire PO6 4RL*

J.M. EWART, *Agricultural Development and Advisory Service, Ministry of Agriculture, Fisheries and Food, Lawnswood, Leeds LS16 5PY*

R.L. FLETCHER, *The Marine Laboratory, Portsmouth Polytechnic, Portsmouth, Hampshire PO11 0DG*

J.R. FURR, *Welsh School of Pharmacy, University of Wales Institute of Science and Technology, PO Box 13, Cathays Park, Cardiff CF1 3XF*

S.E. FURTADO, *Transocean Marine Paint, PO Box 25107, 3001 HC-Rotterdam, The Netherlands*

P.A. GIBBS, *Applied Microbiology Section, Leatherhead Food Research Association, Randalls Road, Leatherhead, Surrey KT22 7RY*

P. GILBERT, *Department of Pharmacy, University of Manchester, Oxford Road, Manchester M13 9PL*

C. GRANT, *Building Research Establishment, Princes Risborough, Aylesbury, Buckinghamshire HP17 9PX*

P.F. HAMLYN, *Shirley Institute, Didsbury, Manchester M20 8RX*

P. HIGDON, *Johnson & Johnson Ltd, Portsmouth, Hampshire PO6 4RL*

E.A. HILDITCH, *Cuprinol Ltd., Adderwell, Frome, Somerset BA11 1NL*

[1] Now: Mrs F.S. Thompson, ICARDA, PO Box 5466, Aleppo, Syria

W.B. Hugo, *Department of Pharmacy, The University, Nottingham NG7 2RD*

J.S. King, *AFRC Institute of Food Research, Reading Laboratory, Shinfield, Reading RG2 9AT*

L.A. Mabbitt, *AFRC Institute of Food Research, Reading Laboratory, Shinfield, Reading RG2 9AT*

B.J. McCarthy, *Wool Industry Research Association, WIRA House, West Park Ring Road, Leeds LS16 6QL*

F. Mendes, *Cuprinol Ltd., Adderwell, Frome, Somerset BA11 1NL*

P. Neaves, *Milk Marketing Board, Microbiology Laboratory, Technical Division, Thames Ditton, Surrey KT7 0EL*

G.J. Nychas, *Sactouri 27 Str., Chalandri, 15232, Athens, Greece*

D.W. Pimbley, *Agricultural Development and Advisory Service, Ministry of Agriculture, Fisheries and Food, Lawnswood, Leeds LS16 5PY*

D.T. Rogers, *Welsh School of Pharmacy, University of Wales Institute of Science and Technology, PO Box 13, Cathays Park, Cardiff CF1 3XF*

B.F. Sagar, *Shirley Institute, Didsbury, Manchester M20 8RX*

A.P. Sharma, *Department of Brewing and Biological Sciences, Heriot-Watt University, Chambers Street, Edinburgh EH1 1HX*

R.J.S. Shaw, *Regional Quality Control Laboratory, Norfolk and Norwich Hospital, Norwich, Norfolk NR1 3SR*

I.J. Stewart, *Department of Brewing and Biological Sciences, Heriot-Watt University, Chambers Street, Edinburgh EH1 1HX*

J.E. Todd, *Shirley Institute, Didsbury, Manchester M20 8RX*

D.S. Wales, *Shirley Institute, Didsbury, Manchester M20 8RX*

L.F.J. Woods, *Applied Microbiology Section, Leatherhead Food Research Association, Randalls Road, Leatherhead, Surrey KT22 7RY*

Naomi Wright, *Department of Pharmacy, University of Manchester, Oxford Road, Manchester M13 9PL*

Preface

This volume, number 22 in the Society for Applied Bacteriology Technical Series, is based on the demonstrations at a meeting of the Society at Chelsea College, University of London, on 17th October 1984.

Once man attempted to store animal products or plant materials he had to contend with the decomposing potential of a vast array of bacteria and fungi; so too the first tentative efforts to use materials from these sources for clothes or shelter would have brought him into conflict with such organisms. In unusual situations, of course, microbial attack on vulnerable materials may have been held in check by physico-chemical features of the environment. Archaeological exploration of the Somerset levels provides an example. The early settlers (*ca.* 4000 years ago) used brush wood to make hurdles that formed crude roads between settlements. From extensive knowledge of peat, we can now offer an explanation of the utility of these early attempts to aid travel. The anoxic, acid and nitrogen-poor peat would have acted as a preservative system for the hurdles.

In general, however, very few environments offer such protection and a need exists for the protection, by physical or more commonly chemical means, of natural materials used in a diverse range of enterprises. Indeed novel problems associated with microbial attack of materials are commonly experienced by enterprises that take man and his technologies into previously unexplored environments. Thus corrosion problems are faced by the oil industry in its quest to recover oil from beneath the sea bed. The sulphate-reducing bacteria can grow on and cause corrosion of the metal of oil rigs and pipelines.

The demonstration meeting upon which this volume is based was planned so that an overview of the current methods for the preservation of a broad spectrum of materials could be obtained by those whose daily work is concerned with some aspect or another of preservation. This volume provides a record of the meeting and offers also a very extensive bibliography on the scattered literature dealing with preservation.

We wish to thank all the demonstrators for the great efforts that they took in the presentation of their exhibits on the 17th October 1984 and in the preparation of their chapters in this book. Their efforts on the day were greatly helped by the local organizer, Dr Sally Bloomfield. We wish to thank also Dr Susan Passmore, the Society's Meetings Secretary, for administering the whole enterprise with such care and efficiency.

R.G. Board
M.C. Allwood
J.G. Banks

Curing Salts in Meat

L. F. J. Woods, P. Neaves and P. A. Gibbs
Applied Microbiology Section, Leatherhead Food Research Association, Surrey, UK

Although the modern process by which pork is cured to produce bacon bears little resemblance to the ancient art of curing, the active chemical ingredients of the process—sodium chloride and sodium nitrite—are the same. The inclusion of polyphosphates is a recent improvement in the curing process. In general, microbiological research into the curing process has been confined to the empirical determination of the amounts of the curing salts and of other factors which are required to produce a safe product, with particular reference to *Clostridium botulinum*. During the course of such research, workers have discovered interactive effects between these various factors resulting in increased inhibition of growth of *Cl. botulinum*, e.g., sodium chloride and sodium nitrite (Baird-Parker & Baillie 1974), pH and sodium chloride and sodium nitrite (Roberts & Ingram 1973) and complex interactions between sodium chloride, sodium nitrite, incubation temperature and inoculum size have been observed (Roberts *et al.* 1976). Such multifactorial experiments have allowed the development of mathematical models which can be used to predict the probability of growth and toxin production by *Cl. botulinum* in a product whose parameters fall within those of the experiment (Roberts *et al.* 1981). However, the prediction of growth of *Cl. botulinum* in a new product which does not resemble that used in such experiments is made difficult by a lack of knowledge about the mechanism of action of the various curing salts.

The broad aim of the work described here was to investigate the mechanism by which curing salts inhibit the growth of micro-organisms, particularly *Cl. botulinum*. In simplistic terms, this inhibition must derive from the modification of those metabolic pathways which are essential to the life of the organism. These are, firstly, the ability to obtain a suitable substrate as a source of energy; secondly, to convert that substrate into useful energy; thirdly, to harness that energy to the synthesis of new substances, which results in growth. As we shall see, one curing salt (nitrite) appears to have a specific site of action in the conversion of a substrate to useful energy, whereas a second curing salt (sodium chloride) has a more generalized effect

Preservatives in the Food,
Pharmaceutical and Environmental
Industries

0–632–01727–9

on the cell, inhibiting both the uptake of substrate and, indirectly, the synthesis of new substances.

Certain classes of cured meats receive a heat process. Some of these canned, pasteurized products are shelf-stable at ambient temperature and have an excellent record of freedom from growth of *Cl. botulinum.* The inhibitory parameters which prevent growth are known to be due to the presence of sodium chloride and sodium nitrite (Yesair & Cameron 1942). It is also known that although the thermal process is too low to cause appreciable spore destruction, it increases the sensitivity of clostridial spores to the effects of the curing salts (Roberts & Ingram 1966). In such products as these, therefore, it is important to consider not only the inhibitory action of the curing salts *per se* but also their effect in combination with the heat process.

Sodium Chloride

Sodium chloride inhibits microbial growth by increasing the ionic strength of the aqueous phase of the cured meat. A general response to such an osmotic stress by non-halophilic bacteria is the intracellular accumulation of compensating solutes—usually glutamate or the related compounds, proline and γ-aminobutyrate, as described by Measures (1975). With regard to the energy metabolism of the cell, such an accumulation of amino acids during osmoregulation, whether by *de novo* synthesis or by active transport, represents a drain on the energy resources of the organism and would adversely affect growth. The production of glutamate from α-ketoglutarate is the first step in the synthesis of these compounds and it is catalyzed by the enzyme glutamate dehydrogenase. Measures (1975) found that this enzyme in cell lysates of *Escherichia coli, Klebsiella aerogenes, Bacillus subtilis* and *Staphylococcus aureus* was stimulated by potassium ions. He proposed that this stimulation was a general mechanism for the accumulation of compensating solutes, since the intracellular concentration of potassium ions increases during osmotic stress. Because of the importance of this enzyme in osmoregulation, it was decided to investigate the properties of glutamate dehydrogenase from *Clostridium sporogenes*, with particular reference to the effect of potassium ions. The effect of growth in the presence of sodium chloride was examined by measuring the maintenance energy requirement during continuous culture of *Cl. sporogenes.*

Glutamate dehydrogenase

To investigate the effect of potassium ions on glutamate dehydrogenase from *Cl. sporogenes*, the organism was grown in Nutrient Broth (Oxoid) containing 1% (w/v) glucose and harvested by centrifugation as described by Woods *et al.* (1981). Cell-free extracts were prepared by sonication of washed-cell suspensions in 0.1 M-Tris/

HCl buffer (pH 8.1) using a Branson Sonifier Model B-30 equipped with a microtip (Ultrasonics Limited, Shipley, West Yorkshire) for 10 min with the machine set to 65% power and 50% pulsed. Cell debris was removed by centrifugation at 10 000 g_{av} for 15 min at 4 °C. The activity of glutamate dehydrogenase (L-glutamate:NAD oxidoreductase, EC 1.4.1.2) was measured in the supernatant fluids by oxidation of NADH in the presence of α-ketoglutarate and ammonium chloride according to the method of Winnacker & Barker (1970). Apparent Michaelis constants (Km′) were determined for each substrate from double reciprocal plots obtained by measuring initial rates of NADH oxidation at saturating concentrations of the non-varying substrate. The effect of sodium and potassium chlorides on glutamate dehydrogenase activity was determined under standard assay conditions, which consisted of (final concentrations): 0.1 M-Tris/HCl (pH 8.1); 0.1 mM-NADH; 180 mM-ammonium chloride and 5 mM-α-ketoglutarate. Protein concentrations were typically 10 μg/ml.

Measurement of the apparent Michaelis constants for each of the three substrates of the enzyme in cell-free extracts from *Cl. sporogenes* gave results as shown in Table 1. Investigation of the effect of growth in the presence of 3.5% (w/v) sodium chloride instead of the 0.5% (w/v) of the control showed that the kinetic parameters of the enzyme were unaffected by adaptation.

TABLE 1. *Comparison of apparent Michaelis constants for glutamate dehydrogenase from* Clostridium sporogenes *adapted to 0.5 and 3.5% (w/v) sodium chloride*

	Apparent k_m (mM)	
Substrate	0.5%	3.5%
α-Ketoglutarate	1.80	1.50
NH_4Cl	67.00	63.00
NADH	0.03	0.03

In an attempt to extend the findings of Measures (1975), in which glutamate dehydrogenase from several bacteria but excluding clostridia, was stimulated by potassium ions, the activity of the enzyme from *Cl. sporogenes* in the presence of various concentrations of potassium chloride was measured. The results presented in Fig. 1 show that potassium chloride inhibited glutamate dehydrogenase and that this inhibition was equal, on a molar basis, to that of sodium chloride. Thus it would seem that stimulation by potassium ions during osmotic stress is not a mechanism for increased glutamate synthesis in *Cl. sporogenes*. Various other mechanisms for increased synthesis of compensating solutes have been proposed; for example a

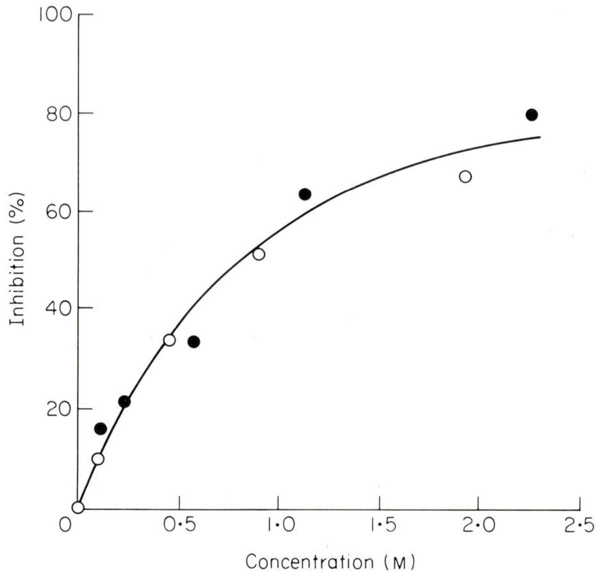

FIG. 1. Inhibition of glutamate dehydrogenase by potassium chloride (○) and sodium chloride (●).

marked pH dependence of the enzyme was measured in *Aerobacter aerogenes* (Tempest *et al.* 1970) and an alternative route for glutamate synthesis via the enzyme glutamine (amide):2-oxoglutarate aminotransferase oxidoreductase (NADP) was shown in this organism (Meers *et al.* 1970). However, glutamate dehydrogenase from *Cl. sporogenes* did not have a marked pH dependence (Fig. 2) and no glutamine (amide):2-oxoglutarate aminotransferase oxidoreductase activity could be detected in cell-free extracts of *Cl. sporogenes.* These results suggest that as amino acid accumulation in *Cl. sporogenes* does not occur by stimulation of glutamate dehydrogenase by potassium ions, the increased activity is likely to be due to *de novo* synthesis.

A closer examination of cells of *Cl. sporogenes* after growth at two sodium chloride concentrations (0.5%, 3.5% w/v) has revealed that an increase in the activity of glutamate dehydrogenase can be measured in cells adapted to the higher concentration. Microscopic examination also showed that there was a decrease in length, but not width, of these adapted cells compared with the control, resulting in a decrease in cell volume. Thus, if the measured increase in glutamate dehydrogenase is expressed as per unit cell volume, i.e. as a concentration, the increase is in fact even greater (Table 2).

Estimates of cell volumes were made from scanning electron micrographs taken after cells were washed in 0.1 M-phosphate buffer (pH 6.5) containing 0.5 or 3.5%

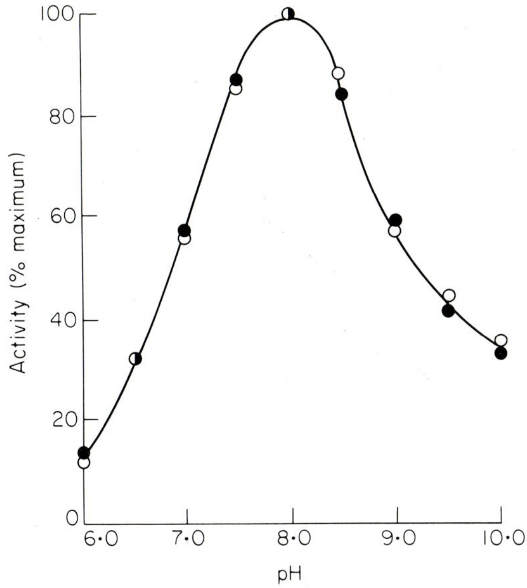

FIG. 2. The pH-activity profile of glutamate dehydrogenase from *Clostridium sporogenes* adapted to 0.5% (●) and 3.5% (○) sodium chloride.

TABLE 2. *Effect of adaptation to sodium chloride on glutamate dehydrogenase (GDH) activity and cell volume of* Clostridium sporogenes

	GDH			GDH	
NaCl (%)	(μmol/min/mg protein)	% control	Cell vol (μm³)	mol/min/ml cell vol	% control
0.5	4.74	100	2.5	1.4×10^4	100
3.5	7.75	164	1.8	3.2×10^4	229

(w/v) sodium chloride according to growth conditions. Samples, deposited on grids, were fixed in glutaraldehyde (2.5% w/v in buffer) followed by 1% (w/v) osmium tetroxide. After being washed with buffer, samples were dehydrated, vacuum-dried on a cover-slip, sputter-coated with gold and examined in a Cambridge SIIA scanning electron microscope (Cambridge Scientific Instruments, England) at 20 kV in the secondary emission mode. Measurements of length (l) and width (w) were taken from contact prints of negatives at 5000× magnification after tilt correction. Cell volumes were calculated by assuming that cells consisted of a cylinder of length ($l-w$) and width w, plus two hemispheres of diameter w.

Measurements taken from the micrographs of thirty-five cells, at each sodium chloride concentration, showed that there was a significant decrease in length ($P < 0.001$) but no significant change in width when data were compared using Student's t-test. This decrease in cell culture (Table 2) on adaptation to sodium chloride is an interesting effect, which may reflect changes in the composition, and hence functionality, of the cell membrane.

Energetics of adaptation

The increased amounts of glutamate dehydrogenase which have been measured in sodium chloride-adapted cells suggest that there would be an increased maintenance energy requirement by these cells compared with controls. We have taken various approaches to examine the energy balance of the cells. Measurement, for example, of the rate of glucose uptake by cells adapted to various sodium chloride concentrations were made, as well as estimations of the intracellular concentrations of ATP. Cells of *Cl. sporogenes* were grown in Nutrient Broth containing 1% (w/v) glucose and sodium chloride, 0.5 or 3.5% (w/v), for 22 h at 37 °C. Cells were harvested by centrifugation and re-suspended in fresh medium containing glucose (0.1% w/v) to a level of approximately 10^9 cells/ml. The rate of glucose disappearance from the medium was measured as described by Woods *et al.* (1981). The concentration of ethanol was also measured, as results from preliminary experiments suggested that this metabolite was the major product of metabolism under these conditions (Table 3). Simultaneous measurement of ATP showed that the total ATP per cell was reduced by adaptation but this was due to the reduced cell volume. The actual mean intracellular concentration of ATP was similar in both control and adapted cells. Although changes in adenylate charge cannot be ruled out since total adenosine phosphate concentrations were not measured, this result suggests a homeostatic mechanism maintaining intracellular ATP.

The energetics of this adaptation to growth in the presence of sodium chloride were investigated further by the use of a simple continuous culture system.

TABLE 3. *Effect of adaptation to sodium chloride on rate of glucose uptake, conversion to ethanol and concentration of intracellular ATP for cells of* Clostridium sporogenes

NaCl (%)	Cell volume (μm^3)	Glucose uptake fmol/h/cell	Glucose uptake mmol/h/ml cell volume	Intracellular ATP fmol × 10^{-3}/cell	Intracellular ATP mmol/l cell vol.	Ethanol production fmol/h/cell
0.5	2.5	0.97	0.39	1.02	0.41	1.78
3.5	1.8	1.45	0.81	0.85	0.47	2.65

Anaerobically, *Cl. sporogenes* can grow in a fermentation system lacking many of the normal control features for maintaining such parameters as pO_2. The system consisted of a Quickfit VF culture vessel (nominal capacity 500 ml) with an MAF 1/75 multi-socket/flange lid in a water bath at 37 °C. Inlet and outlet medium flow rates were controlled by HR Flow Inducers (Watson–Marlow Ltd., Cornwall, England). Culture liquid was maintained at a constant volume of 200 ml and sparged with oxygen-free dinitrogen (British Oxygen Company) at a rate of 10 l/h through a sintered glass tube, porosity O (Gallenkamp, London). Inlet and outlet gas was sterilized by passage through hydrophobic membrane filters (Gelman Sciences Inc., Ann Arbor, Michigan, USA). Observed growth yields were calculated by dividing cell yield by glucose utilization for a given volume of medium. Glucose utilization was determined from the difference in glucose concentration between fresh and spent media. Cell yield was determined by centrifuging 100 ml of culture at 2500 g_{av} for 10 min, washing the cells three times in distilled water and evaporating to dryness at 80 °C for 20 h on a watch glass. Adaptation to 2.5% (w/v) sodium chloride was chosen as this did not affect growth rate to the same extent as higher concentrations. Observed growth yields (Y) were measured at various dilution rates (D) at two sodium chloride concentrations and graphs of $1/Y$ against $1/D$ (Fig. 3) were used to determine true growth yield and maintenance coefficient (Table 4) by the method of Pirt (1965). These results demonstrate that there is an

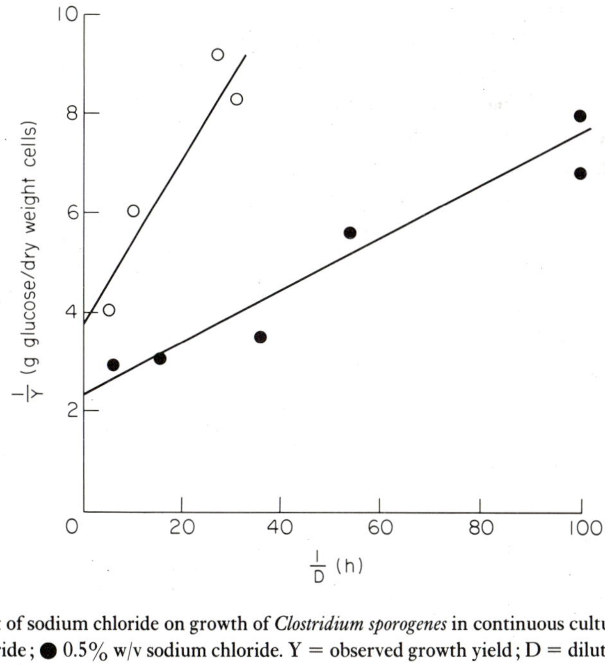

FIG. 3. Effect of sodium chloride on growth of *Clostridium sporogenes* in continuous culture (○ 2.5% w/v sodium chloride; ● 0.5% w/v sodium chloride. Y = observed growth yield; D = dilution rate).

TABLE 4. *Growth parameters of* Clostridium sporogenes *during continuous culture at two sodium chloride concentrations*

Sodium chloride (% w/v)	Maintenance coefficient (g glucose/g dry wt/h)	True growth yield (g dry wt/g glucose)	Correlation coefficient of linear regression
0.5	0.052	0.44	0.97
2.5	0.166	0.27	0.93

increase in maintenance energy and a decrease in growth yield during adaptation. Such an increase in maintenance energy caused by growth-inhibiting substances was proposed by Stouthamer & Bettenhaussen (1973) and, in our system, part of this increased energy requirement would be the need for adapting cells to synthesise glutamate and other osmoregulatory compounds by means of *de novo* synthesis of glutamate dehydrogenase and associated enzymes.

Polyphosphates

Polyphosphates have antimicrobial properties as well as an ability to improve the organoleptic properties of cured meats. The data have been reviewed by Tompkin (1983). Iron sequestration by polyphosphate was proposed as the inhibitory mechanism in canned cured-meat systems.

We have been investigating the effect of polyphosphates on the growth of a range of bacteria, with particular reference to *Cl. sporogenes*, in a simple system measuring culture turbidity. In such a system the inhibitory properties of polyphosphates can be easily demonstrated (Fig. 4). We have observed, however, that in this system the degree of inhibition caused by the polyphosphate is dependent on inoculum size, i.e. the number of bacterial cells present in inoculated culture. The inoculation dependence is shown in Fig. 5. This effect is not due to the presence of compounds in the inoculum since it was not altered by addition of spent medium. The reaction between the cells and polyphosphate is reversible since the inhibitory effect can be diluted out after pre-incubation with polyphosphate. A similar inoculum dependence was observed by Soracco & Pope (1983) in their investigation into the bacteriocidal activity of bis(tributyltin)oxide on *Legionella pneumophila* and they concluded that such results showed the presence of a site of action in or on the cells. Certainly the involvement of the cell surface (p. 255) is implied in the mode of action of polyphosphates by the differing susceptibilities shown by Gram-positive and Gram-negative organisms (Post *et al.* 1963; Tompkin 1983).

We have used a nephelometric system to examine the structure-activity

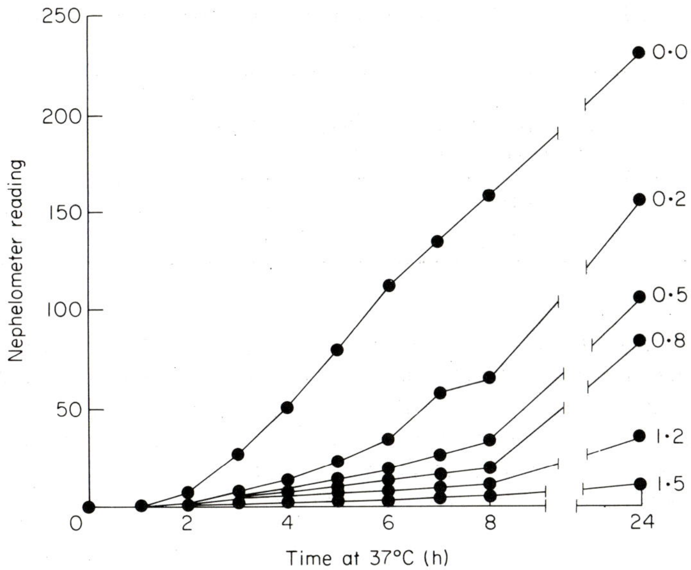

FIG. 4. Effect of various concentrations of sodium tripolyphosphate (% w/v) on the growth of *Clostridium sporogenes.*

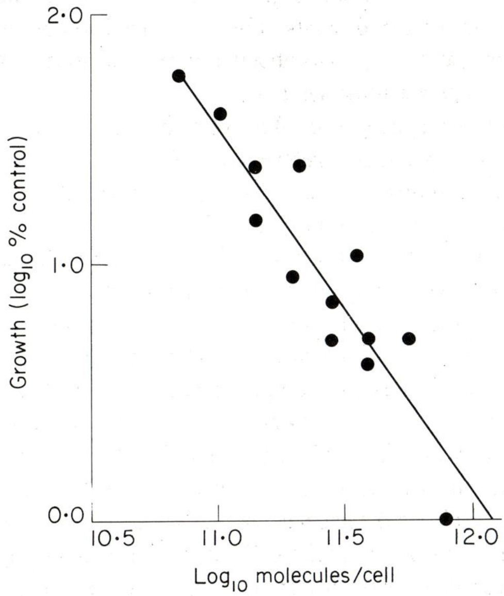

FIG. 5. Effect of sodium tripolyphosphate (amount per cell) on growth inhibition of *Clostridium sporogenes* (linear regression; r = 0.92).

relationship of the polyphosphates and the inhibition of growth. To compensate for the inoculum-dependence described previously, experiments are performed at three inoculum levels and percentage inhibition relative to the control calculated. We then used these results to construct a line similar to that shown in Fig. 5 and the degree of inhibition at a given cell density (usually $\log_{10}$ cell/ml = 7.5) interpolated. As the activity of the individual polyphosphates differs and thus different concentrations of the compounds are required to obtain useful information, this interpolated value of growth inhibition is then calculated for a polyphosphate concentration of 1% (w/v). Although such a system gives results on an arbitrary scale which exceeds 100%, it does allow comparison of those derived on different occasions and with different polyphosphates. These results, which are on a weight basis, can be converted subsequently to a molar basis by using the molecular weight of the polyphosphate based on the formula $Na_{n+2}\,P_n\,O_{3n+1}$. It is evident from Fig. 6 that, on a molar basis, the inhibition of growth of *Cl. sporogenes* increases with increasing chain length, whereas on a weight basis there is a maximum effect at a chain length of approximately 20. The data of Irani & Morgenthaler (1963) are shown also; these indicate that the inhibition by polyphosphate does not follow iron-chelating activity, implying that iron chelation is not the mechanism of action of these compounds.

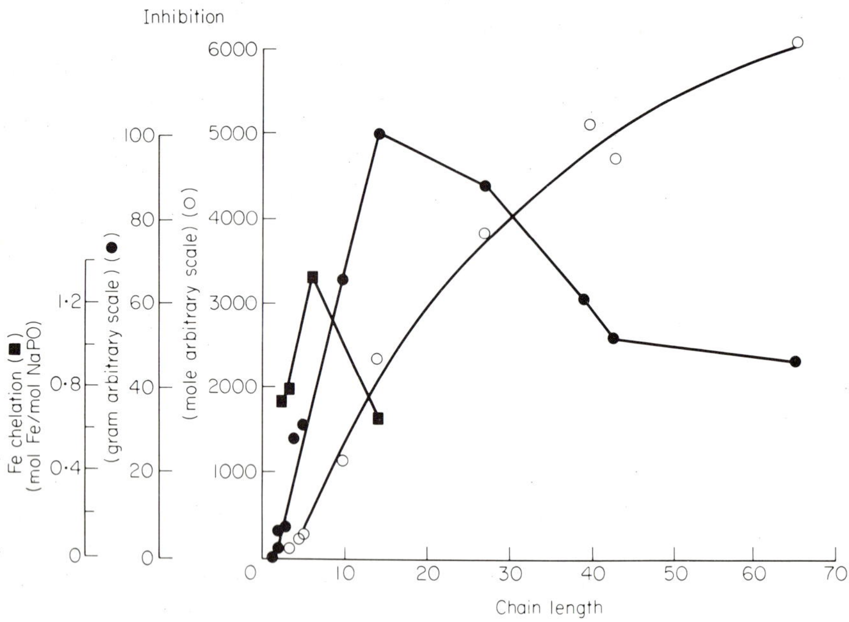

FIG. 6. Effect of different chain-length polyphosphates on growth of *Clostridium sporogenes* (○ mole; ● gram; ■ Fe chelation).

The rate of hydrolysis of various polyphosphates was investigated by paper chromatography according to the method of Kollof (1961). The solvent system was modified and consisted of 140 ml dioxan (BDH, chromatography grade), 50 ml distilled water, 1 ml ammonia (Fisons, SG 0.88) and 13.5 g trichloroacetic acid (Fisons). Whatman 31ET chromatography paper was used. This system gave good separation of polyphosphates up to chain length 10 (hexametaphosphate) and indicated that hydrolysis of long chain polyphosphates was slow in aqueous solution. This evidence suggests that our observations are due to long chain-length compounds rather than their shorter hydrolysis products.

The mode of action of polyphosphates remains elusive. Our results suggest an involvement of the cell wall or membrane, there being apparently some 10^{12} sites at which polyphosphates can bind to the cell. Whether, once bound, these molecules have an effect on specific proteins in the cell membrane such that transport, for example, of substrate is inhibited or a more generalized effect on the cell is not known. We have shown that polyphosphates such as sodium tripolyphosphate can act as inhibitors of enzymes which use phosphorylated substrates, e.g. phosphofructokinase from *Cl. sporogenes* (Fig. 7). Whether or not these compounds can gain access to the interior of the cell *in vivo* and inhibit cellular processes is not known. Phosphofructokinase activity was measured according to the method of Ling *et al.* (1955). Concentrations of fructose-6-phosphate and ATP were varied up to 5 mM and 1 mM, respectively. Other components of the system (final concentrations) were Tris/HCl, pH 8.0 (0.1 mol/l), $MgCl_2$ (10 mmol/l), cysteine hydrochloride (6.7 mmol/l), NAD (0.12 mmol/l), sodium arsenate (3.3 mmol/l), rabbit muscle aldolase (Sigma Type IV, 2.6 units/ml) and rabbit muscle glyceraldehyde-3-phosphate dehydrogenase (20.7 units/ml). A crude phosphofructokinase preparation was made from *Cl. sporogenes* by ammonium sulphate precipitation of a cell-free extract as described by Woods *et al.* (1981) after removal of ferredoxin by DEAE-cellulose chromatography. The results were analysed graphically using double-reciprocal (Lineweaver–Burk) plots and derivative plots as described by Mahler & Cordes (1971).

Sodium Nitrite

Nitrite appears to inhibit at least one specific enzyme in the pathway for the breakdown of glucose by *Cl. sporogenes* (Woods *et al.* 1981) and *Cl. botulinum* (Woods & Wood 1982). This enzyme forms part of the phosphoroclastic system which converts pyruvate to acetate and is an important site of substrate level phosphorylation in these organisms. This system contains two components which possess non-haem iron and which are therefore likely sites of inhibition by nitrite *via* nitric oxide. We have suggested one of these components—pyruvate : ferredoxin oxidoreductase—as the most likely site of action (Woods *et al.* 1981). The formation

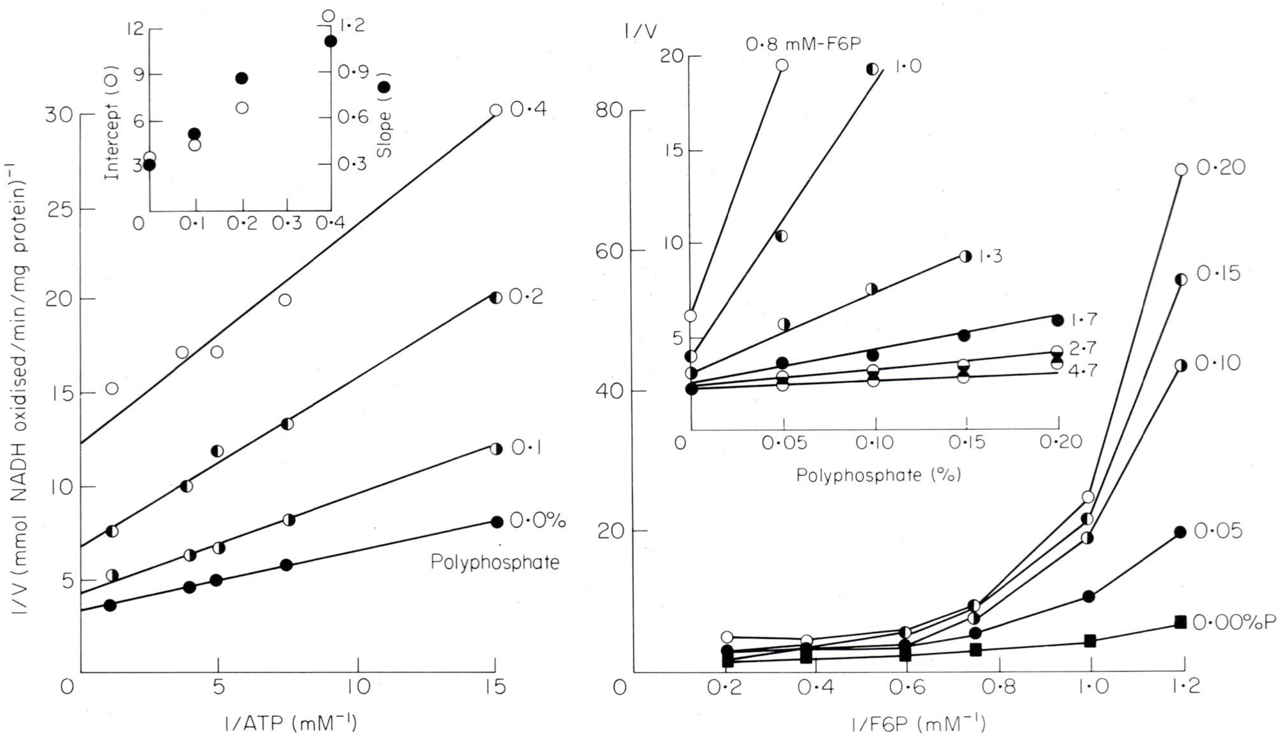

FIG. 7. Lineweaver-Burk plots and derivative plots of phosphofructokinase activity from *Clostridium sporogenes.*

of iron-nitric oxide complexes has been demonstrated recently by direct measurement using electron-spin resonance spectroscopy of preparations of *Cl. botulinum* in the presence of nitrite (Reddy *et al.* 1983). Such observations as these indicate that nitrite exerts its inhibitory effect by altering the chemical composition of sensitive components of bacterial cells. There is some evidence also of a less sensitive site elsewhere in glucose metabolism or transport in *Cl. sporogenes* (Woods 1982). Presumably in different organisms other sulphur-containing enzymes or co-factors are affected by nitrite. O'Leary & Solberg (1976), for example, showed that nitrite inhibited the sulphydryl enzymes glyceraldehyde-3-phosphate dehydrogenase and aldolase in *Cl. perfringens* while Yarbrough *et al.* (1980) demonstrated that nitrite can inhibit a number of different biological functions in aerobic bacteria including active transport, oxygen uptake and oxidative phosphorylation, all of which depend on iron-sulphur centres present in cytochromes. These effects occur generally at higher concentrations of nitrite than those necessary for the inhibition of *Cl. sporogenes* and *Cl. botulinum,* indicating how sensitive the phosphoroclastic system of the latter organisms is to inhibition by nitrite.

Effect of Heat on Spores

It is well known that the thermal process used to pasteurize canned cured meat is inadequate. It does not ensure the stability of such products. This stability results from both the low incidence of clostridial spores and interactions between the curing salts and the heat process on the spores. These interactions between sodium chloride, nitrite, pH value and thermal process produce a system which is more inhibitory to the spores than is any one factor alone (Roberts & Ingram 1973). It is the presence of curing salts after rather than during heating which is important (Roberts & Ingram 1966), since heat damage sensitizes spores to the presence of the salts during the germination and growth of the spore (Jarvis *et al.* 1976).

Investigations of the heat resistance of clostridial spores are generally based on the assumption that there is a logarithmic reduction in viable spore number with time, at a constant temperature, and that the rate of change of reduction with temperature is also constant. There is also some evidence, however, of non-linearity of survivor curves in the literature although linearity is generally assumed since calculation of heat-process parameters is more straightforward on this basis—for a review of this topic see Cerf (1977). In view of the uncertainty concerning the thermal death kinetics of clostridial spores when subsequently exposed to curing salts, we decided to investigate the apparent heat resistance of spores of *Cl. sprorogenes* as measured in a culture medium with pH value and concentrations of sodium chloride and of nitrite typical of a canned cured meat.

Spores of *Cl. sporogenes* were obtained by subculturing from actively growing broth cultures into freshly steamed and cooled Robertson's Cooked Meat Broth

(Southern Group Laboratories, London) and incubated for 18 h at 37 °C. A 5 ml sample of this culture was transferred to freshly steamed and cooled sporulation medium (6% w/v Trypticase (BBL), pH 7; 500 ml), and incubated at 37 °C for 4 d, at which time a high proportion (>80%) of phase-bright spores could be observed microscopically. The cultures were centrifuged at 18 000 g_{av} for 20 min and the spores resuspended in sterile distilled water (pH 7). The spores were washed five times by being resuspended in distilled water. They were stored at 4 °C until required.

Spores were heated in Reinforced Clostridial Medium (RCM) prepared from individual ingredients (Hirsch & Grinsted 1954). It contained 0.5% (w/v) sodium chloride but no agar. The pH of the medium was adjusted such that the desired pH was achieved after autoclaving (121 °C for 15 min). The heating medium was inoculated in bulk with spores to give a concentration of approximately 10^7 colony-forming units/ml and dispensed aseptically in 20 ml amounts in sterile glass scintillation vials fitted with serum caps. The vials were submerged in a heating bath and removed at suitable times according to the temperature of the bath for estimation of survivors. The heating bath contained concentrated calcium chloride solution which elevated the boiling point to above 125°C. This allowed experiments to be made without resort to oil baths or autoclaves. Viable counts of the spore suspension at 'time zero' were estimated by heat activation at 80 °C for 10 min. Survivors were estimated in duplicate—pour plates were prepared with 0.5 ml of suitable dilutions and RCM containing 1.2% (w/v) agar (RCA). The culture was incubated anaerobically at 37 °C for 6 d. All colonies were counted, and weighted mean numbers of colony-forming units/ml were calculated by the method of Farmiloe *et al.* (1954). Survivor curves were plotted as $\log_{10}$ percentage survivors against heating time at a specific temperature, the decimal reduction time (*D* value) being the reciprocal of the gradient.

The marked effect that the pH of the enumeration medium has on the apparent heat resistance of the spores (Fig. 8) was one of the most striking results of our investigations. This effect is due presumably to an increased sensitization of heat-damaged spores to a relatively mild environmental stress. Such an effect is not well documented in the literature. Jarvis *et al.* (1976) obtained similar apparent D values for *Cl. botulinum* type B grown on blood agar in the range pH 6.0 to 7.0 while Roberts & Ingram (1966) observed an interaction between pH and sodium chloride and nitrite concentrations rather than an effect of pH alone. In general, media used to recover heated spores are at the optimum pH for growth (*ca.* 7.0) in order to achieve maximum recovery rather than at lower values as a model for the effect of the pH of a typical meat product. Clearly, this effect of pH on the recovery of heated spores represents a significant contribution to the observed safety of pasteurized cured products.

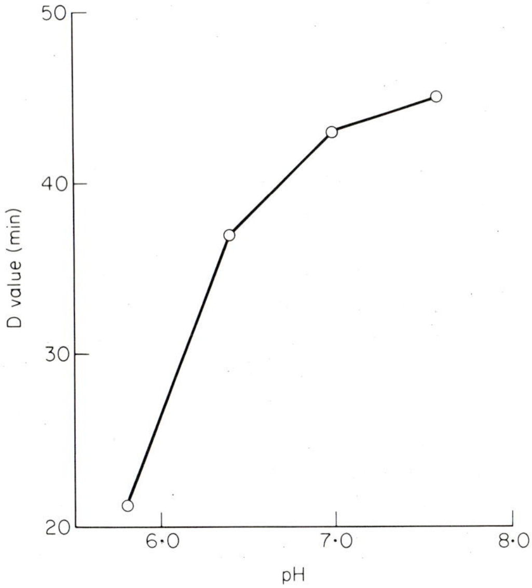

FIG. 8. Effect of pH on D value of *Clostridium sporogenes.*

References

BAIRD-PARKER, A. C. & BAILLIE, M. A. H. 1974 Inhibition of *Clostridium botulinum* by nitrite and sodium chloride. In *Proceedings of the International Symposium on Nitrite in Meat Products*, eds. Krol, B. & Tinbergen, B. J. pp. 77–90. Wageningen, Netherlands: Centre for Agricultural Publishing and Documentation.

CERF, O. 1977 Tailing of survival curves of bacterial spores. *Journal of Applied Bacteriology* **42**, 1–19.

FARMILOE, F. J., CORNFORD, S. J., COPPOCK, J. B. M. & INGRAM, M. 1954 The survival of *Bacillus subtilis* spores in the baking of bread. *Journal of the Science of Food and Agriculture* **5**, 292–304.

HIRSCH, A. & GRINSTED, E. 1954 Methods for the growth and enumeration of anaerobic spore-formers from cheese. *Journal of Dairy Research* **21**, 101–110.

IRANI, R. R. & MORGENTHALER, W. W. 1963 Iron sequestration by polyphosphates. *Journal of the American Oil Chemists Society* **40**, 283–285.

JARVIS, B., RHODES, A. C., KING, S. E. & PATEL, M. 1976 Sensitization of heat-damaged spores of *Clostridium botulinum* type B to sodium chloride and sodium nitrite. *Journal of Food Technology* **11**, 41–50.

KOLLOF, R. H. 1961 The rapid trace analysis of phosphate mixtures by a paper chromatographic, densitometric procedure. *Analytical Chemistry* **33**, 373–379.

LING, K-H., BYRNE, W. L. & LARDY, H. 1955 Phosphohexokinase. *Methods in Enzymology* **1**, 306–310.

MAHLER, H. R. & CORDES, E. H. 1971 *Biological Chemistry* 2nd edn, pp. 295–299. Philadelphia: Harper & Row.

MEASURES, J. C. 1975 Role of amino acids in osmoregulation of non-halophilic bacteria. *Nature*, London **257**, 398–400.

MEERS, J. L., TEMPEST, D. W. & BROWN, C. M. 1970 Glutamine (amide): 2-oxoglutarate aminotransferase oxidoreductase (NADP): an enzyme involved in the synthesis of glutamate by some bacteria. *Journal of General Microbiology* **64**, 187–194.

O'LEARY, V. & SOLBERG, M. 1976 Effect of sodium nitrite inhibition on intracellular thiol groups and on the activity of certain glycolytic enzymes in *Clostridium perfringens*. *Applied and Environmental Microbiology* **31**, 208–212.

PIRT, S. J. 1965 The maintenance energy of bacteria in growing cultures. *Proceedings of the Royal Society B163*, 224–231.

POST, F. J., KRISHNAMURTY, G. B. & FLANAGAN, M. D. 1963 Influence of sodium hexametaphosphate on selected bacteria. *Applied Microbiology* **11**, 430–435.

REDDY, D., LANCASTER, J. R. & CORNFORTH, D. P. 1983 Nitrite inhibition of *Clostridium botulinum*: electron spin resonance detection of iron-nitric oxide complexes. *Science* **221**, 769–770.

ROBERTS, T. A. & INGRAM, M. 1966 The effect of sodium chloride, potassium nitrate and sodium nitrite on the recovery of heated bacterial spores. *Journal of Food Technology* **1**, 147–163.

ROBERTS, T. A. & INGRAM, M. 1973 Inhibition of growth of *Clostridium botulinum* at different pH values by sodium chloride and sodium nitrite. *Journal of Food Technology* **8**, 469–475.

ROBERTS, T. A., JARVIS, B. & RHODES, A. C. 1976 Inhibition of *Clostridium botulinum* by curing salts in pasteurised pork slurry. *Journal of Food Technology* **11**, 25–40.

ROBERTS, T. A., GIBSON, A. M. & ROBINSON, A. 1981 Prediction of toxin production by *Clostridium botulinum* in pasteurised pork slurry. *Journal of Food Technology* **16**, 337–355.

SORACCO, R. J. & POPE, D. H. 1983 Bacteriostatic and bacteriocidal modes of action of bis(tributyltin)oxide on *Legionella pneumophila*. *Applied and Environmental Microbiology* **45**, 48–57.

STOUTHAMER, A. H. & BETTENHAUSSEN, C. 1973 Utilisation of energy for growth and maintenance in continuous and batch cultures of micro-organisms. *Biochimica et Biophysica Acta* **301**, 53–70.

TEMPEST, D. W., MEERS, J. L. & BROWN, C. M. 1970 Influence of environment on the content and composition of microbial free amino acid pools. *Journal of General Microbiology* **64**, 171–185.

TOMPKIN, R. B. 1983 Indirect antimicrobial effects in foods: phosphates. *Journal of Food Safety* **6**, 13–27.

WINNACKER, E. L. & BARKER, H. A. 1970 Purification and properties of a NAD-dependent glutamate dehydrogenase from *Clostridium* SB4. *Biochimica et Biophysica Acta* **212**, 225–242.

WOODS, L. F. J. 1982 The mode of action of curing salts. *Chemistry and Industry* 975–977.

WOODS, L. F. J., WOOD, J. M. & GIBBS, P. A. 1981 The involvement of nitric oxide in the inhibition of the phosphoroclastic system of *Clostridium sporogenes* by sodium nitrite. *Journal of General Microbiology* **125**, 399–406.

WOODS, L. F. J. & WOOD, J. M. 1982 A note on the effect of nitrite inhibition on the metabolism of *Clostridium botulinum*. *Journal of Applied Bacteriology* **52**, 109–110.

YARBROUGH, J. M., RAKE, J. B. & EAGON, R. G. 1980 Bacterial inhibitory effects of nitrite: inhibition of active transport, but not group translocation and of intracellular enzymes. *Applied and Environmental Microbiology* **39**, 831–834.

YESAIR, J. & CAMERON, E. J. 1942 Inhibitive effect of curing agents on anaerobic spores. *Canner* **94**, 89–92.

Sulphite Preservation of Meat Products

J. G. Banks, G. J. Nychas and R. G. Board

School of Biological Sciences, University of Bath, Avon UK

Background

Metabisulphite or sulphite is added to British style sausages and some other carbohydrate-containing meat products to delay microbial spoilage. The content of preservative in the product has been determined traditionally by collection of the sulphur dioxide (SO_2) released from a sample suspended in boiling acid (Monier-Williams 1927). This procedure led to a widely held view that SO_2 *per se* was the antimicrobial moiety. In practice, the pH of such products (6.8–6.2) would not be expected to cause significant SO_2 formation from metabisulphite and sulphite (Fig. 1; Ingram 1948; King *et al.* 1981) and it is more likely that bisulphite (HSO_3^-) and sulphite (SO_3^{2-}) are the active agents (Hammond & Carr 1976). Vas & Ingram (1949) were probably the first food microbiologists to recognize that the pH of a food product was an important determinant of the efficacy of the preservative and they published a set of curves showing the proportions of the various molecular species at different H-ion concentrations based on the value of 5×10^{-6} for K_2 (2nd dissociation constant). Although these curves have been reproduced widely in the literature e.g. Rehm & Wittman (1962), King *et al.* (1981) were of the opinion that they were in error by two full pH units (7.2 *vs* 5.3) at the pK_2 point. But even with the recalculated percentage (Table 1) of each molecular species of undissociated sulphurous acid (H_2SO_3) in the range of pH 7.5–5.0, based on a K_1 of 1.7×10^{-2} and K_2 of 6.31×10^{-8} (pK_1 = pH 1.77 and pK_2 = pH 7.2), SO_2 *per se* forms a negligible proportion of the preservative in meat products.

The absolute content of SO_3^{2-}/HSO_3^- in sausages diminishes slowly with time (Brown 1977) probably as a consequence of oxidation to SO_4^{2-} (Abel 1913; Fridovitch & Handler 1961). As the loss in minced meat is very much greater (Nychas, unpublished observations), it would appear that the pseudo-emulsion state of sausage protects against sulphite oxidation. Many substances in foods bind reversibly with sulphite *viz* pyruvate (Burroughs & Sparks 1963), acetaldehyde and ketones (Joslyn & Braverman 1954), glucose and maltose (Ingram & Vas 1950*a*, *b*),

Preservatives in the Food,
Pharmaceutical and Environmental
Industries

0-632-01727-9

TABLE 1. *Percentage distribution of molecular species of sulphurous acid as a function of pH**

pH	H_2SO_3 (%)	HSO_3^- (%)	SO_3^{2-} (%)
2	37.03	62.97	0.0
3	5.56	94.43	0.006
4	0.59	99.35	0.063
5	0.058	99.31	0.63
6	0.006	94.15	5.84
7	0.002	61.30	38.70

* Adapted from King *et al.* (1981).

and amines (Joslyn & Braverman 1954). The addition products show little antimicrobial activity (Neuberg 1929; Rehm 1964). Recent observations in this laboratory (Dalton, unpublished observations) have shown that acetaldehyde produced by yeasts is the main cause of the progressive binding of sulphite which occurs in stored sausages.

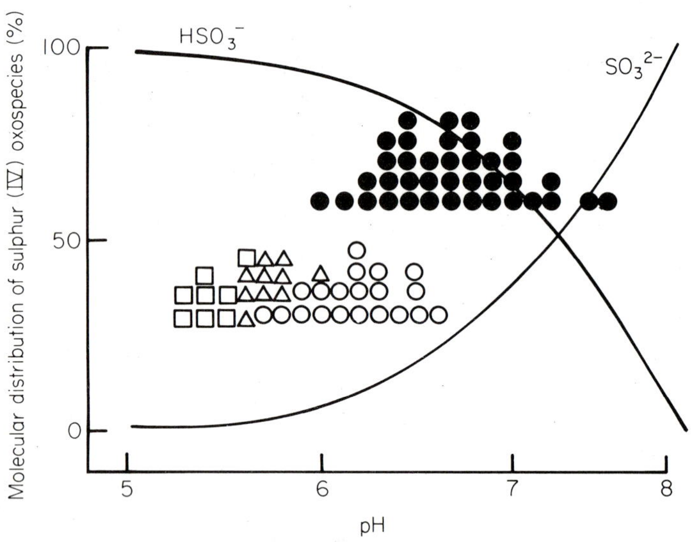

FIG. 1. Influence of pH on the molecular distribution of sulphur (IV) oxospecies and pH of freshly-manufactured or stored sulphited sausage. A sample (20 g) of sausage was homogenized in distilled water (180 ml) in a Colworth Stomacher (Seward, London) for 60 s. The pH of the homogenate was measured with the glass pH electrode (Russell). ●, pH of sulphited sausage at point of manufacture; open symbols, lowest pH attained with storage at: 4 °C, ○; 10 °C, △ or 22 °C, □. (Adapted from Banks & Board 1983.)

Lafontaine and his co-workers (1955) were perhaps the first to note that the addition of sulphite to minced meat resulted in the inhibition of the growth of the microbial contaminants. This situation was dependent not only on the concentration of sulphite but also on the temperature of storage and the initial concentration of micro-organisms. Thus a concentration of 300 μg of sulphite/g caused bacteriostasis at 4 °C but not at 20 °C whereas higher concentrations (1000 and 3000 μg/g) were effective at both temperatures. These workers observed also that the addition of sulphite favoured the growth of facultative anaerobes and that the efficacy of preservation was inversely related to the initial size of the microbial population. Broth studies with isolates from minced meat corroborated the above observations. Krol & Moerman (1959/60) confirmed the findings of Lafontaine *et al.* (1955) and reported also that microbial growth, as indicated by total viable counts, was restricted for up to 6 d at refrigeration temperatures in sulphited (300 μg/g) minced meat balls; a concentration of 900 μg/g resulted in the death of micro-organisms. Even at the former concentration of sulphite, the Enterobacteriaceae were particularly sensitive with upwards of a 90% kill during chilled storage for 6 d. Fournaud *et al.* (1971) observed that sulphite was a most effective preservative against the Enterobacteriaceae and microbial contaminants in general in the minced pork ingredients of saucissons. Dyett & Shelley (1962, 1965, 1966) commented on the inhibition of growth of the coliforms, and, in particular, *Salmonella* spp. in sulphited pork or beef sausages. Three observations in the last mentioned report are of particular importance: firstly that with incubation at temperatures below 22 °C, sulphite was most potent against the 'coli-aerogenes' group and other Gram-negative bacteria; secondly that the addition of sulphite resulted initially in a kill of the numerically dominant micro-organisms, and thirdly that it retarded the growth rate of the microbial association thereafter.

Despite the high incidence (Roberts *et al.* 1975) of contamination of sausage with salmonellas, the level is low (Banks & Board 1983). Thus it has been surmised (Dyett & Shelley 1966) that the good public health record of sausages is associated with the widespread use of sulphite. Such a view has been supported by Banks and Board (1982*a*) who noted that at storage temperatures below 25 °C, *Salmonella virchow* inoculated into sulphited sausage failed to grow; it did so in sausage without sulphite at 15, 20 or 25 °C. Banks & Board (1982*a*) demonstrated also that changes in the size of populations of Enterobacteriaceae provide an index of the preservative role of sulphite such that members of this group multiplied only when the concentration of the free form was less than 40 μg/g sausage. They noted also that the preservative influenced the composition of the Enterobacteriaceae flora. Thus *Enterobacter cloacae*, *Serratia liquefaciens* and *Escherichia coli* flourished in sulphited sausage when the content of free preservative had fallen below the critical threshold for growth, whereas *Enterobacter agglomerans* did so only in the unsulphited product. According to Christian (1963) sulphite, at a concentration of 3.5 grains/lb, extended the shelf life of minced beef two- to three-fold with chilled storage. In accord with the observations of Dyett & Shelley (1962) and Krol & Moerman (1959/60), Christian noted that sulphite increased the 'lag' period and suppressed the rate of growth of the dominant micro-organisms. There was evidence also that the preservative exerted a selective action on the microbial association. Thus with storage of unsulphited minced beef, Gram-negative bacilli predominated, whereas in the presence of sulphite, short Gram-positive bacilli were numerically dominant. Gardner (1968) could not reproduce all the findings of Christian (1963) with respect to sulphite preservation of vacuum-packed baconburgers. It should be recognized, however, that the principal raw materials of the product contained nitrite and nitrate which probably affected the composition and behaviour of the microbial association as well as negating the antimicrobial effect of sulphite (Tompkin *et al.* 1980). Gardner (1968) found that sulphite prevented an acid drift with storage and that shelf-life extension was related directly to temperature of storage.

An investigation of the influence of sodium sulphite on *Escherichia coli*, *Salmonella typhimurium* and micro-organisms indigenous to raw minced meat (Moerman *et al.* 1966) revealed that a concentration of 0.03% (w/w) at 15 °C caused bacteriostasis of the 2 named organisms but was less effective against the others. With incubation at 20 °C or 25 °C, the presence of the preservative had a negligible effect on the rate or extent of growth of any of the micro-organisms sought. Moerman *et al.* (1966) demonstrated also a significant decrease in the concentration of total sulphite over the storage period. Thus with incubation for 20 h at 37 °C, only 35% was recovered whereas after 41 h at 25 °C, 60% was still present. The utility of sulphite in the preservation of chilled raw mince was confirmed by Mulder (1969) who noted that, with storage at 7 °C, sulphite prevented microbial growth for upwards of 7 d.

Following an extensive survey (Dowdell & Board 1968) of British fresh sausages, Dowdell & Board (1971) defined the composition of the 'microbial association' of the product. The dominant members, yeasts, homofermentative *Lactobacillus* spp. and *Microbacterium thermosphactum* (*Brochothrix thermosphacta*, Gardner 1981), grew in sulphited culture media whereas pseudomonads and coliforms isolated from fresh sausage did not. The elective nature of sulphite for a restricted microbial consortium was first proposed by Hurst (1972) and confirmed by others (Brown 1977; Banks & Board 1981).

Tyson (1976) noted that polyphosphate enhanced the antimicrobial action of sulphite against lactobacilli, pseudomonads and yeasts in sausage meat. The preservative alone enhanced the rate and extent of growth of yeasts but, through suppressing those of the lactobacilli, extended the shelf life of sausage meat. She demonstrated that pseudomonads grew only when the concentration of active (free) sulphite had fallen below a critical level. Brown (1977) was of the opinion that sulphite in sausage 'steered' the microbial fermentation which was dominated by *Brochothrix thermosphacta*, and proposed that free sulphite was bound (rendered inactive) by unidentified compounds produced by the developing microflora as well as by meat and rusk. Recent findings (Dalton & Board, unpublished observations) have shown that some (e.g. *Debaryomyces hansenii*, *Candida lipolytica*, *Pichia membranaefaciens* and *Torulopsis candida*) but not all (e.g. *Cryptococcus albidus* var. *albidus* and *Rhodotorula rubra*) yeasts bound sulphite added to minced pork. All of the sulphite-binding yeasts produced acetaldehyde in amounts sufficient to account for upwards of 90% of the binding of sulphite.

The above review of the limited literature concerned with the nature of sulphite preservation of neutral-pH meat products has revealed several features. It has shown firstly that the preservative is effective against micro-organisms only when present in the free (unbound) form. Secondly, that free sulphite is most potent at low temperatures and, as a consequence of binding, its potential to inhibit microbial growth decreases with time. Thirdly, that it elects for a Gram-positive bacterial yeast microflora, which sours the product (Fig. 1), through inhibition of Gram-negative micro-organisms, particularly the Enterobacteriaceace.

Assay of Free and Total Sulphite

Monier-Williams (1927) steam distillation method, modified by Shipton (1959)

A sample (20 g) of sulphited meat is placed in a three-necked round bottom flask (Fig. 2) (250 ml) and purged with dinitrogen (20 $cm^3\ s^{-1}$ for 5 min) before the flask

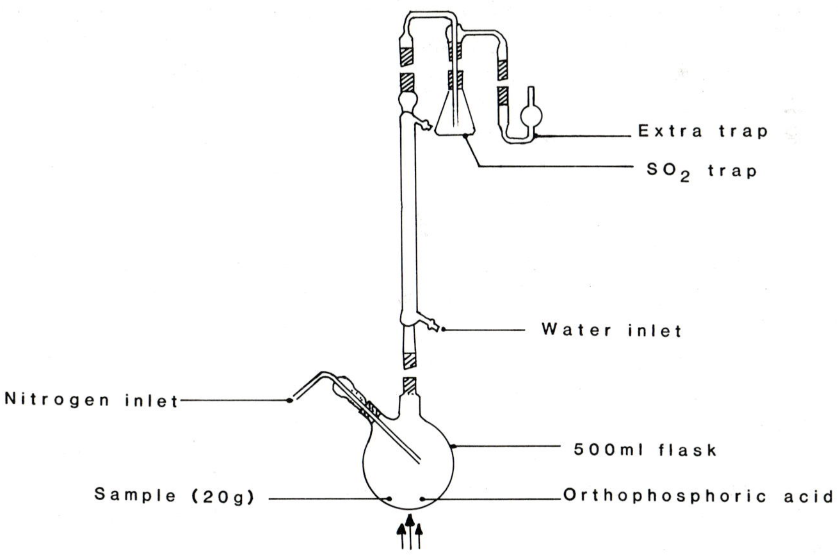

FIG. 2. The Shipton's steam distillation apparatus for assay of total sulphur (IV) oxospecies in sulphited meat products. (Adapted from Shipton 1959)

is sealed and acidified with orthophosphoric acid (2 mol/1). Heat is applied and the sample distilled for 60 min. Hydrogen peroxide (3% v/v, 25 ml) containing Tashiro's indicator (2 vol. methyl red, 1% w/v and 1 vol methylene blue, 0.1% w/v in 95% w/v ethanol) is used to trap the released SO_2 (Burroughs & Sparks 1963). Before distillation is begun, the trapping liquid is titrated to the first appearance of a green colour from the neutral end-point (grey, pH 4.5) and, after distillation, is titrated back to the green colour with freshly prepared sodium hydroxide (0.01 mol/l).

Electrometric ion-selective probe method

The theory and use of gas-sensing membrane probes for the determination of SO_2 in solution has been discussed by Bailey & Riley (1975). Probes are available commercially from E.I.L. or Orion. Such probes incorporate an SO_2-permeable membrane so that a hydrogen-sensing electrode is separated from the test solution by a buffered aqueous solution of hydrogen sulphite. The concentration of SO_2 is usually maximized by poising the test solution below pH 1.0 before analysis. The equilibria associated with the hydrolysis of SO_2 in aqueous solution (Fig. 1) can be summarized (King *et al.* 1981):

$$SO_2 + H_2O \rightleftharpoons H_2O \cdot SO_2 \quad (1)$$

$$H_2O \cdot SO_2 + H_2O \rightleftharpoons HSO_3^- + H_3O^+ \quad (2)$$

$$HSO_3^- + H_2O \rightleftharpoons SO_3^{2-} + H_3O^+ \quad (3)$$

The following methods for the use of the probe with sulphited meats are based largely on the work of Brown (1977).

Procedure

A sample (5 g) of sulphited meat is added to deoxygenated and chilled (5 °C) distilled water (45 ml) and homogenized in a Colworth Stomacher (Seward, London) for 60 s. The homogenate is centrifuged at 35 000 **g** for 10 min. When the set speed is achieved, the compressor is regulated to cool at 2 °C. Chilled centrifugation separates a fatty layer from an aqueous one and a solid base comprising starch and meat particles. The aqueous layer is decanted and, to reduce SO_2 loss, stored in a stoppered volumetric flask (50 ml) in an ice bath before analysis.

Total sulphite determination

To dissociate sulphite addition compounds, equal volumes of sodium hydroxide (1 mol/1) and the sample solution are mixed for 30 min at 25 °C in a shaken Erlenmeyer flask (100 ml). A sample (25 ml) of this alkalized solution is placed in an Erlenmeyer flask (50 ml) containing a magnetic stirring bar and an equal volume of sulphuric acid (1 mol/1) added. The flask is put in an ice bath on a stirrer bed and the SO_2 probe (Fig. 3) immersed to a depth of 1.5 cm. The peak deflection (mV) after 2 min is noted directly on a pH meter (monovalent cation setting) or from the chart of a pen recorder attached to it (Fig. 4).

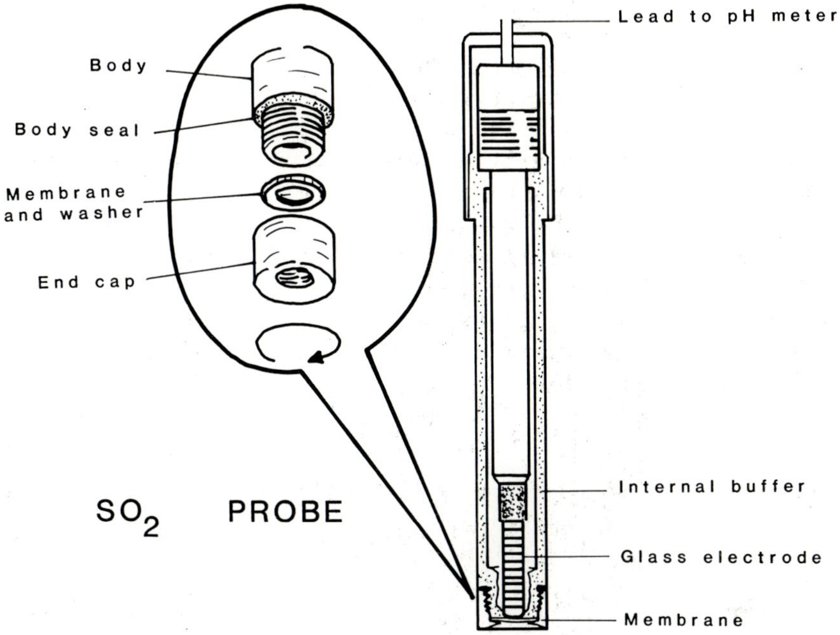

FIG. 3. An exploded view of an SO_2 ion-selective electrode. (Taken from Brown 1977.)

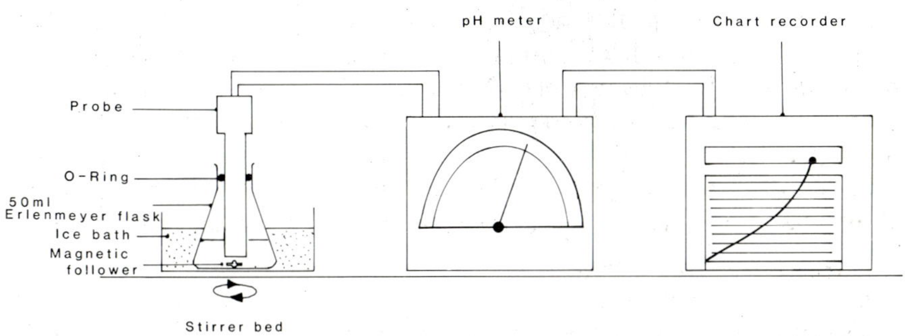

FIG. 4. Apparatus used for the electrometric determination of free or total sulphur (IV) oxospecies in meat serum.

Free sulphite determination

Analysis is done immediately after sample preparation. No alkalization procedure is used and acidification with sulphuric acid releases the free SO_2 which is then estimated as described for 'total sulphite' (see before).

Two technical points need to be stressed. Firstly, chilled centrifugation of homogenized samples is essential because fat tends to coat the membrane of the probe. This results in loss of sensitivity and a progressive lowering of the estimations of SO_2 concentration, and a short membrane life. The poor recovery (cf. spectrophotometric method) of both free and total SO_2 in samples needs to be stressed also. The SO_2 level of the aqueous phase only is measured; the SO_2 trapped in the fat or solid phases of the centrifuged meat homogenate is omitted from the analysis. Secondly, at very low concentrations (< 50 μg SO_2/ml) the peak deflection after 2 min is often difficult to resolve because of (a) a sluggish response by the pH meter, and (b) interference from unidentified factors. At high concentrations (400 g SO_2/ml) the peak deflection after 2 min is measured easily but the recovery period (dialysis against distilled water) required to remove excess SO_2 from the internal buffer solution of the probe is seldom less than 10 min. Under optimal conditions the probe can detect as little as 5 μg SO_2/ml and the relationship between the logarithm of SO_2 concentration and the electrometric deflection (mV) is linear up to 5×10^{-3} mol/l $Na_2S_2O_5$ (Fig. 5).

Spectrophotometric method

The sulphite ion reacts with many organic disulphides to displace a thiol anion forming the organic thiosulphate, commonly referred to as a 'Bunte' salt (4). A large excess of sulphite ion is normally required to estimate quantitatively the disulphide and the thiol anion is usually titrated with silver nitrate (Cecil & McPhee 1954).

$$RSSR + SO_3^{2-} \rightleftharpoons RSSO_3^{-} + RS^{-} \quad (4)$$

Some disulphides e.g. 5,5′ dithiobis (2 nitrobenzoic acid)—DTNB or 'Ellman's reagent'—and 4,4′ dithiodipyridine (Grassetti & Murray 1967) are used also for the determination of thiol groups. The disulphide is reduced to the corresponding thiol which absorbs at a different wavelength from the disulphide. As the disulphide is in excess, a mixed disulphide is probably formed:

$$R'SSR' + RSH \rightleftharpoons R'SSR + R'SH \quad (5)$$

With Ellman's reagent, the disulphide absorbs maximally at 324 nm and the 5 mercapto-2-nitrobenzoic acid at 412 nm. When the concentration of the disulphide is *ca.* 5×10^{-5} mol/l, a mole ratio plot of SO_3^{2-} indicates a stoichiometric reaction, 1 mole of SO_3^{2-} reacting with 1 mole of DTNB, see Humphrey *et al.* (1970). The accentuated inflection seen in this plot suggests that no intermediates are formed and that the 5 mercapto-2-nitrobenzoate moiety is produced in each reaction (Fig. 6).

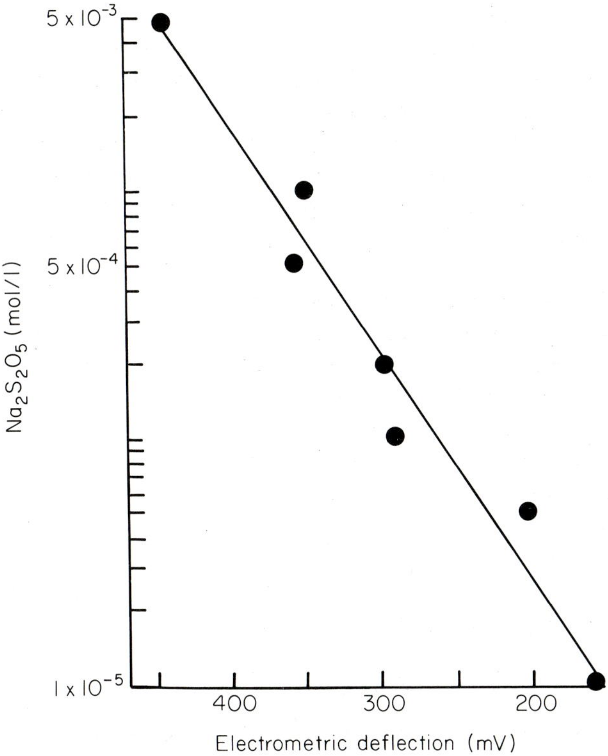

FIG. 5. The influence of sulphite concentration on the electrometric deflection measured by the SO_2 ion-selective probe.

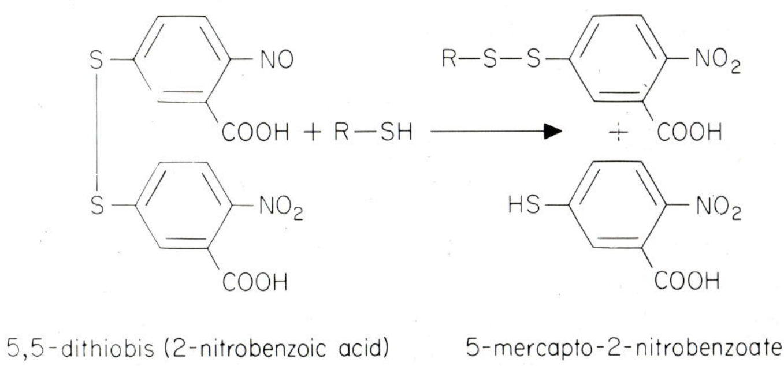

FIG. 6. The reaction of 5,5′ dithiobis (2-nitrobenzoic acid) 'Ellmans' reagent with moieties such as the sulphite ion to form 5-mercapto-2-nitrobenzoic acid. Unreacted DTNB is pale yellow whereas 5-mercapto-2-nitrobenzoate is a dark yellow colour, absorbing maximally at 412 nm.

Procedure (see also Banks & Board 1982*b*)

A sample (5 g) of sulphited meat is added to chilled, deoxygenated distilled water (25 ml) in a screw-capped bottle (100 ml) containing glass beads, and shaken 60 times through an arc of 0.5 m for 30 s. The homogenate is stored in an ice bath before analysis.

Free sulphite determination

The homogenate is transferred (with beads) to a 3 neck distillation flask (Q&Q 250/35/22A) containing a magnetic stirring bar (Fig. 7). One side neck is connected to a supply of dinitrogen (20 cm^3/s) *via* a 'drawn out' Drechsel bottle head (Q&Q

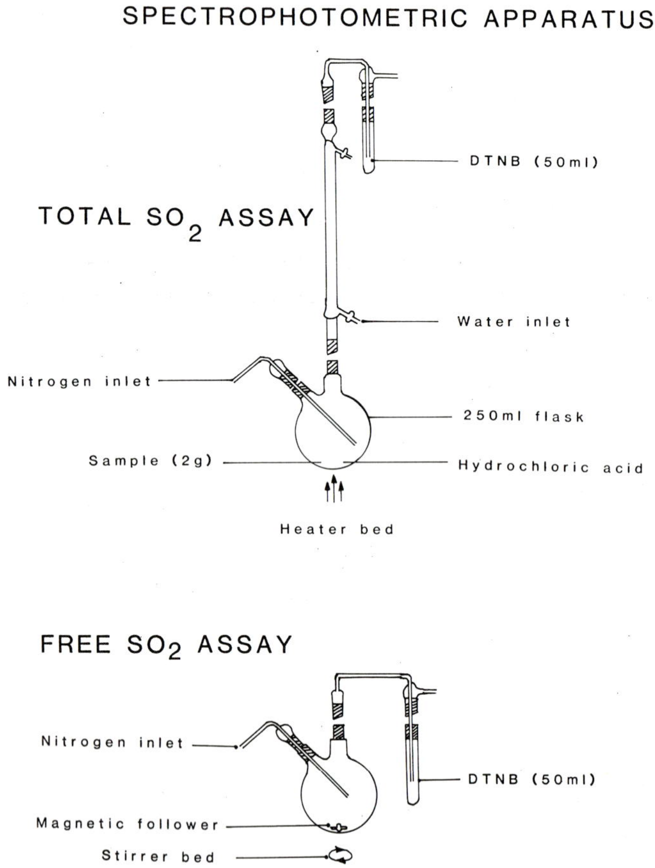

FIG. 7. Apparatus used for the spectrophotometric determination of total and free sulphur (IV) oxospecies in meat homogenates.

MF/27/3/13). A 'Y' junction adapter (Q&Q MF 102B) with taps is placed on the centre neck of the flask and it is attached to two Dreschel bottle heads arranged in parallel. The dip tubes of the bottle heads are situated beneath the surface of a solution of DTNB (2.5×10^{-3} mol/l, 50 ml) dissolved in phosphate buffer (pH 8.0; 1.6×10^{-2} mol/l Na_2HPO_4 and 1.8×10^{-3} mol/l KH_2PO_4) containing ethanol (10% v/v) in boiling tubes (Q&Q MF 24/3). The mixture is stirred and the reaction allowed to continue for 15 min after the addition of hydrochloric acid (2 mol/l, 50 ml) to the meat homogenate. The free SO_2 is carried over with the dinitrogen and bubbled through the DTNB solutions; it reacts with DTNB to form a stable deep yellow complex. To avoid saturation of DTNB and loss of SO_2 from the system, the flow of SO_2 in the carrier gas is diverted by means of the 'Y' junction adaptor. DTNB is mixed with an equal volume of phosphate buffer (pH 8.0) and the absorbance measured at 412 nm on a spectrophotometer (Pye Unicam SP6-550 UV/VIS)—see Wedzicha & Bindra (1980) for additional details.

Total sulphite determination

The procedure described above is followed but heat is applied to the distillation flask throughout the 15 min reaction period following HCl addition. Furthermore, the free and bound SO_2 driven off by heat and carried in the dinitrogen is passed through a vertical condenser (Q&Q CI/13) before collection in DTNB (Fig. 7). At a wavelength of 412 nm and a light path of 1 cm, the limit of detection is *ca.* 0.04 µg SO_2/ml and a straight line plot is obtained up to an absorbance of *ca.* 2 units (Fig. 8).

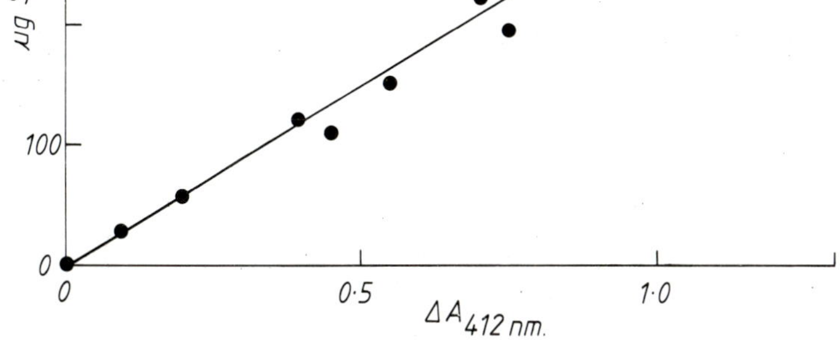

FIG. 8. Influence of sulphite concentration on the absorbance of 5-mercapto-2-nitrobenzoate at 412 nm.

Enzymatic method

The reagents for this method are available in kit form from the Boehringer Company, Park Royal, Sussex. In the presence of oxygen, sulphite is oxidized by sulphite oxidase (SO_2—OD) to sulphate (6):

$$SO_3^{2-} + O_2 + H_2O \xrightarrow{SO_2\text{-OD}} SO_4^{2-} + H_2O_2 \qquad (6)$$

The hydrogen peroxide is reduced by NADH-peroxidase/NADH: hydrogen peroxide oxidoreductase (EC 1.11.1.1-NADH-POD) in the presence of reduced nicotinamide—adenine dinucleotide (NADH) (7):

$$H_2O + NADH + H^+ \xrightarrow{NADH\text{-POD}} 2H_2O + NAD^+ \qquad (7)$$

The amount of NADH consumed in (7) is equivalent to the amount of free sulphite or chemically bound sulphite. NADH is determined by absorbance at 334, 340, or 365 nm.

Procedure

Total SO_2 determination. A sample (10 g) of sulphited meat is homogenized in 90 ml of chilled (5 °C) deoxygenated water in a Colworth Stomacher (Seward, London) for 60 s. The homogenate is centrifuged (4500 **g**) for 15 min and stored in an ice bath before analysis. NADH (0.4 mg) is dissolved in 1 ml of triethanolamine buffer (pH 8.0) in a glass or plastic cuvette. A sample (0.1 ml) of the aqueous phase of the centrifuged meat homogenate is placed in the test cuvette. As the SO_2 concentration in the cuvette should be between 3 and 30 μg (Fig. 9); the sample solution may require dilution. Samples of distilled water (1.9 ml) and 0.01 ml of NADH-POD (activity = 3U) are added to the cuvette and the contents mixed. A reagent blank containing all the above as well as 0.1 ml distilled water in place of the 0.1 ml of sulphited sample is prepared also. The absorbance (A_1) at 340, 365 Hg or 334 Hg nm of the reagent blank and test cuvettes is read after incubation for 5 min at 25 °C against air or water. The reaction is started by addition of 0.05 ml of SO_2—OD (activity = 4 U; measured as oxygen consumption, 1 mol/min). The contents of the cuvettes are mixed, incubated for 30 min at 25 °C and the absorbance (A_2) noted. The absorbance differences ($A_1 - A_2$) for the test sample (A_s) and the reagent blank (A_B) are calculated, the absorbance of the latter being subtracted from the former (A_S–A_B) to give the true difference (A).

According to the general formula for calculation of the concentration, the equation is:

$$c = \frac{V \times \text{mw}}{d \times v \times 1000} \times A \text{ (g/l)}$$

where V, final volume in cuvette (ml); v, test sample volume (ml); mw, molecular weight of substance to be assayed; d, light path (cm), and A, absorption coefficient of NADH at:

340 nm	= 6.3	1 mmol/cm
Hg 365 nm	= 3.4	1 mmol/cm
Hg 334 nm	= 6.18	1 mmol/cm

Thus for SO_2:

$$c = \frac{3.06 \times 64.06}{1 \times 0.1 \times 10\,000} \times A = 1.96 \times A\,[\text{g } SO_2/1]$$

If the sample was diluted during preparation, the result must be multiplied by the dilution factor (Fig. 9).

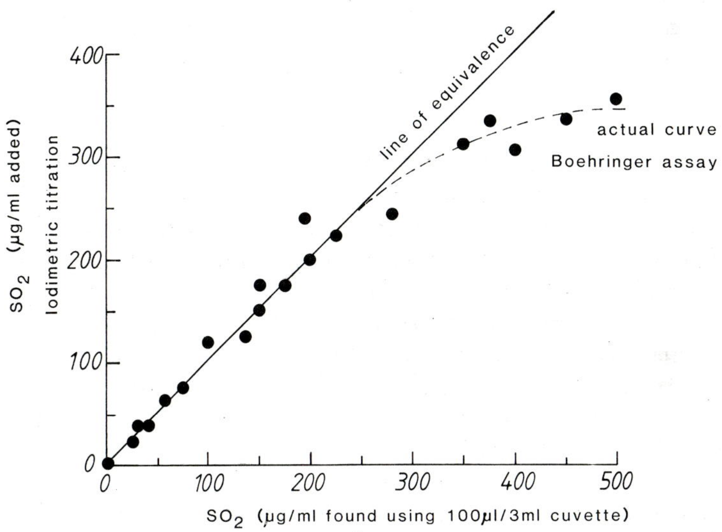

FIG. 9. Comparison of determination of total sulphur (IV) oxospecies by iodimetric titration and by the enzymatic method. It should be noted that there is excellent agreement between the two methods until the concentration of SO_2 in the test cuvette exceeds 300 µg/ml. As the manufacturers (Boehringer) recommend a dilution of the test sample above this concentration, no deviation from the line of equivalence should be detected in routine assay.

Standard sulphite solutions

Sulphite solutions, which are made up from potassium metabisulphite, should contain ethylenediaminetetraacetic acid (EDTA, 1×10^{-4} mol/l) to prevent oxidation. The solutions are standardized iodimetrically by volumetric analysis.

Determination of Sulphite Binding Compounds

*Agar plate method**

The ability of metabolizing micro-organisms, or solutions of chemicals, to sequester free sulphite can be tested with a nutrient agar medium containing a Schiffs base indicator decolourized by sulphite. The medium contains (g/l): glucose, 10; mycological peptone, 10; sodium sulphite, 2.5; Lab M agar, 12, and 0.4 g of basic fuchsin in 4 ml ethanol. The complete medium is sterilized at 121 °C for 15 min, cooled to 60 °C and poured (20 ml) into 9 cm, single-vented Petri dishes (Sterilin). When set, the Petri dishes are wrapped in aluminium foil to avoid photodecomposition of ingredients in the nutrient agar and stored at 5 °C. The medium is dried (37 °C, 60 min) before inoculation. A sample (0.01 ml) of a suspension of micro-organisms in broth or in a meat homogenate is spread over the dried surface of the medium. The Petri dishes are inverted, wrapped in aluminium foil and incubated aerobically until colonies visible to the naked eye are formed. Sulphite binding is manifested by an intense red zone around the colonies; the uninoculated medium has a pale pink hue. Alternative methods for plate inoculation may be used. These include streaking and multipoint inoculation. The indicator base of sodium sulphite and basic fuchsin may also be incorporated directly into selective isolation media.

Acetaldehyde determinations

Enzymatic method

A sample (1 g) of sulphited meat or meat product is deproteinized in perchloric acid (70% w/v, 9 ml) at 5 °C for 15 min. The samples are tested immediately or stored at −20 °C before assay. In the latter case the samples are thawed and then mixed and centrifuged (3000 g, 10 min.) as in the manner used to prepare a freshly taken sample. The supernatant (6 ml) is added to potassium phosphate (0.7 mol/l, 3 ml) at 5 °C. The pH is adjusted to 7.5 with sodium hydroxide (2 mol/l) and centrifuged (3000 **g**, 10 min.). The supernatant is stored in a screw-capped bottle at 5 °C.

The concentration of acetaldehyde is proportional to the change in the extinction at 340 nm arising from NADH oxidation (8):

$$CH_3CHO + NADH + H^+ \xrightarrow{\text{Alcohol dehydrogenase}} C_2H_5OH + NAD^+ \qquad (8)$$

A sample (2 ml) of deproteinized supernatant is placed in a glass or plastic cuvette. A freshly prepared solution (0.2 ml) of β-reduced nicotinamide adenine dinucleotide (1.1 mmol/l) in sodium bicarbonate (1% w/v) is added. The cuvette is

* The authors wish to thank Mrs V Dillon for details of the agar plate method used to demonstrate Sulphite binding.

sealed with Parafilm tape (Gallenkamp), the contents mixed and the absorbance (ΔA_1) measured at 340 nm against air. The reaction is started by addition of yeast alcohol dehydrogenase (0.02 ml) in ammonium sulphite (8.2 mol/l). The cuvette is re-sealed, mixed and incubated for 15 min. at 25 °C and the absorbance (ΔA_2) noted.

The concentration of acetaldehyde (c) as μg/ml is calculated from the equation (9):

$$c = \Delta A_2 - \Delta A_1 \times \text{dilution factor} \times 7.86 \qquad (9)$$

Steam distillation

A sample (20 ml) of culture supernatant is heated for 60 min in a Markham steam-distillation apparatus inclined at an angle of 45 ° to the horizontal. Acetaldehyde is trapped after passage down a vertical water condenser in 10 ml of a phosphate (pH 7.0) buffered sodium bisulphite (0.5%, w/v) /EDTA solution. The distillate is acidified with hydrochloric acid (2 mol/l) and excess bisulphite removed by titration with 0.1 and 0.01 *N* iodine to a pale blue end point using starch (1% w/v) as indicator. The bisulphite bound to acetaldehyde is freed by addition of sodium bicarbonate (5 g) and its concentration determined by iodimetric titration as before.

References

ABEL, E. 1913 Inaktivierung eines nicht-Kolloiden, anorganischen katalysators durch temperatursteigerung. *Monatsh* **34**, 1361–1391.

BAILEY, P. L. & RILEY, M. 1975 Performance characteristics of gas-sensing membrane probes. *Analyst* **100**, 145–156.

BANKS, J. G. 1983 *Sulphite preservation of British fresh sausage.* PhD Thesis, University of Bath, UK.

BANKS, J. G. & BOARD, R. G. 1981 Sulphite—the elective agent for the microbial associations in British fresh sausages. *Journal of Applied Bacteriology* **51**, ix.

BANKS, J. G. & BOARD, R. G. 1982*a* Sulphite-inhibition of Enterobacteriacaeae—including *Salmonella*—in British fresh sausages and in culture systems. *Journal of Food Protection* **45**, 1292–1297.

BANKS, J. G. & BOARD, R. G. 1982*b* Comparison of methods for the determination of free and bound sulphur dioxide in stored British fresh sausages. *Journal of the Science of Food and Agriculture* **33**, 197–203.

BANKS, J. G. & BOARD, R. G. 1983 The incidence and level of contamination of British fresh sausages and ingredients with salmonellas. *Journal of Hygiene* **90**, 213–223.

BROWN, M. H. 1977 *The microbiology of the British fresh sausage.* PhD Thesis, University of Bath, UK.

BURROUGHS, L. F. & SPARKS, A. H. 1963 The determination of total SO_2 in cider. *Analyst* **88**, 304–309.

CECIL, R. & MCPHEE, J. R. 1954 The estimation of thiols and disulphides by potentiometric titration with silver nitrate. *Biochemical Journal* **59**, 234–240.

CHRISTIAN, J. H. B. 1963 The preservation of minced beef with sulphur dioxide. *CSIRO Food Preservation Quarterly* **23**, 30–36.

DOWDELL, M. J. & BOARD, R. G. 1968 A microbiological survey of British fresh sausages. *Journal of Applied Bacteriology* **31**, 378–396.

DOWDELL, M. J. & BOARD, R. G. 1971 The microbial associations in British fresh sausages. *Journal of Applied Bacteriology* **34**, 317–337.

DYETT, E. J. & SHELLEY, D. 1962 Microbiology of British fresh sausage. *Proceedings of the 1st International Congress of Food Science and Technology* p. 54.

DYETT, E. J. & SHELLEY, D. 1965 Microbiology of the British fresh sausage. *Proceedings of the 1st International Congress of Food Science and Technology* **2**, 393.

DYETT, E. J. & SHELLEY, D. 1966 Effects of sulphite preservation in British fresh sausage. *Journal of Applied Bacteriology* **29**, 439–446.

FOURNAUD, J., JACQUET, B. & POUJARDIEU, B. 1971 Development of microflora of sausage meat depending on preservatives used and method of storage. *Industrielle Alimentation et Agriculture* **88**, 15–26.

FRIDOVITCH, I. & HANDLER, P. 1961 Detection of free radicals generated during enzyme oxidations by the initiation of sulphite oxidation. *Journal of Biological Chemistry* **236**, 1836–1840.

GARDNER, G. A. 1968 Effects of pasteurization or added sulphite on the microbiology of stored vacuum packed baconburgers. *Journal of Applied Bacteriology* **31**, 462–478.

GARDNER, G. A. 1981 *Brochothrix thermosphacta* (*Microbacterium thermosphactum*) in the spoilage of meats: A review. In *Psychrotrophic micro-organisms in spoilage and pathogenicity*, ed. Roberts, T. A., Hobbs, G., Christian, J. H. B. & Skovgaard, N., pp. 139–173. London: Academic Press.

GRASSETTI, D. R. & MURRAY, J. F. 1967 Determination of sulfhydral groups with 2,2′ or 4,4′ dithiopyridine. *Archives of Biochemistry and Biophysics* **119**, 41–49.

HAMMOND, S. M. & CARR, J. G. 1976 The antimicrobial activity of sulphur dioxide—with particular reference to fermented and non-fermented fruit juices. In *Inhibition and inactivation of vegetative microbes*, ed. Skinner, F. A. & Hugo, W. B., pp 86–110. London: Academic Press.

HUMPHREY, R. E., WARD, M. H. & HINZE, W. L. 1970 Spectrophotometric determination of sulphite with 4,4′ dithiopyridine and 5,5′ dithiobis-(2-nitrobenzoic acid). *Analytical Chemistry* **42**, 698–702.

HURST, J. 1972 *The selection of the microbial association of British fresh sausage.* MSc Thesis. University of Bath, UK.

INGRAM, M. 1948 Germicidal effects of free and combined sulphur dioxide. *Journal of the Society of Chemical Industry* **67**, 18–21.

INGRAM, M. & VAS, K. 1950*a* Combination of sulphur dioxide with concentrated orange juice I. Equilibrium states. *Journal of the Science of Food and Agriculture* **1**, 21–27.

INGRAM, M. & VAS, K. 1950*b* Combination of sulphur dioxide with concentrated orange juice. II. The rate of the reaction. *Journal of the Science of Food and Agriculture* **1**, 63–66.

JOSLYN, M. A. & BRAVERMAN, J. B. S. 1954 The chemistry and technology of the pretreatment and preservation of fruit and vegetable products with sulphur dioxide and sulfites. *Advances in Food Research* **5**, 97–160.

KING, A. D., PONTING, J. D., FANSHUCK, D. W., JACKSON, R. & MIHARA, K. 1981 Factors affecting the death of yeasts by sulphur dioxide. *Journal of Food Protection* **44**, 92–97.

KROL, B. & MOERMAN, P. C. 1959/60 Effect of sulphite on the colour and bacteriological condition of mince. *Conserva* **8**, 1–11.

LAFONTAINE, A., GOBLET, J. & DE MAYER-CLEEMPOEL, S. 1955 Action bacteriologique des sulfites sur les viandes. *Archives of the Belge Medical Society* **13**, 288–292.

MOERMAN, P. C., KROL, G., KAMPELMACHER, E. H. & VAN NOORLE JANSEN, L. M. 1966 Invloed van natriumsulfiet op de groei van salmonellae en *E. coli* in rauw gehakt. *Conserva* **14**, 172–175.

MONIER-WILLIAMS, G. W. 1927 Determination of sulphur dioxide in foods. *Reports of Public Health and Medicinal Subjects. Ministry of Health, London* **43**, 1–56.

MULDER, S. J. 1969 Bewaren van gehakt bij lage temperaturen. *Koeltechniek* **62**, 228–230.

NEUBERG, C. 1929 Uber das verhalten des glucose-schwefligsauren natriums zu hefe und damit zugammenhangende fragen. *Biochemica Zeitschrift* **212**, 477–517.

REHM, H. J. 1964 The antimicrobial action of sulphurous acid. In *Microbial Inhibitors in Foods*, ed. Molin, N. Uppsala: Almquist and Wiksells.

REHM, H. J. & WITTMAN, H. 1962 Beitrag zur kenntis der antimikrobiellen wirkung der schwefligen saure. I. Ubersicht uber einflussnehmende faktoren auf die antimicrobielle wirkung der schwefligen saure. *Zeitschrift fur Lebensmittel Untersuchung und Forschung* **118**, 413–425.

ROBERTS, D., BOAG, K., HALL, M. L. M. & SHIPP, C. R. 1975 The isolation of salmonellas from British fresh sausages and sausage meat. *Journal of Hygiene* **75**, 173–184.

SHIPTON, J. 1959 Estimation of sulphur dioxide in foods. *Food Preservation Quarterly* **14**, 54.

TOMPKIN, R. B., CHRISTIANSEN, L. N. & SHAPARIS, A. B. 1980 Antibotulinal efficacy of sulphur dioxide in meat. *Applied and Environmental Microbiology* **39**, 1096–1099.

TYSON, C. B. 1976 *A study of sulphite, sorbate and polyphosphate as preservatives of pork sausage meat.* BSc Project Report 302, University of Bath, UK.

VAS, K. & INGRAM, M. 1949 Preservation of fruit juices with less sulphur dioxide. *Food Manufacture* **24**, 414–416.

WEDZICHA, N. L. & BINDRA, N. J. 1980 A rapid method for the determination of sulphur dioxide in dehydrated vegetables. *Journal of the Science of Food and Agriculture* **31**, 286–288.

The Use of Carbon Dioxide for the Preservation of Milk

J. S. KING AND L. A. MABBITT

AFRC Institute of Food Research, Reading Laboratory, Shinfield, Reading, U.K.

In the UK the widespread use of farm bulk tanks which cool milk to <4 °C before collection, together with insulated road tankers and subsequent storage in insulated silos has, in general, eliminated spoilage of stored milk by lactic acid bacteria. Under these conditions, however, the microflora slowly becomes dominated by psychrotrophs so that bacterial growth often reaches unacceptable levels after only 3 d storage. Subsequent pasteurization of the milk, while destroying heat-labile microorganisms, does not destroy the heat-stable proteolytic and lipolytic enzymes produced by the psychrotrophic bacteria. These enzymes may subsequently adversely affect the quality of milk products such as cream, ice cream mix, butter and cheese.

Any system which reduces the growth of psychrotrophs in refrigerated stored milk could be useful to the dairy industry. It would permit longer storage times, thereby giving greater flexibility of working, and smooth the seasonal surges in milk production. The only acceptable methods currently available for delaying the onset of spoilage by psychrotrophs are thermization of milk on arrival at the dairy, or storing and transporting milk at lower temperatures. The addition of carbon dioxide (CO_2) offers a possible alternative.

The inhibitory properties of CO_2 have been exploited for the preservation of a wide range of foods including fruit, vegetables, processed meats and bacon, fresh beef, pork, poultry, fish and eggs (Clark & Takacs 1980). More recently CO_2 has been used also to extend the storage life of milk products such as cheese, yoghurt quarq (Corinth 1982) and liquid milk (Shipe *et al.* 1982; King & Mabbitt 1982).

Preservation of milk with CO_2 was first proposed in the 1920s. It is unfortunate that the early investigations by Valley & Rettger (1927) were confined to attempts to control spoilage by *Streptococcus lactis* in milk at ambient temperature (20–23 °C). This initiative could not be expected to be successful. Since then no interest has been shown in milk preservation with CO_2 although the inhibitory effects of this gas on Gram-negative bacteria have been known for a long time (Haines 1933; Coyne 1933).

Preservatives in the Food,
Pharmaceutical and Environmental
Industries

0–632–01727–9

After preliminary experiments showed that CO_2 dissolved in refrigerated raw milk retarded the rate of growth of *Pseudomonas* spp. for 2–3 d, further experiments were made to examine the effect of various concentrations of dissolved CO_2 on spoilage of milk stored at temperatures representative of those used in practice, and to evaluate its possible usefulness to the dairy industry.

Materials and Methods

Determination of the CO_2 content of milk

Two methods of determining the CO_2 content of milk were used. For accurate determinations the concentration of CO_2 was measured by gas chromatography analysis.

Samples (5 ml) of milk were transferred to rubber sealed, evacuated 20 ml vials ('Hypo-Vials', Pierce and Warriner (UK) Ltd., Chester) by hypodermic syringe and 1 ml H_2SO_4 (0.5) M injected. After shaking for 2 min to release the dissolved gas, the CO_2 concentration in the headspace was measured by gas chromatography using silica gel (80–100 mesh) in glass columns (id 6 mm, length 0.45 m) at 100 °C, with a Unicam 104 (Pye Unicam Ltd, Cambridge) and a katharometer detector (bridge current 90 mA). Nitrogen (50 cm^3/min) was used as a carrier gas. A standard curve, which was prepared with solutions of $NaHCO_3$ of known concentrations, was used to determine the CO_2 content of milk samples.

A faster but less accurate estimate was made by measuring the pH of milk. The CO_2 concentration was determined from a graph showing the relationship between pH and dissolved CO_2 concentration as determined by gas chromatography analysis (Fig. 1). This method was found to be particularly useful for determining the progress of carbonation of the large volumes of milk used for cheese making.

Bacterial counting methods

The total count (TC) and psychrotroph count (PC) were determined on milk agar after 3 d at 30 °C and 10 d at 6 °C respectively. The methods of Law *et al.* (1979) were used.

Milk

All milks were mixtures of evening and morning milk collected aseptically from farm tanks or from milk tankers on arrival at the dairy. Samples were stored at 4 °C until used, which was normally within 6 h of collection.

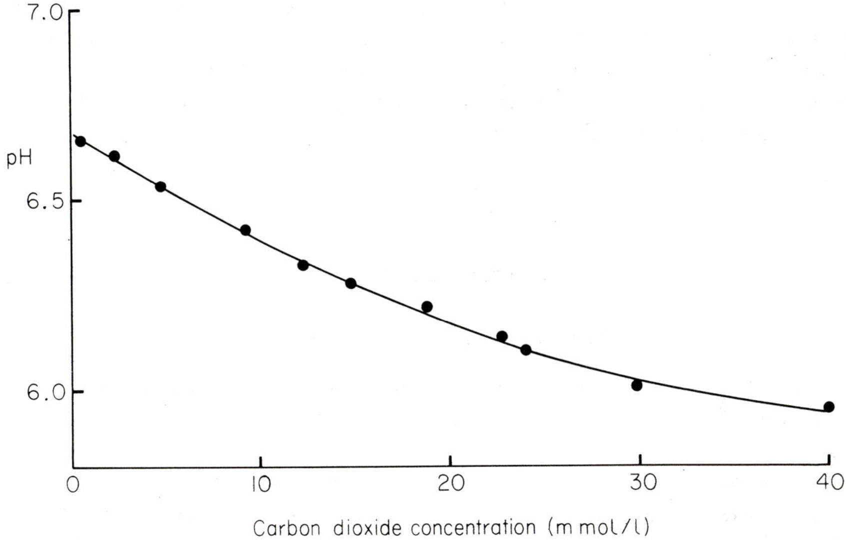

FIG. 1. The relationship between dissolved CO_2 concentration and pH of raw milk at 4 °C.

Stock cultures

These were maintained on Nutrient agar slopes and cultured in Nutrient Broth (Oxoid No. 2) overnight at 22 °C.

Buffered Nutrient Broth

This was prepared by aseptically adding sodium phosphate buffer to Nutrient Broth to a final concentration of 0.025 M (pH 6.6). It gave a buffering capacity similar to that of fresh milk. This broth was used in experiments in which the rate of growth was determined by changes in opacity.

Incorporation of CO_2 into milk and broth

In the laboratory CO_2 was incorporated into milk by vigorous stirring in an atmosphere of CO_2 or by shaking 100 ml of milk in sealed conical flasks in which the headspace (approx. 90 ml) was filled with CO_2 at atmospheric pressure. After CO_2 determination by gas chromatography, the CO_2 concentration was adjusted when necessary by further treatment or by dilution with untreated milk. Large volumes of milk (450 l) were carbonated with a 'Ytron' mixer pump (Mason & Morton Ltd., Hemel Hempstead, Herts.) which had a gas-injection nozzle in the mixer head. The apparatus enabled 450 l milk to be carbonated to 30 mmol/l CO_2 in approximately 40 min.

Incubation

Milks under test were kept in constant temperature (± 0.5 °C) water baths at 4, 7 or 10 °C.

Results

The maximum concentration of CO_2 used in these experiments was 30 mmol/l. Preliminary experiments showed that the acidity associated with concentrations in excess of 30 mmol/l caused instability of the milk protein and 'bittiness'.

The effect of CO_2 concentration and storage temperature

Figure 2 shows the effects of CO_2 concentration and storage temperature on the growth rate of a natural flora of psychrotrophs in bulk milk. The effects on milk from two tankers are shown. One contained good quality (low count) milk produced under hygienic conditions, and the other poorer quality milk. For both milks the inhibitory effects were greatest at the highest concentration of CO_2 and at the lowest temperature. In all cases the degree of inhibition was dependent on the initial numbers of psychrotrophic bacteria in the milk. As the temperature of ex-farm milk rises during transport and storage at the creamery, the results obtained at 7 °C are probably the most relevant to consider in a commercial context. At this temperature a concentration of 30 mmol/l CO_2 increased the time taken for the TC to reach 10^6 cfu/ml by 1.2 d for the poor milk and by >3 d for the better quality milk. This level is considered by many to be the safe upper limit for processing milk.

Farm bulk milk

In order to verify the findings of the first experiment the effects of 20–30 mmol/l CO_2 on the TC of 15 samples of ex-farm milk were examined after storage at 7 °C for 4 d (Fig. 3). Each sample was treated by shaking in a sealed conical flask with CO_2. The final concentration of CO_2 varied between 19 and 28.5 mmol/l.

The psychrotrophic component of the total flora in the 15 samples of milk varied from $<1\%$ to $>90\%$. Consequently there was only a poor correlation between the initial and final TC of the stored milks ($r = 0.34$ and $r = 0.7$ for the untreated and CO_2-treated milks respectively). The correlation coefficients for the initial PC and the final TC of the same samples were $r = 0.89$ and $r = 0.80$. The geometric mean increase in cfu/ml during 4 d storage for CO_2-treated milk samples

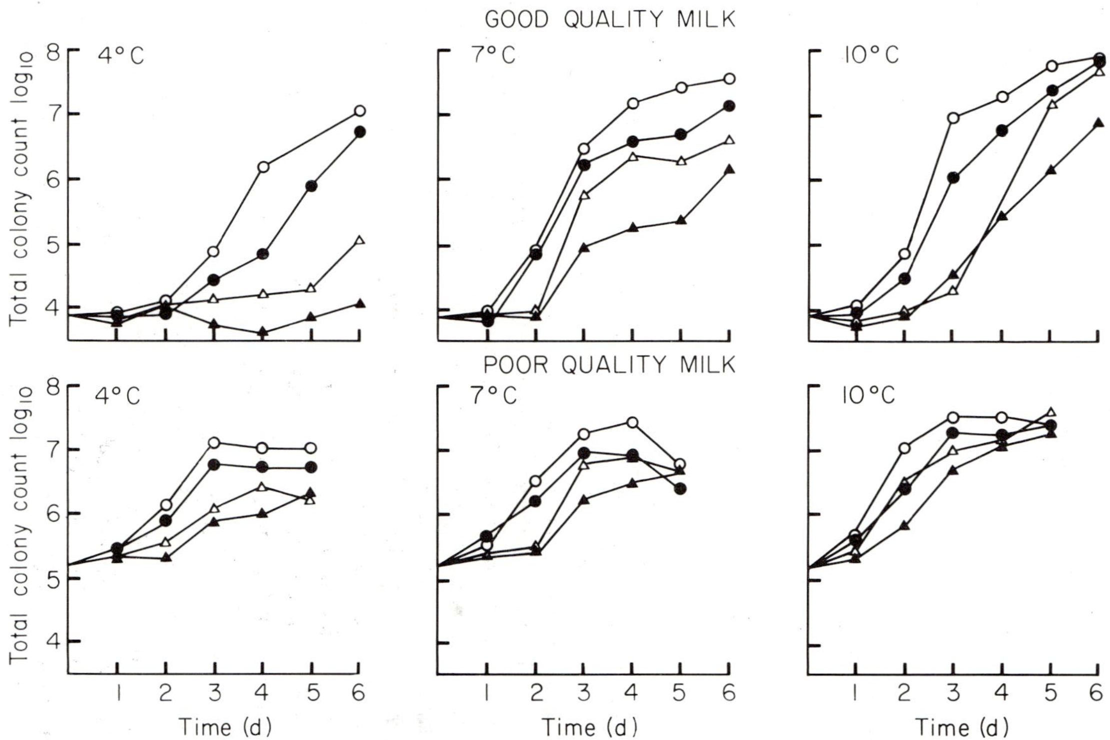

FIG. 2. Effect of CO_2 and storage temperature on the growth of psychrotrophs in good and poor quality milk. ○, untreated milk; ●, 10 mmol/l CO_2; △, 20 mmol/l CO_2; ▲, 30 mmol/l CO_2.

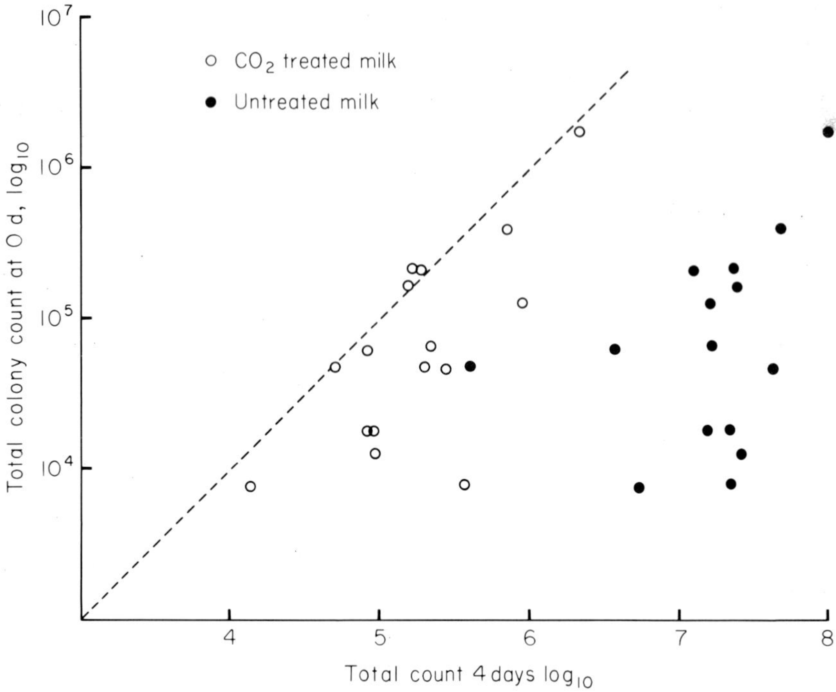

FIG. 3. The effect of CO_2 on the multiplication of bacteria in raw milk stored at 7 °C. ○, CO_2 treated milk; ●, untreated milk.

was 47.8×10^3 cfu/ml; for untreated milk the increase was 10.75×10^6 cfu/ml during the same period.

With the exception of one sample which had a TC of 1.7×10^6 cfu/ml at the start of the experiment, the TC of all CO_2-treated milks remained below 10^6 cfu/ml for 4 d while the TC of 12/15 untreated samples rose to $> 10^7$ cfu/ml.

The effects of CO_2 on cheese starter bacteria

It is known that *Strep. cremoris* is particularly resistant to the effects of CO_2 when grown in liquid culture (Enfors & Molin 1980). It has been observed also that in gas-packed meat, CO_2 causes microbial selection. This results in a predominance of lactic acid bacteria. These organisms are considerably less 'damaging' to meat than are *Pseudomonas* spp.

The indications were therefore that it should be possible to use CO_2-treated milk for cheese making without affecting the growth of the starter cultures and therefore without need to remove the gas. The rates of multiplication and acid production of a selection of cheese starter cultures in pasteurized milk were measured in the presence and absence of CO_2, 30 mmol/l. Representative results obtained with *Strep. cremoris*, *Strep. lactis* and *Strep. lactis* var. *diacetilactis* are shown in Fig. 4. No significant difference was found between the growth rates of these bacteria in the treated and untreated milks. These results suggest that cheese making is unlikely to be perturbed by interactions between the starter bacteria and CO_2. The gas might have undesirable effects on curd formation.

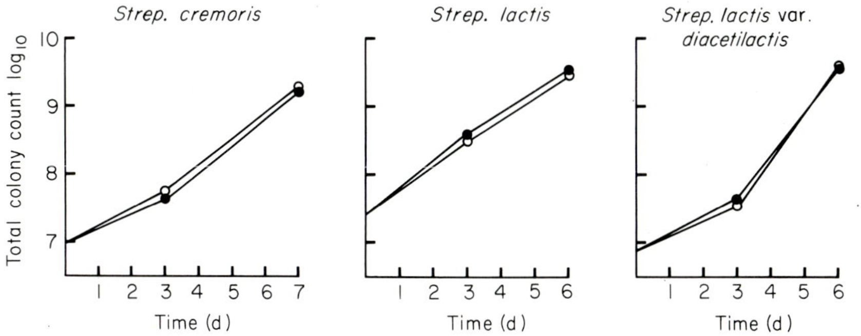

FIG. 4. The effect of CO_2 (30 mmol/l) on the growth rate of three cheese starter streptococci in milk at 30 °C. ○, untreated milk; ●, 30 mmol/l CO_2.

The use of CO_2 treated milk in milk products

Cheese

CO_2-treated milk has been used to make Cheddar cheese; the established procedures of the National Institute for Research in Dairying Experimental Dairy (Chapman & Burnett 1972) were used. In a typical experiment milk containing 30 mmol/l CO_2 was stored for 3 d at 4 °C before HTST pasteurization. Cheese made from this milk was compared with a control cheese made from fresh milk from the same farm.

The dissolved CO_2 in the experimental milk enhanced the normal development of acidity due to starter activity. Consequently clot formation was accelerated and only half of the normal quantity of rennet was required to produce a satisfactory curd. In the 100 l vats used in these experiments, the combination of agitation and warming of the milk to 39 °C resulted in the gradual loss of CO_2 to the atmosphere. Indeed by the time the curd and whey were separated (approx. $3\frac{1}{2}$ h after adding the starter), no CO_2 was detectable in the whey by gas chromatography analysis and the titratable acidity of the whey was the same in both vats. The cheese making process then proceeded normally. The cheeses were vacuum packed and, after maturation for 6 months, submitted to a trained taste panel. No significant differences between the treatments in either flavour or texture were noted.

Yoghurt

Yoghurt made from skim milk containing 29 mmol/l CO_2 was found to contain only a trace (<5 mmol/l) of CO_2 after incubation at 42 °C for 5 h. Although the pH of the carbonated milk at the time of adding the starter culture was 6.1, while that of the untreated milk was 6.5, the acidities of the final products were very similar, pH 4.2 and 4.15 respectively. The taste and consistency of the two batches of yoghurt were indistinguishable.

Cream

Fresh unpasteurized double cream (containing 40% fat) was warmed to 30 °C to reduce its viscosity before carbonation. This was done by stirring the cream in an atmosphere of CO_2 until the dissolved CO_2 level reached 30 mmol/l. Part of the treated cream was then diluted with fresh cream to give concentrations of 20 and 10 mmol/l CO_2.

The treated creams together with untreated controls were dispensed into 25 ml screw capped bottles for storage at 4, 7 or 10 °C ± 0.5 °C. One bottle of each treatment was removed daily for TC determinations.

Figure 5 shows that the effect of CO_2 on stored cream and stored milk (cf., Fig. 1) is comparable. Thus at 7 °C, 30 mmol/l CO_2 increased the time taken for the TC to reach 10^6 cfu/cm^3 from 2.5 to 4 d; at 10 °C the delay was limited to 1 d. As with stored raw milk the benefits of CO_2 increased with increased CO_2 concentration and were inversely related to storage temperature.

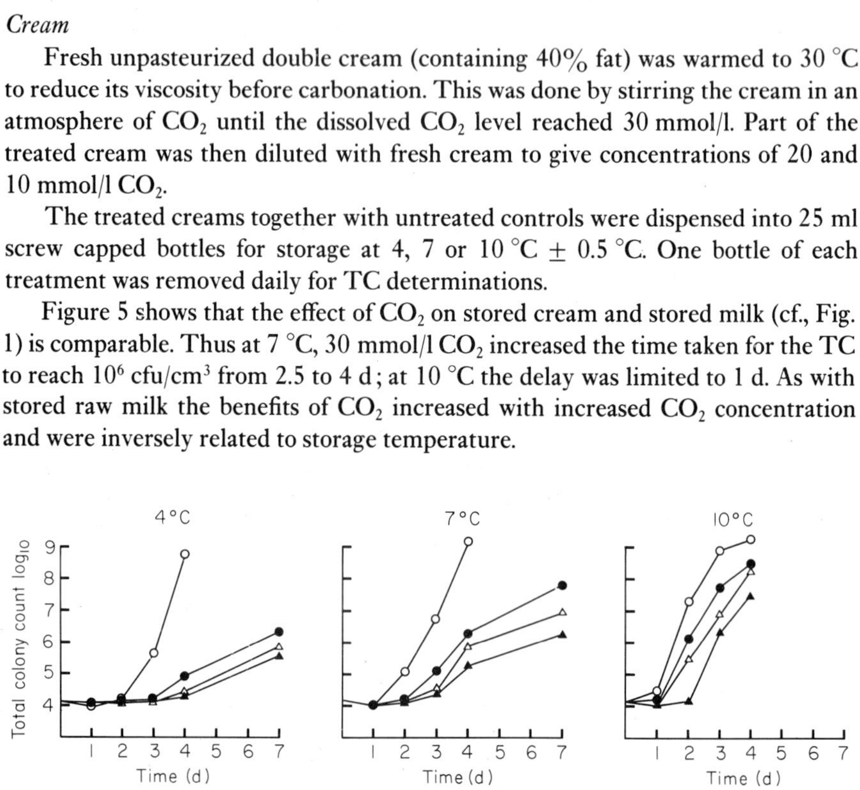

FIG. 5. The effect of CO_2 concentration and storage temperature on the growth of psychrotrophs in cream. ○, untreated; ●, 10 mmol/l CO_2; △, 20 mmol/l CO_2; ▲, 30 mmol/l CO_2.

Conclusions

These experiments have demonstrated that the addition of CO_2 to refrigerated milk can double the time for which it can be safely stored without spoilage by psychrotrophs. It has also been shown that such treatment has no adverse effect on milk products. Although the removal of the gas from manufacturing milk is not necessary, CO_2 can be removed easily should the milk be required for liquid consumption.

Carbon dioxide is completely safe and free from aesthetic, toxicological and practical objections to its use. It is a normal component of milk (approximately 5% v/v, as milk leaves the udder) and it is already consumed in large quantities in drinks ranging from beer and lemonade to kefir. It is also plentiful and inexpensive.

Whether or not CO_2 preservation is adopted by the dairy industry will depend ultimately on cost and convenience. Before a final decision is taken it will be necessary to compare these with the cost of the alternatives viz. refrigeration to lower temperatures, e.g. 1–2 °C has been shown to retard spoilage to a similar degree, or thermization.

References

CHAPMAN, H. R. & BURNETT, J. 1972 Seasonal changes in physical properties of milk for cheesemaking. *Dairy Industries* **37**, 207–211.

CLARK, D. S. & TAKACS, J. 1980 Gases as preservatives. In *Microbial ecology of foods Vol. 1. Factors affecting life and death of micro-organisms*, ed. Silliker, J. H. & Elliot, R. P. pp. 170–189. London: Academic Press.

CORINTH, H. G. 1982 Carbon dioxide. Its uses in the Dairy Industry. *Deutsche Molkerei Zeitung* **103**, 942–952.

COYNE, F. P. 1933 The effect of carbon dioxide on bacterial growth. *Proceedings of the Royal Society London Series B* **113**, 196–217.

ENFORS, S. O. & MOLIN, G. 1980 Effect of high concentrations of carbon dioxide on growth rate of *Pseudomonas fragi, Bacillus cereus* and *Streptococcus cremoris*. *Journal of Applied Bacteriology* **48**, 409–416.

HAINES, R. B. 1933 The influence of carbon dioxide on the rate of multiplication of certain bacteria as judged by viable counts. *Journal of the Society for Chemical Industry* **52**, 13T–17T.

KING, J. S. & MABBITT, L. A. 1982 Preservation of raw milk by the addition of carbon dioxide. *Journal of Dairy Research* **49**, 439–447.

LAW, B. A., COUSINS, C. M., SHARPE, M. E. & DAVIES, F. L. 1979 Psychrotrophs and their effect on milk and dairy products. In *Cold tolerant microbes in spoilage and the environment*, ed. Russell, A. D. & Fuller, R. Society for Applied Bacteriology Technical Series No. 13. pp. 137–152. London: Academic Press.

SHIPE, W. F., SENYK, G. F., ADLER, E. J. & LEDFORD, R. A. 1982 The effect of infusion of carbon dioxide on the bacterial growth in fluid milk. Proceedings of the 77th Annual Meeting of the American Dairy Science Association ADSA Annual Meeting and Divisional Abstract Supplement 1 *Journal of Dairy Science* **65**, 77.

VALLEY, G. & RETTGER, L. F. 1927 The influence of carbon dioxide on bacteria. *Journal of Bacteriology* **14**, 101–137.

Preservation of Milk Samples

J. M. Ewart

Ministry of Agriculture, Fisheries and Food (MAFF), Agricultural Development and Advisory Service (ADAS), Government Buildings, Lawnswood, Leeds, UK

Increased interest in the microbiological quality of raw milk has resulted in the examination of greater numbers of on-farm samples. There is an inevitable delay between sampling and testing, particularly when samples are transported long distances for centralized testing. Samples are usually refrigerated during transport in containers with ice packs but because of the great variability in the types of organisms present in raw milk, microbiological changes during refrigerated transport are difficult to predict. An alternative approach to sample preservation is to use chemical inhibitors which induce bacterial stasis. Such inhibitors should not interfere with subsequent testing.

Refrigerated Transport of Samples

Mechanical refrigeration is seldom used for sample transport although suitable equipment, capable of operating on a 12 V DC (road vehicle electrics) supply, is both inexpensive and readily available. The usual arrangement is a well insulated robust box that will accommodate sample tubes together with plastic containers of 1.5% (v/v) ethylene glycol which is frozen before use.

Cooled transport containers

A number of factors influence the performance of insulated containers: surface to volume ratio, insulation thickness and thermal resistance, temperature differential across the insulation, sample volume, and initial temperature and volume of frozen glycol. A good example of a practical design which accommodates these various criteria and meets a specific performance requirement is the Milk Marketing Board (MMB) central testing box (Fig. 1) which was developed by MMB, Technical

Preservatives in the Food,
Pharmaceutical and Environmental
Industries

0–632–01727–9

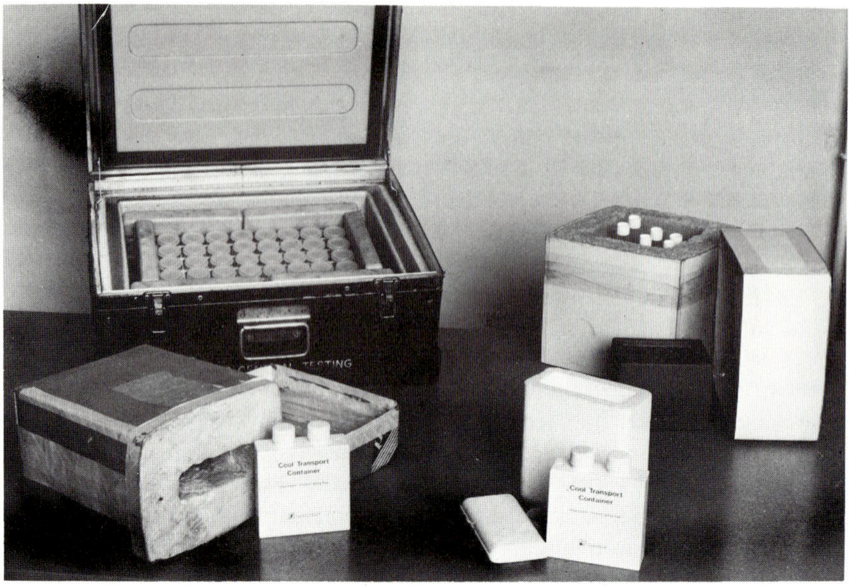

FIG. 1. Insulated containers for cooled transport of milk samples. Clockwise from top left: MMB 'Central Testing' box; MAFF 'Development' container for 6 samples; commercial postal pack for 2 samples; commercial pack with modified (50 mm) insulation.

Division, Thames Ditton, Surrey, KT7. This container, which is 44 × 37 × 23 cm, holds up to 40 samples of 25–30 ml, in plastic pots, together with eight ice packs of approximately 450 ml volume. Polyurethane insulation (4 cm thick) is used. Samples are maintained below 2 °C for up to 36 h and below 4 °C for up to 48 h at summer ambient temperatures within these containers.

Containers for small numbers of samples

A number of insulated packs for the transport of small volumes and small numbers of samples are available commercially. These are intended mainly for transmission by post; the consequent constraint on size may compromise insulation thickness. It is also possible to make containers to individual requirements. Polyurethane foam is the ideal insulator for sample containers, because it has a number of advantages over other materials, especially its superior insulating properties (Table 1).

Polyurethane foam is rigid and does not require mechanical support as do glass fibre and granular materials. It does, however, require some physical protection against damage during handling. Containers formed in polyurethane may be readily constructed to any reasonable size and wall thickness using two-part resins (CGT Glassfibre, Unit B3, Cross Green Way, Cross Green Industrial Estate, Leeds 9).

TABLE 1. *Characteristics of various insulating materials*

Type	Thermal conductivity (k) (W/m °C)	Properties
Polyurethane	0.023	rigid, formable, self supporting
Glassfibre	0.037	fibrous, non-rigid
Polystyrene	0.037	rigid, preformed, self-supporting
Cork	0.042	semirigid, preformed (sheet)
Perlite	0.042	granular, non rigid
Vermiculite	0.065	granular, non rigid

These are mixed and poured into a cardboard or hardboard mould. A silicone mould-release agent is necessary unless it is intended that the foam should bond, as it will do naturally, with the container wall. As heat is produced during foaming, care should be taken in selecting materials for preparation of the resins and the moulds. Polyurethane foam has a density of 0.016 g/ml and approximately 95% closed-cell content. Containers (Fig. 1) with various specifications (Table 2) have been evaluated by the Agriculture Development and Advisory Service (ADAS) and their performances are shown in Fig. 2. The effects of increased insulation thickness and surface-to-volume ratio can be seen from these data and should be considered when designing insulated packs for specific applications.

Microbial Changes During Cooled Transport

The microbial flora of milk samples is subjected to a number of influences during transport in refrigerated containers. Movement may break up clumps of organisms,

TABLE 2. *Specifications of insulated transport containers*

Description	Sample capacity (ml)	Insulation/ thickness (mm)	Volume of 1.5% (v/v) glycol (ml)	Overall dimensions (cm)
Commercial*	2 × 12	Polystyrene 15	250	14 × 7 × 17
Modified Commercial	2 × 12	Polyurethane 50	250	20 × 16 × 28
Development†	6 × 12	Polyurethane 50	600	24 × 22 × 28

*'Cool transport container': Walter Sarstedt (UK) Ltd, 66 Boston Road, Beaumont Leys, Leicester, LE4 1AW, UK.

†Based on 'Kryorack 6': Alpha Laboratories Ltd, 40 Parham Drive, Eastleigh, Hants, SO5 4UU, UK.

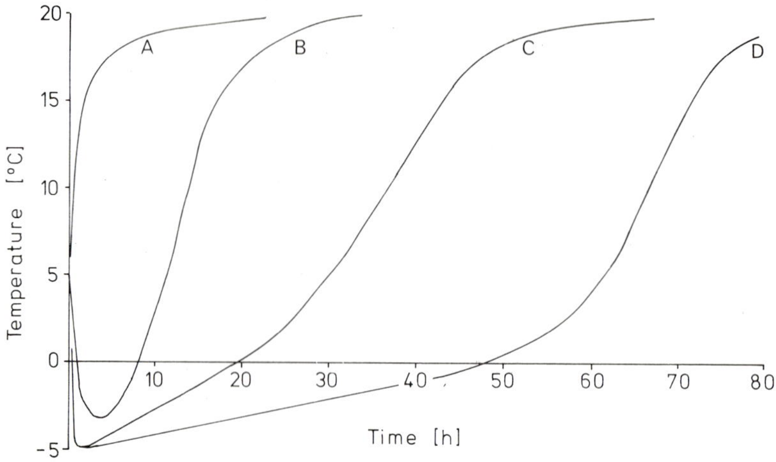

FIG. 2. Temperature rise in 10 ml milk samples stored in various types of insulated container at 20 °C. A, control, no insulation; B, commercial pack for 2 samples; C, commercial pack with modified (50 mm) insulation; D, MAFF 'development' container for 6 samples.

temporary freezing of samples can occur—especially in smaller packs (Fig. 2), the number of some organisms will increase, while that of others may fall. Proportions of different types of organisms in the milk will vary. For example if the main source of bacteria is infected udders, Gram-positive types may predominate whereas organisms arising from milking equipment may be thermoduric or psychrotrophic.

Microbiology of cooled milk

Psychrotrophic organisms in milk, of which some 60–90% are identified with the genus *Pseudomonas* (Kikuchi & Matsui 1974), are of greatest importance during cooled transport and storage. Bloquel & Veillet-Poncet (1980) found that at 4 °C for 4 d, mesophilic counts increased 1.5 times whereas psychrotrophic counts increased ten-fold. The nature of the microflora also changed during storage; 82% of the initial mesophilic flora and 41% of the psychrotrophic organisms were Gram-positive. After storage the percentages were 57 and 12 respectively. The proportion of fluorescent pseudomonads also increased during storage, as did numbers of members of the Enterobacteriaceae.

Our investigation of cooled milk during storage has indicated that changes may be more complex than indicated above. Samples of raw milk were counted by dilution plating when received at the laboratory on the day of sampling, and after 24 and 48 h storage at 4 °C. Regression analysis gave the following equation ($n = 48$):

$$\text{At 24 h } xy = 0.70 + 0.84\, x \ (r = 0.86)$$
$$\text{At 48 h } xy = 0.25 + 1.01\, x \ (r = 0.81)$$
(x = log count before storage, y = log count after storage)

The regression line describing the 24 h results had a slope < 1.0 and a positive intercept, suggesting that bacterial numbers underwent a net decrease (death exceeded growth) when the initial count was high. At low initial counts the growth of some organisms was greater than death of others. After 48 h a net increase in count was observed at all values of initial count (slope = 1.01), but further examination of the 48 h results indicates that variability in change in plate count was still present.

On the basis that an individual plate count has a typical plating and counting error of 15%, the results were arranged into groups accordingly (±30%), and the figures in Table 3 were obtained. Sixteen of the 48 samples examined were within 30% of the initial count. This group had the lowest average initial count and a mixed flora with a predominance of neither Gram-positive (mainly *Staphylococcus* and *Micrococcus* spp.) nor Gram-negative (mainly *Pseudomonas* spp.).

In samples where plate counts rose or fell by more than 30% the predominant floras were pseudomonads and Gram-positive types respectively. Large numbers of psychotrophs in a milk sample are likely to result in an increase in plate count during cool storage. As noted by Kikuchi & Matsúi (1974) these are mainly pseudomonads. Similarly, some of the Gram-positive species may be associated with mastitis. The fall in their numbers will be due in part to environmental factors and, in addition, the presence of natural antibacterial activity in raw milk, such as the lactoperoxidase system (Kneifel & Foissy 1982; Björck 1982). In small transport

TABLE 3. *Bacterial flora of milk after storage at 4 °C for 48 h, grouped according to change in plate count*

Change in plate count	Number of samples	Mean initial count	Mean final count	Predominant flora on representative plates (60 isolates per group) (%)
30% Increase	27	4.7×10^4	2.8×10^5	Fluorescent pseudomonads, 44 Non-fluorescent pseudomonads, 41 Gram-positive, 15
Within ±30%	16	2.6×10^4	2.5×10^4	Fluorescent pseudomonads, 4 Non-fluorescent pseudomonads, 33 Gram-positive, 53 Enterobacteriacae, 10
30% fall	5	9.7×10^4	6.1×10^4	Non-fluorescent pseudomonads, 5 Gram-positive, 95

packs, where freezing and slow thawing can take place, as much as 87% of some species will be destroyed (Gebre–Eyziabher *et al.* 1982). These apparent deficiencies of cooled storage/transport should not be exaggerated however. For general requirements of a hygiene test, valid results will be obtained for the majority of milk samples, where the count is less than 50 000/ml.

Chemical Preservation

A number of bacteriostatic chemicals were tested for milk preservation by Pettipher & Rodrigues (1982). Formaldehyde (0.2% v/v) and Bronopol (0.5 mg/ml) permitted the growth of yeasts and moulds whilst a mixture of boric acid, glycerol and sodium propionate (Heeschen *et al.* 1969; von Grove & Terplan 1978) inhibited bacteria and moulds, but permitted the growth of yeasts. A combination of potassium sorbate and nystatin was found by Pettipher & Rodrigues (1982) to give adequate control of both yeasts and moulds, and resulted in minimal change in bacterial count during storage at room temperature (20–22 °C). The bacteriostatic effect of this preservative combination is due mainly to the inclusion of boric acid, although the mechanism of its action is unclear. It is surmised that boric acid is antagonistic to vitamin and phosphate metabolism, and to the activities of some dehydrogenases (Heeschen *et al.* 1969; Lueck 1980). Final concentrations in the milk sample of the mixture used by Pettipher & Rodrigues (1982) were as follows (mg/ml): boric acid, 8.3; glycerol, 1.7; potassium sorbate, 0.12; and nystatin (Sigma), 40 units/ml. Pettipher (1983) reported a good correlation ($r = 0.80$) between colony counts before storage and counts by the Direct Epifluorescent Filter Technique (DEFT) after 3 d at 20–22 °C. The slope of regression was similar to that for DEFT and colony counts without storage. Similar results were obtained by Dutton (personal communication). The preservative mixture has been found to have no effect on somatic cell counts and automatic fat and lactose determinations; protein determinations, however, suffer interference from boric acid (Pettipher & Rodrigues 1982). Tests for antibiotic residues, based on bacterial growth, could not be done unless the preservative was neutralized.

Use of Preservatives

It might be expected that samples containing a bacteriostatic preservative would be unsuitable for bacterial counting by dilution plating methods. As the preservative is not lethal if it is diluted below active concentration, it should not interfere with bacterial growth. When samples are plated, the overall dilution can be ten times the nominal dilution as a result of the addition of agar. Samples giving counts of 500 or more will thus be diluted at least 100 times in a ‘countable’ plate; at such a concentration the preservative will have a negligible effect. A number of trials have been carried out by ADAS to examine the effect of the boric acid–glycerol–potassium sorbate–nystatin (BGSN) preservative system on samples transported by post at ambient temperatures. The BGSN mixture used contained (g/100 ml): boric acid, 5; glycerol, 1; potassium sorbate, 0.075, and nystatin 2400 units. Nystatin is typically supplied at a concentration of 100 M units/g. The solution was filter-sterilized, stored under refrigeration and, if not used, discarded after 30 d. In practice it was convenient to dispense the solution in 2 ml amounts in screw-capped plastic sample tubes which allowed the addition of 10 ml of the sample to give the final concentrations noted above. Counts on preserved samples were multiplied by a factor of 1.2 to correct for the volume of preservative added. Preservative in sample tubes can be freeze dried, thus extending storage life beyond 2 months. Samples of raw milk were split. One portion was counted by dilution plating. The other was treated with preservative, taken by road to a post office 25 miles from the laboratory and posted back. On return (generally within 24 h) samples were again dilution-plated. This trial was conducted in late July/early August.

Correlation between the two sets of counts was good ($r = 0.97$) as was the regression relationship: $y = 0.029 + 0.977\,x$, $n = 69$ (where $y =$ log count of preserved sample after 24 h and $x =$ log initial count). Pettipher & Rodrigues (1982) found that plate counts diminished slightly during BGSN preservation and storage (intercept $= -0.75$, $r = 0.85$). Their results were based on fewer samples ($n = 30$) which were not transported during storage. Pettipher & Rodrigues (1982) found also that the BGSN preservative did not cause any large changes in types of bacteria present although micrococci were more sensitive than other groups. They also showed that milk refrigerated for 72 h before preservation, with a resultant initial count $> 10^6$/ml, had a higher count after preservation.

Conclusions

Both refrigeration and bacteriostatic chemicals are effective methods of preservation for milk samples before bacteriological examination. For routine quality control or where delays between sampling and testing are short, refrigeration in cooled, insulated containers is a satisfactory method, although precautions to prevent freezing of the sample are desirable. When more critical examination of the microflora is contemplated, particularly when delays between sampling and testing are anticipated, the BGSN preservative system is likely to prove satisfactory. Use of the preservative eliminates the need for insulated containers and ice packs, and substantially reduces costs when samples are transported by post.

References

Björck, L. 1982 Activitation of the lactoperoxidase system as a means of preventing bacterial deterioration of raw milk. *Kieler Milchwertschaftliche Forschungsberichte* **34**, 5–11.

Bloquel, R. & Veillet-Poncet, L. 1980 Changes in the bacterial flora of grade A raw milk during refrigerated storage. *Lait* **60**, 474–486.

Gebre-Eyziabher, A., Thomson, B. & Blankenagel, G. 1982 Destruction of micro-organisms during thawing of skim milk. *Journal of Food Protection* **45**, 125–126.

Heeschen, W., Reichmuth, J., Tolle, A. & Zeidler, H. 1969 Preservation of milk samples for bacteriologic and cytologic examinations and examinations for inhibitors. *Milchwissenschaft* **24**, 729–734.

Kikuchi, M. & Matsui, Y. 1974 Change in the bacterial count and bacterial flora in bulk-cooled milk. *Japanese Journal of Zootechnical Science* **45**, 592–596.

Kneifel, N. & Foissy, H. 1982 Antibacterial activity of the lactoperoxidase system on coliform strains from raw milk. *Proceedings of the XXI International Dairy Congress* **1**, 320–321. Mir Publishers, Moscow, USSR.

Lueck, E. 1980 *Antimicrobial food additives; characteristics, uses and effects*, p. 75. Berlin: Springer-Verlag.

Pettipher, G. L. 1983 *The Direct Epifluorescent Filter Technique*, pp. 164–168. Chichester: John Wiley.

Pettipher, G. L. & Rodrigues, U. M. 1982 A bacteriostatic mixture of milk samples and its effect on bacteriological, cytological and chemical compositional analysis. *Journal of Applied Bacteriology* **52**, 259–265.

von Grove, H. H. & Terplan, G. 1978 Zur Konservierung von Milchproben für die bakteriologische Untersuchung. *Archiv für Lebensmittelhygiene* **29**, 81–84.

Preserving Stored Crops with Urea

Fiona S. Carlile*

Agricultural Development and Advisory Service, Ministry of Agriculture, Fisheries and Food, Kenton Bar, Newcastle upon Tyne, UK.

A variety of crops such as hay (Ghate & Bilanski 1979), silage (Huber *et al.* 1968) and high moisture grain (Bothast *et al.* 1975) have been treated with urea. The primary aim of such treatment has been to increase the crude protein content (Ghate & Bilanski 1979), although in some cases an improvement in digestibility has also been observed (Hadjipanayiotou 1984). These improvements are particularly important where the cellulose content of a crop is high, making crude protein content and digestibility too low to meet the maintenance requirements of ruminants (Herrera-Saldana *et al.* 1982). Treatment of whole barley grain with urea reduces the cost of processing and also improves the quality of animals fed with the treated material (Ørskov *et al.* 1974*a*). Urea has also been shown to have a preservative effect which is thought to be due to its breakdown by urease in treated crops, producing ammonia and carbon dioxide (Ghate & Bilanski 1979). The released ammonia acts as the preservative. Such treatment allows the harvesting of crops with a high moisture content, and reduces drying and processing costs.

Urea is readily available as both a feed additive and as a fertilizer and can be applied, using machinery already available, in solid or liquid form depending on the crop to be treated. The treatment rate is usually calculated on a percentage of the dry matter (DM) of the crop, commonly at levels at or below 6%. This avoids toxicity in stock fed treated material (Price 1981). A number of laboratory—and farm—scale trials have been carried out by the Agricultural Development and Advisory Service (ADAS) on different crops and the microbial flora examined before and during storage.

Microbiological Examination

All samples were examined by the same technique and 40 g of sample were mixed in peptone saline (containing, sodium chloride 0.85, peptone 0.1 g/l, in distilled

* Now Mrs F. S. Thomson, ICARDA, PO Box 5466, Aleppo, Syria.

Preservatives in the Food,
Pharmaceutical and Environmental
Industries

0–632–01727–9

water) for 2 min using a Colworth 400 stomacher. Serial tenfold dilutions were prepared in peptone saline. The following media were used to enumerate micro-organisms.

OAES agar (Kaufman *et al.* 1965)

This medium was used for the enumeration of moulds and yeasts, using the pour plate method with 1 ml of each dilution. Plates were incubated at 25 °C for 5–7 d (isolation of field fungi) and at 37 °C for 3–4 d (isolation of storage fungi). Before use the following filter-sterilized antibiotics were added to each 100 ml of the basal medium: 1 ml of 0.5% streptomycin, and 2 ml of 0.25% chloramphenicol.

Half-strength Nutrient Agar

Nutrient Agar (Oxoid) was prepared at half strength and supplemented with 0.05% (w/v) actidione dissolved in acetone and water and filter-sterilized. The medium was used to isolate thermophilic actinomycetes (farmer's lung organisms). Petri dishes containing the medium were prepared using the spread plate method and incubated at 55 °C for 48 h.

Nutrient Agar

A total bacterial count was made on Nutrient Agar (Oxoid) containing actidione (0.05%, w/v). Plates were prepared by the pour plate method and incubated at 30 °C for 5 d.

pH measurements were made on 1:10 dilutions of the samples in distilled water.

Barley Grain

Farm trial 1

As part of a trial to investigate the preservative effect of sodium hydroxide on whole barley grains, quantities of grain were also treated with urea and with a mixture of urea and sodium hydroxide. Solutions of both chemicals were applied to 700 kg batches of grain resulting in a range of treatments and moisture contents viz: (a) sodium hydroxide (% of DM), 0; urea (% of DM), 2.0 and moisture contents (%), 15, 20, 25 or30; (b) sodium hydroxide, 3.5; urea 2.0; moisture content 15, 20, 25 or 30. A solution is preferable to powder for treating grain as it avoids the problem of grains and urea separating and causing rejection by stock (Ørskov *et al.* 1974*a*). The sodium hydroxide was applied with a commercial applicator and the urea with a complete diet feeder. After treatment the grain was stored for 38 weeks in unsealed polypropylene 'minibulk' bags of one tonne capacity which were placed on wooden pallets under cover. Samples were taken at monthly intervals for microbiological examination.

Microbial counts (Fig. 1) were low throughout storage for both treatments at all moisture contents. This indicated effective preservation. The addition of urea to sodium hydroxide treated grain appeared to prevent the aerobic deterioration observed previously with sodium hydroxide treatment alone (Perks 1981).

TABLE 1. *pH readings of urea treated barley throughout storage*

	Date of Sampling			
Treatment	6.1.81	9.2.81	2.3.81	31.3.81
20/1.5	7.49	8.20	8.42	8.60
20/2.5	7.44	8.14	8.45	8.60
20/3.5	7.59	8.11	8.53	8.60
25/1.5	7.58	8.08	8.65	8.70
25/2.5	7.51	8.19	8.57	8.70
25/3.5	8.00	8.41	8.89	8.70
30/1.5	7.70	8.07	8.55	8.70
30/2.5	8.10	7.61	8.68	8.90
30/3.5	8.54	8.18	8.90	8.90

Laboratory trial

Samples (100 kg) of barley grain at 20, 25 and 30% moisture contents were each treated with 1.5, 2.5 and 3.5% urea on a dry matter (DM) basis (Table 1). The urea was applied with a watering can, and a spade used to mix the grain. The treated grain was stored in uncovered plastic drums in a heated outbuilding. Samples were taken monthly for microbiological examination.

The results (Fig. 2) indicate a general pattern. At each moisture content there were lower counts for bacteria, moulds and actinomycetes with increasing urea concentrations. For a particular concentration of urea, the microbial counts were also lower at higher moisture contents. This is in agreement with the findings of Ørskov *et al.* (1979) who also suggested that at low moisture contents (17%) microbial numbers may have been increased by the addition of urea.

The effect of moisture content on preservation by urea may be a result of the stoichiometry of the reaction for urea breakdown, a reaction requiring water.

$$(NH_2)_2CO + H_2O \rightarrow 2NH_3 + CO_2$$

Increasing moisture content would tend to favour urea breakdown with more ammonia released and hence better preservation. pH readings (Table 1) were higher

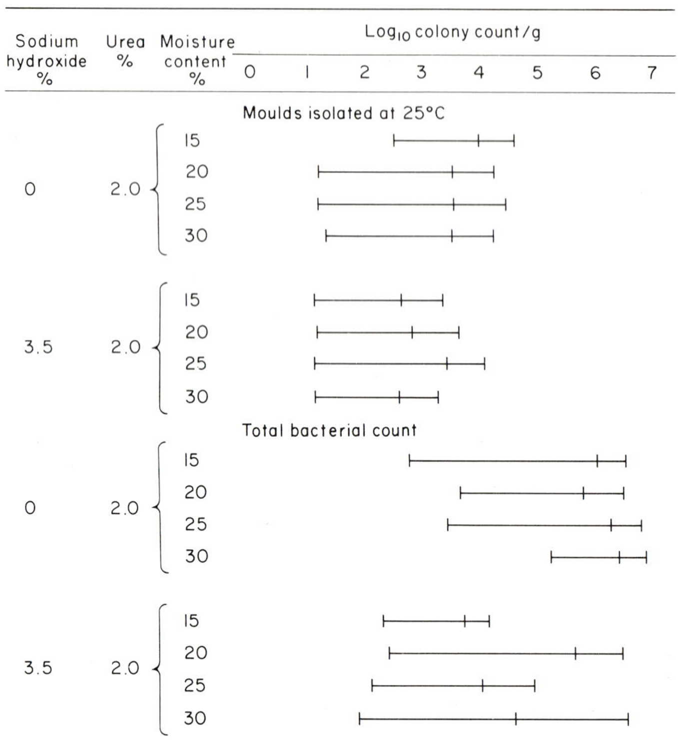

FIG. 1. Microbial counts of urea and sodium hydroxide treated barley. Horizontal lines represent range of colony counts, vertical bars represent average colony counts.

at higher levels of urea treatment and at greater moisture contents, suggesting increased release of ammonia under these circumstances. Alternatively, the ammonia produced from urea breakdown may go into solution in the water surrounding the grains. At higher moisture contents this would reduce the loss of ammonia by volatilization, resulting in an improved preservative effect.

Farm trial 2

On a farm scale approximately 10 tonnes of barley at 25% moisture content were treated with 2% urea (in powder form) on a DM basis, using a Berwyn applicator. The grain was stored, uncovered, on a concrete floor in a shed for 20 weeks. Samples were taken for microbiological examination before and immediately after treatment, and at monthly intervals during storage. Table 2 indicates that low microbial numbers were maintained throughout storage. However, the heap started to heat

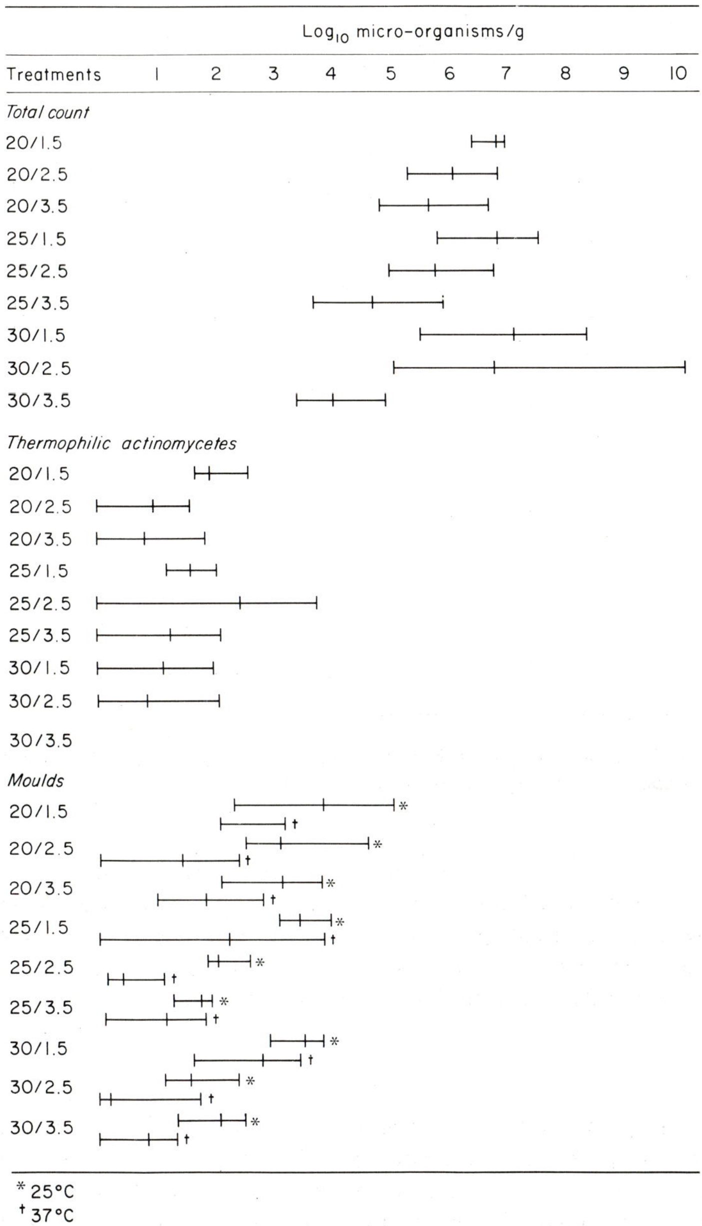

FIG. 2. Urea treated grain; microbial counts at various moisture contents and treatment rates. Horizontal lines represent range of colony counts, vertical bars represent average colony counts.

TABLE 2. *Urea treated barley, Trial 3; Farm scale*

Sample	Weeks after treatment	Colony count ($\log_{10}$) per gram of grain: Moulds 25 °C	Moulds 37 °C	Yeasts 25 °C	Yeasts 37 °C	Thermophilic actinomycetes 55 °C	Total bacteria 30 °C
Untreated	0	5.294	NI*	4.108	NI	1.301	7.845
Treated	0	2.602	NI	NI	NI	1.301	6.157
Surface	6	4.217	3.176	NI	NI	2.173	7.164
Subsurface	6	1.903	1.00	NI	NI	2.114	4.623
Hotspot 1	6	2.176	1.954	NI	NI	NI	7.409
2	6	2.079	1.845	NI	NI	NI	6.349
3	6	3.826	3.061	NI	NI	1.301	7.568
Surface	12	3.779	2.380	1.00	2.230	2.903	7.170
Subsurface	12	2.114	1.954	NI	NI	2.301	7.919
Hotspot	12	1.903	NI	NI	NI	1.602	5.397
Surface	18	3.114	1.778	NI	2.322	2.973	7.434
Hotspot	18	2.230	1.301	NI	NI	2.663	6.863
Surface	22	2.322	1.903	NI	NI	2.146	6.398
Subsurface	22	2.447	2.255	NI	NI	1.602	6.255

* NI = Not Isolated.

early in the storage period. This was found subsequently to be due to a mite infestation. Areas not affected by heating were well preserved.

Whole Crop Barley

In 1979 and 1980 ADAS carried out trials with sodium hydroxide treatment of whole crop barley (i.e., the straw and grain were not separated). Despite an improvement in the nutritional value, several problems were encountered. The treated crop deteriorated on exposure to air and the high sodium content resulted in increased urination by animals fed this material. The sodium hydroxide also proved hazardous to the handlers. Following the relative success of treating grain with urea and the problems associated with sodium hydroxide, a series of trials was set up to examine the preservation of whole crop barley with urea.

Trial 1

A 'simulated whole crop' consisting of grain at 19.5% moisture content (MC) and straw at 20.6% MC was prepared. The MC of the mixture was adjusted to 55% by soaking the two constituents in water for 72 h. For each of the treatments (control = untreated; urea = 6.5% on a DM basis, applied as a solution), 7 kg of straw and 7 kg of grain were mixed. Control and treated materials were stored in sealed plastic bags in a heated outbuilding. Samples were taken for microbiological examination on the day of treatment and after 3 months' storage.

Table 3 shows that although the untreated crop sampled on the day of treatment had high mould counts their numbers were reduced during storage. With untreated material, a layer of mould was noted at the top of the material, which smelled sweet, when bags were opened. The treated crop was not visibly mouldy and the mould counts were low after storage. Thus urea appeared to be an effective preservative. It needs to be stressed that, as the bags were sealed, the bulk of the material would be anoxic and therefore less likely to deteriorate.

Trial 2

Approximately two tonnes of whole crop barley were treated in the swath with a solution of urea (applied with a watering can) at a concentration resulting in 4% urea (on a DM basis) in the treated material. The treated material was made into big bales immediately after treatment. The bales were stored in individual plastic bags. Of the seven bales used in the trial, four were treated and three untreated. Two of the treated bales were sealed in the bags, one bag was unsealed and the fourth was treated with urea in the bag before sealing. All the untreated bales were placed in sealed bags. Samples were taken for microbiological examination on the day of treatment and after approximately 6 weeks' storage.

TABLE 3. *Urea treatment of simulated whole crop barley*

			Colony count ($\log_{10}$) per gram of sample					
			Moulds		Yeasts		Thermophilic actinomycetes	Total bacteria
Sample	Treatment	pH	25 °C	37 °C	25 °C	37 °C	55 °C	30 °C
Taken day of treatment	Control	8.82	5.079	4.143	5.987	5.820	2.881	9.623
After 3 months' storage	Control	4.71	3.991	3.544	5.009	3.477	2.146	7.114
Taken day of treatment	Urea 6.5%	9.09	4.643	3.778	4.940	1.00	2.833	9.505
After 3 months' storage	Urea 6.5%	8.88	2.000	1.301	1.00	1.00	2.146	8.613

The untreated crop (Table 4) was found to have undergone spoilage, despite being stored in sealed bags, as shown by high counts of all groups of micro-organisms. It may be that because of the nature of the material, pockets of air were trapped in the bales and small holes in the bags may have allowed air to enter and hence mould growth to occur. The treated crop appeared to have been well preserved. It had low counts of all groups of micro-organisms, even where bags were left unsealed.

Straw

On a farm scale, straw at 20–25% moisture content was treated with a solution of urea to achieve 5% urea in the treated crop. The solution was sprayed over big

TABLE 4. *Urea treated whole crop barley—farm scale trial*

			Colony count ($\log_{10}$) per gram of crop					
			Moulds		Yeasts		Thermophilic actinomycetes	Total bacteria
Bale no.	Sample	Treatment	25 °C	37 °C	25 °C	37 °C	55 °C	30 °C
–	Before treatment	Untreated	6.127	2.255	NI*	1.000	2.255	9.629
–	Taken on day of treatment	Treated	6.276	2.079	NI	NI	2.079	9.702
1†	A	Treated/	1.903	NI	NI	NI	1.602	6.679
	B	Bag closed	NI	1.301	NI	NI	1.903	7.272
2	A	Treated/	1.903	1.602	NI	NI	2.204	4.756
	B	Bag closed	1.778	1.301	NI	NI	1.602	7.510
3	A	Treated/	1.778	2.505	NI	NI	2.204	7.593
	B	Bag open	NI	1.301	NI	NI	1.602	9.655
4	A	Untreated/	8.188	7.973	NI	NI	8.00	8.544
	B	Bag closed	7.778	7.225	5.732	NI	5.488	8.292
5	A	Untreated/	5.505	4.250	6.496	5.963	1.602	8.732
	B	Closed	3.250	3.334	6.342	6.491	3.510	8.041
6	A	Untreated/	3.418	2.964	3.477	NI	2.556	7.833
	B	Closed	2.377	1.903	NI	NI	5.057	6.580
7	A	Treated 'in bag'/	1.903	1.903	NI	NI	1.903	6.114
	B	Closed	7.246	7.283	NI	NI	NI	9.100

* NI = Not Isolated.

† 2 samples per bale.

bales of straw which were then placed in individual plastic bags, sealed and stored outdoors. Samples were taken approximately monthly for microbiological examination.

Results (Table 5) suggest some preservative effect as the numbers of storage fungi (37 °C) and field fungi (25 °C) were insignificant during storage. This preservative effect may have resulted from storage in anaerobic conditions, or from the presence of urea. Numbers of thermophilic actinomycetes increased, however, during storage although small holes were found in some of the bags which would account for their development. It may be that straw in comparison with other crops is too dry for preservation with urea to be as effective.

TABLE 5. *Urea treated straw*

			Colony count ($\log_{10}$) per gram of straw			
		Weeks	Moulds		Thermophilic	Total
Bale		after			actinomycetes	bacteria
number	Treatment	treatment	25 °C	37 °C	55 °C	30 °C
1	Untreated	0	7.960	NI*	3.912	10.137
	5% Urea	7	6.934	5.017	3.459	8.604
		13	6.152	NI	5.000	8.250
2	Untreated	0	6.857	NI	4.272	9.690
	5% Urea	4	5.964	1.301	3.064	8.903
		13	6.152	NI	5.000	8.250
3	Untreated	0	7.436	NI	7.438	9.688
	5% Urea	7	6.524	1.778	NI	9.093
		13	6.225	2.204	2.505	8.380
4	5% Urea	4	6.380	2.301	NI	8.914
		13	5.394	2.643	2.301	8.283

* NI = Not Isolated.

Hay

Hay treated in the swath with urea solution to produce 3.1% urea in the treated crop (fresh weight basis), and stored aerobically under cover, resulted in a measure of preservation (Table 6). Although mould growth was controlled to some extent in the treated material compared with 'wet' and 'dry' controls, the growth of thermophilic actinomycetes was not. It may be that in this particular case, the treatment rate was too low for effective preservation.

TABLE 6. *Urea treated and untreated hay*

Treatment	Weeks after treatment	Colony count ($\log_{10}$) per gram of hay					
		Moulds		Yeasts		Thermophilic actinomycetes	Total bacteria
		25 °C	37 °C	25 °C	37 °C	55 °C	30 °C
Control	0	5.982	1.602	5.869	NI*	3.204	8.964
	10	8.033	8.158	NI	NI	6.017	7.00
	27	7.857	8.041	NI	NI	5.415	9.384
Control	0	6.422	2.531	6.164	NI	6.944	9.006
	10	8.215	6.833	NI	NI	5.00	8.850
	27	7.100	5.833	NI	NI	4.881	7.00
Wet Control	0	6.230	1.699	6.100	NI	4.146	9.438
	10	6.279	4.857	NI	NI	6.358	7.00
	27	7.235	6.064	NI	NI	4.00	9.274
Wet Control	0	6.806	2.623	7.301	NI	4.064	9.312
	10	7.288	5.220	NI	NI	6.787	7.00
	27	6.903	5.270	NI	NI	4.00	7.00
Urea	0	6.100	2.079	6.198	NI	3.204	8.584
	10	3.152	3.140	NI	NI	3.107	8.823
	27	4.093	3.748	NI	NI	5.973	8.935
Urea	0	6.250	1.903	5.255	NI	4.193	9.591
	10	3.297	2.924	NI	NI	5.134	7.785
	27	4.134	3.935	NI	NI	5.107	7.00
Urea	0	4.914	2.681	4.079	NI	4.158	8.107
	10	5.736	3.934	NI	NI	5.944	7.00
	27	6.025	6.079	NI	NI	5.305	8.870

* NI = Not Isolated.

Discussion

It appears that urea can be an effective preservative, especially at high crop moisture contents. The degree of preservation achieved depends on the type of crop, its moisture content and the concentration of urea in the treated crop. Preservation has been observed in both aerobic and anaerobic conditions of storage. The upper limit for urea concentration should be approximately 6% in the treated material. Higher concentrations may cause urea toxicity in stock. Whether urea is applied in solid or liquid form depends on the crop to be treated and how evenly the urea can be applied.

References

BOTHAST, R. J., ADAMS, G. H., HATFIELD, E. E. & LANCASTER, E. B. 1975 Preservation of high moisture corn: A microbiological evaluation. *Journal of Dairy Science* **58**, 386–391.

GHATE, S. R. & BILANSKI, W. K. 1979 Treating high moisture alfalfa with urea. *Transactions of the American Society of Agricultural Engineers* **22**, 204–206.

HADJIPANAYIOTOU, M. 1984 The value of urea-treated straw in diets of lactating goats. *Animal Feed Science and Technology* **11**, 67.

HERRERA-SALDANA, R., CHURCH, D. C. & KELLEMS, R. O. 1982 The effect of ammoniation treatment on intake and nutritive value of wheat straw. *Journal of Animal Science* **54**, 603.

HUBER, J. T., THOMAS, J. W. & EMERY, R. S. 1968 Response of lactating cows fed urea treated corn silage harvested at varying stages of maturity. *Journal of Dairy Science* **51**, 1806–1810.

KAUFMAN, D. D., WILLIAMS, L. E. & SUMNER, C. B. 1963 Effect of plating medium and incubation temperature on growth of fungi in soil-dilution plates. *Canadian Journal of Microbiology* **9**, 741–751.

ØRSKOV, E. R., SMART, R. & MEHREZ, A. Z. 1974*a* A method of including urea in whole grains. *Journal of Agricultural Science* **83**, 299–302.

ØRSKOV, E. R., FRASER, C. & GORDON, J. G. 1974*b* The effect of processing of cereals on rumen fermentation, digestibility, rumination time, and firmness of subcutaneous fat in lambs. *British Journal of Nutrition* **32**, 59–69.

ØRSKOV, E. R., STEWART, C. S. & GREENHALGH, J. F. D. 1979 The effect of sodium hydroxide and urea on some storage properties of moist corn. *Journal of Agricultural Science* **92**, 185–188.

PERKS, D. 1981 Whole crop cereals and grain; developments. In *Upgrading of crops and by-products by chemical or biological treatment.* Proceedings of an ADAS/ARC/CSG Seminar.

PRICE, D. P. 1981 Urea as a feedstuff for beef cattle. *Feedstuffs*, September 14, p. 30.

The Preservation of On-farm Pre-packed Beetroot

D. W. PIMBLEY

*Ministry of Agriculture, Fisheries and Food (MAFF), Agricultural Development and Advisory Service (ADAS), Lawnswood, Leeds, UK**

The expansion in demand for pre-packed vegetables has led to an increase in the numbers of farmers and growers who process and pack their own produce for direct sale through retail outlets. Beetroot is only one of a wide range of vegetables which are now marketed in pre-packs, others include individual lettuces in loose bags and film-wrapped mixed salad packs. There is also a market for semi-cooked vegetables in larger catering sizes. Pre-packing has a number of advantages for the producer, the retailer and the consumer. For the producer there is the attraction of higher profit margins by cutting out the wholesaler, although this has to be balanced against higher costs for labour and capital expenditure. For the retailer and consumer the advantages include ease of handling, a longer shelf life and a reduction in wastage. Organizing a pre-packing operation requires a sizeable capital investment in equipment for grading, processing and packaging; labour costs also have to be taken into consideration.

Producers who mainly supply supermarkets are generally aware of the high quality standards required but others, supplying smaller retail outlets or considering entering the market, may not always be aware of the need for good hygiene, particularly with cooked vegetables. Neglecting good hygiene will invariably lead to poor product quality and microbiologists of the Agricultural Development and Advisory Service (ADAS) are able to identify defects, either by visual observation, or from the results of laboratory tests, and improve product quality.

Hygiene

The manufacture of foods on agricultural premises is covered by the Food Hygiene Regulations (Anon. 1970). These lay down standards relating to the cleanliness of premises, of persons engaged in the handling of foods, the provision of water and washing facilities and the disposal of waste. The regulations are enforced by local

* Address all correspondence c/o The Regional Microbiologist.

Preservatives in the Food,
Pharmaceutical and Environmental
Industries

0–632–01727–9

authorities through the Department of Environmental Health. The role of the ADAS is to provide farmers with general advice on hygiene, cleaning routines, the use of detergents and disinfectants, and the layout of the plant so as to minimize cross-contamination.

The first priority in most advisory cases is to establish good hygienic practice. This frequently leads to an improvement in product quality without the need to resort to additional safeguards, such as preservatives. Sources of contamination in packing sheds include machinery, such as conveyor belts, the raw material, often stored in the same building, the air and the staff. Once the finished product is contaminated the high humidity inside pre-packs is ideal for the growth of micro-organisms. The maintenance of high hygiene standards is thus necessary, not only to comply with the law but also to extend the shelf life of the product.

Preservatives

The use of preservatives in foods is covered by the Preservatives in Foods Regulations (Anon. 1979) and Amendments (Anon. 1980; 1982). These specify the maximum levels of preservatives and the foods in which they are permitted. Other permitted additives include salt and vinegar. Colouring agents and anti-oxidants are covered by other regulations. The preservatives currently permitted in vegetables and vegetable products are listed in Table 1.

If a producer wants to use a preservative which is not listed in the regulations then he must make a submission to the Food Advisory Committee of the Ministry of Agriculture, Fisheries and Food. The submission would normally include evidence of a commercial interest in the use of the preservative, together with

TABLE 1. *Permitted preservatives in vegetable products*

Preservatives	Vegetable	Max level mg/kg
Benzoic acid Methyl 4-hydroxybenzoate Ethyl 4-hydroxybenzoate Propyl 4-hydroxybenzoate	Cooked-pre-packed beetroot	250
Sulphur dioxide	canned garden peas and cauliflower	100
(as sodium sulphite;	raw potatoes	50
sodium metabisulphite;	dehydrated sprouts	2500
sodium hydrogen sulphite;	cabbage	2500
or the calcium salts)	potato	550
	other vegetables	2000

information on the method of use, levels of preservative used, expected shelf life of the product and the way in which the product will be packaged and marketed. The use of preservatives is not normally permitted in products which are described as 'fresh', or which the consumer would expect to be fresh.

Packaging

The three most common methods of pre-packing vegetables are: loose wrapping in bags (lettuce), shrink wrapping (cucumbers), and film wrapping in plastic trays (tomatoes, salad packs etc.). The method of packaging and choice of packaging materials can influence consumer response. An attractively packaged and presented product will attract the eye and help to increase sales, and the choice of packaging material may influence the shelf life of the product. The producer is faced with a range of packaging materials, each with properties which make them suitable for specific uses. Some of the commonly used materials and their properties are listed in Table 2. Properties such as heat sealability, and permeability to gases and water can often be altered by changes in formulation. Materials may be chosen to modify the internal atmosphere of the pack, thereby restricting the growth of micro-organisms which cause spoilage (Browning 1982).

TABLE 2. *Packing materials*

Material	Comments
Polythene	Strong, inexpensive, waterproof
Polyvinyl chloride	Often used for cling or shrink wrapping
Polypropylene	Heat shrinkable, good water vapour barrier, poor gas barrier
Cellulose	Cellophane, which is not heat sealable or moisture proof, has been superceded by special celluloses with thermo-plastic coatings

Labelling

Under EEC regulations pre-packed vegetables must be clearly labelled with the name of the product, the net weight, a complete list of ingredients and a date mark, 'sell by' or 'best before' date (Anon. 1984).

Storage

Storage facilities on farms are often inadequate. Cold stores may be too small for the volume of products in storage. As a result shelf life is reduced and in warm weather the product may spoil before despatch. Rapid cooling and storage of pre-packed vegetables at chill temperatures and in chilled transport will usually extend the product shelf life (Snowden & Ahmed 1981), but only the larger producers can afford the necessary capital investment.

Preservation of Pre-packed Beetroot

Most of the beetroot sold in the UK is processed before sale, either by commercial food companies to be sold in jars, or by the growers themselves, for sale in pre-packs (ADAS unpublished data). Commonly, on-farm packed beetroot is preserved by dipping in acetic acid solution. Under hygienic conditions and with correct storage, the high concentration of acetic acid used should ensure a shelf life of 7–10 days. Spoilage is usually associated with acid tolerant micro-organisms, mostly moulds and yeasts picked up as contaminants along the production line. Some beetroot packers in Lancashire and Humberside have problems with mould and yeast spoilage, particularly during the summer months. In certain cases the beetroot spoiled within 2 to 3 days of packing.

Consumers now want a less acid beetroot so some growers are trying to replace at least part of the acetic acid with other preservatives. A laboratory trial was set up to investigate the use of alternative preservatives for pre-packed beetroot in order to improve the advice given to growers. The main aims were to identify suitable preservatives, and their effective concentration, so as to reduce the acetic acid levels without reducing shelf life and to develop a commercially acceptable method of application.

Choice of preservative

Potassium sorbate is a suitable preservative for acid beetroot because the sorbates are more active at acid pH levels (Kabara 1981) and they are effective against moulds and yeasts (p. 182). When the concentration of acetic acid is reduced then preservatives such as hydroxybenzoates, active at higher pH levels, are required. Mixtures of methyl, ethyl and propyl 4-hydroxybenzoates are more effective than single preservatives and the activity of hydroxybenzoates against micro-organisms increases with increasing molecular chain length (p. 903; Strandskow *et al.* 1966). However, since ethyl and propyl parabenzoate are only sparingly soluble in the presence of acetic acid, only methyl 4-hydroxybenzoate was used in this study.

Under the regulations, the hydroxybenzoates are permitted at levels of up to 250 mg/kg (ppm) in cooked, pre-packed beetroot, whereas sorbate is not currently permitted. Both sorbates and hydroxybenzoates have characteristic tastes which govern the maximum concentrations which may be used, particularly in delicately flavoured foods.

Application of preservatives to beetroot

Preservatives can be applied to foods as a powder or in solution. For beetroot, a dip in an aqueous solution of the preservative was found most satisfactory. The sorbates and hydroxybenzoates are unstable in aqueous solutions, so dips have to be replaced at regular intervals, preferably daily.

Beetroot preparations

The sequence of operations in the preparation of cooked beetroot at a commercial pre-packer is as follows:

Raw beetroot is washed and graded.
Beetroot is steamed for 15–20 min.
Beetroot is peeled and sorted.
Dipped for 5 min in a 5–7% (v/v) solution of acetic acid.
Packed into plastic trays.
The trays are film wrapped, boxed and placed into store.

Preparation of the beetroot followed commercial practice as closely as possible. Initial experiments to find effective preservative concentrations were undertaken using beetroot prepared, washed and packed in the laboratory. At a later stage beetroot was taken from the pre-packers line and treated with preservative.

Experimental work

At least six replicate packs were prepared for each treatment in the course of the experimental work, some of which was done in the laboratory and some on the premises of a producer.

In the laboratory

Three treatments were examined:

Treatment A: beetroots dipped for 5 min in 5% (v/v) acetic acid and for 2 min in 0, 0.1 or 5% (w/v) potassium sorbate.

Treatment B: beetroots dipped for 2 min in 0, 0.1, 1.0 or 5.0% (w/v) methyl 4-hydroxybenzoate alone.

Treatment C: beetroots dipped for 5 min in 0, 0.1, 1.0 or 5.0% (v/v) acetic acid and for 2 min in 1% (w/v) potassium sorbate.

At a packing station

Beetroots were taken from the packing line after the acetic dip. The pre-packer aimed for an acetic acid concentration of *ca.* 5–7% (v/v). Two treatments (2 and 3) were examined and compared with a control (treatment 1) which contained no additional preservative. In treatment 2 the beetroots were dipped for 2 min in 1% (w/v) potassium sorbate, and in treatment 3, for 2 min in 1% (w/v) methyl 4-hydroxybenzoate.

Samples of treatments A, B and C were inoculated with a suspension containing yeasts and spores of a *Penicillium* spp. isolated from mouldy pre-packed beetroot.

Storage

Beetroots prepared in the laboratory were stored at 20 °C. Those prepared at the packing station were stored at ambient temperature, as this is the normal commercial practice.

Examination

The following examination was carried out:
Visual examination for spoilage at daily intervals; flavour and appearance of unstored samples; total count on Plate Count Agar (Oxoid) at 30 °C to enumerate bacteria, yeasts and moulds; and the pH on selected samples was measured.

Results and Discussion

Visual spoilage

Beetroots prepared in the laboratory according to commercial practice showed no spoilage after 21 d at 20 °C. This confirmed that good hygiene can produce a product with a long shelf life without the need for additional preservatives, even at a relatively high storage temperature. It was necessary therefore to inoculate the beetroot with a suspension of spores of *Penicillium* spp. and yeasts to reproduce the levels of contamination at a commercial packing station. The inoculated beetroots went mouldy in 2 to 3 d. This storage period was similar to the shelf life of some batches of commercially packed beetroots. The use of 1% or 5% (w/v) sorbate as a dip in addition to acetic acid, prevented mould development for more than 21 d (Table 3).

TABLE 3. *The effect of sorbate dip concentration on the time to visual spoilage of acetic acid dipped pre-packed beetroot*

Sorbate concentration % (w/v)	Days to spoilage at 20 °C	
	with inoculum*	without inoculum
0	2	21
0.1	3	21+
1.0	21+	21+
5.0	21+	21+

* Suspension of yeast and spores of *Penicillium* spp.

A 1% (w/v) hydroxybenzoate dip prevented the moulding of inoculated beetroots prepared without treatment with acetic acid for 3 d and a 5% (w/v) dip for over 21 d. Without preservative the beetroots went mouldy after only 1 day (Table 4).

TABLE 4. *The effect of hydroxybenzoate dip concentration on the time to visual spoilage of acetic acid-free pre-packed beetroot*

Hydroxybenzoate concentration % (w/v)	Days to spoilage at 20 °C	
	with inoculum*	without inoculum
0	1	2
0.1	1	2
1.0	3	7
5.0	21+	21+

* Suspension of yeast and spores of *Penicillium* spp.

Reducing the acetic acid concentration to 1% (v/v)—approximately one-fifth of the concentration used commercially—had no effect on the time to visual spoilage (over 14 d) when the beetroots were also dipped in 1% (w/v) sorbate (Table 5). Inoculated beetroots dipped in 0.1% (v/v) acetic acid and 1% (w/v) sorbate took 13 d to spoil, compared with only 3 d for beetroots treated with sorbate alone. This suggests a synergistic effect between the sorbate and the acetic acid. After 14 d at 20 °C, the beetroots were beginning to dry out due to loss of moisture through the wrapping film. This would inhibit microbial growth, particularly at the surface, so the samples were discarded at this stage, rather than at 21 d, as in previous

TABLE 5. *The effect of acetic acid dip concentration on the time to visual spoilage of pre-packed beetroot dipped in 1% (w/v) sorbate*

Acetic acid concentration % (w/v)	Days to spoilage at 20 °C	
	with inoculum*	without inoculum
0	3	5
0.1	13	13
1.0	14+	14+
5.0	14+	14+

* Suspension of yeast and spores of *Penicillium* spp.

experiments (Tables 3 & 4). The shelf life of the beetroot in terms of flavour and texture is therefore likely to be less than the 21 d obtained by visual examination.

Beetroots from the packing station which were treated with either sorbate or hydroxybenzoate spoiled more rapidly than beetroots prepared in the laboratory with the same level of preservative (Table 6). This was due partly to differences in storage temperatures. The beetroot was stored at summer ambient temperatures of up to 25 °C and with 3–4 whole beetroots per pack the humidity levels remained high for a longer period. Sorbate was more effective than hydroxybenzoate in preventing moulding of the pre-packed beetroot.

TABLE 6. *The effect of sorbate and hydroxybenzoate dips on the time to visual spoilage of pre-packed beetroot prepared at a packing station*

Treatment	Days to visual spoilage at ambient
Acetic acid dip only*	3
Acetic acid dip plus 1% (w/v) sorbate dip	10
Acetic acid dip plus 1% (w/v) hydroxybenzoate dip	7

* 5–7 (v/v) acetic acid.

Total viable counts

Samples of uninoculated beetroots from Treatments B (hydroxybenzoate only) and C (acetic acid plus sorbate) were analysed for total viable counts. The higher the concentration of preservative, the lower the total viable counts during storage (Tables 7 & 8). This was due to direct inhibitory effects of the preservatives on micro-organisms and also a reduction in pH with acetic acid (Table 9).

TABLE 7. *The effect of hydroxybenzoate dip concentration on the total viable count during storage at 20 °C of acid-free pre-packed beetroot*

Hydroxybenzoate concentration	Log_{10} total viable counts g^{-1}			
% (w/v)	3	6	9	13 days
0	3.3	7.9	9.8	NT*
0.1	4.4	7.8	8.1	NT
1.0	1.3	3.1	5.9	NT
5.0	1.3	3.0	2.8	2.1

* Not tested.

TABLE 8. *The effect of acetic acid dip concentration on the total viable count during storage at 20 °C of pre-packed beetroot dipped in 1% (w/v) sorbate*

Acetic acid concentration % (w/v)	Log_{10} total viable counts g^{-1}			
	3	6	9	13 days
0	2.0	6.7	8.7	3.1
0.1	2.6	4.7	5.5	6.8
1.0	2.3	4.1	5.4	8.7
5.0	1.0	1.9	3.2	4.0

TABLE 9. *The effect of acetic acid dip concentration on the pH of sorbate dipped pre-packed beetroot during storage at 20 °C*

Acetic acid concentration % (v/v)		pH (days)	
	3	6	13
0	5.61	5.96	4.53
0.1	5.31	5.14	4.95
1.0	4.95	5.03	4.20
5.0	4.32	4.42	4.47

Flavour and appearance

Sorbate and hydroxybenzoate were detectable 'by taste' on beetroot dipped in 5% (w/v) solutions of these preservatives. Beetroots dipped in 1% (w/v) solutions could not be distinguished from untreated controls. The less acetic acid there is in the beetroots the more likely it is that the other preservatives would be detectable by taste. Sorbic acid, which is less strongly flavoured than the hydroxybenzoates, might therefore be more suitable for use in beetroots containing lower concentrations of acetic acid, although it was not evaluated. A temporary bleaching effect was noted at the highest concentration of preservative used (5% w/v). Normal colour gradually returned after 1 to 2 h and beetroots treated with preservative were then indistinguishable from untreated beetroots.

Conclusions

The sale of pre-packed vegetables direct to retailers can be profitable, but growers must bear in mind the need for extra facilities and for careful hygiene if a product of high quality and with a long shelf life is to be produced. Growers must also be aware of and comply with the regulations governing hygiene, labelling and the use of preservatives and other food additives. Work with pre-packed beetroots has shown that preservatives, such as sorbate and parabenzoates, can be used to extend shelf life. They should not however be used as a substitute for strict hygiene and the correct storage of the product. The use of sorbate will enable growers to reduce the acetic acid concentration without affecting shelf life, thereby satisfying consumer demand for a less acid product.

References

ANON. 1970 *The Food Hygiene (General) Regulations SI 1172.* London: HMSO.

ANON. 1979 *The Preservatives in Food Regulations SI 752* London: HMSO.

ANON. 1980 *The Preservatives in Food (Amendment) Regulations SI 931.* London: HMSO.

ANON. 1982 *The Preservatives in Food (Amendment) Regulations SI 15.* London: HMSO.

ANON. 1984 *Look at the Label.* Ministry of Agriculture, Fisheries and Food. London: HMSO.

BROWNING, T. H. 1982 *The Preservation and Marketing of Fruits and Vegetables in Forms Other than Canned and Frozen.* Technical Memorandum 295. Chipping Campden: Campden Food Preservation Association.

KABARA, J. J. 1981 Food Grade Chemicals for Use in Designing Food Preservation Systems. *Journal of Food Protection* **44**, 633–647.

STRANDSKOV, F. B., ZILLIOTTO, H. L., BRESCIA, I. A. & BOCKELMAN, J. A. 1966 The inhibition of microbial Growth in Package Beer. *American Society of Brewing Chemists Proceedings 1965*, 129–134.

SNOWDEN, ANNA L. & AHMED, AWAD, H. M. 1981 *The Storage and Transport of Fresh Fruit and Vegetables.* London: National Institute of Fresh Produce.

Preservatives for use in the Wool Textile Industry

B. J. McCarthy

Wool Industry Research Association, Wira House, West Park Ring Road, Leeds, U.K.

The growth of micro-organisms on wool textile materials is a phenomenon which has been observed and recorded throughout human history. The coat (fleece) derived from sheep, initially of limited domestic importance, now forms the basis of a major international industry. Wool is almost entirely composed of cystine-containing proteins, the keratins. The natural insolubility of wool is attributed to the presence of interchain disulphide linkages formed by post-synthetic oxidation of pairs of cysteinyl residues (Fraser & Gillespie 1976). Although relatively resistant to attack in its natural state, microbial growth on wool or woollen materials may result in a variety of interrelated spoilage phenomena (e.g. odour production, pigmentation, loss of fibre strength) all of which may contribute to significant financial and material losses during wool processing (Wehrner 1902; Osborn 1912; Georqiewics 1924; McCarthy 1983*a*). Wool is made more susceptible to microbial biodeterioration by mechanical and chemical damage incurred during textile production and finishing. These cause an increased solubility and fibre disruption. Relatively mild microbial growth may render wool unusable because of superficial staining and characteristic musty odours associated with infected material, before any apparent loss of strength is detectable. Biodeterioration of wool results in a decrease in the value of the material and a reduction in the ability of the product to fulfill the function for which it was intended (Eggins 1983; Eggins & Oxley 1980). Losses due to biodeterioration in the UK industry have been estimated between (UK) £8 and £10 million per annum, including direct losses, re-processing and replacement costs (McCarthy 1983*a*). Chemical or environmental prevention of spoilage is required during the working life of the substrate, but should not prevent its ultimate degradation (Ilyichev 1979). In the present paper, the selection, application and efficacy of biocides will be discussed, with special reference to microbiological methods for assessing biocide potency.

Preservatives in the Food, Pharmaceutical and Environmental Industries

0–632–01727–9

Biocides for Wool Products

A diverse range of chemicals — antimicrobial agents or biocides — are applied to wool and wool blends at various stages of manufacture to prevent microbial degradation (known in the trade as 'mildew'). Table 1 gives a representative list of the major biocides which may be applied to wool or products blended with wool. The list is incomplete and ephemeral; traditional textile biocides may be withdrawn from the market and new or existing ones introduced from other industries. Little published material is available for guidance in the areas of biocide selection and methods of application (British Standard 2087. Anon. 1981a), or to record quantities of biocides used specifically in the UK woollen and worsted industries. Although the overall demand for biocides may be limited (in terms of quantities of biocides used annually), there is undoubtedly a requirement for specific biocides for wool protection. Biocides may be supplied directly for application as a protective treatment, or incorporated into a textile additive or auxiliary. The major biocides currently in use include: pentachlorophenyl laurate, dichlorophen, *o*-phenylphenol, pentachlorophenol, and tributyltin oxide.

TABLE 1. *List of biocides used in the Wool Textile Industry*

Active ingredient	Trade name	Supplier
Copper naphthenate	Nuodex	Durham Chemicals
Copper 8-hydroxyquinolinolate	Mystox 8	Catomance
Zinc dimethyl dithiocarbamate		Robinson Bros.
Zinc naphthenate	Nuodex	Durham Chemicals
Zinc 8-quinolinolate		Ward Blenkinsop
Tri-n-butyltin oxide		Schering
o-Phenyl phenol	Preventol O	Bayer
Sodium *o*-phenyl phenate	Mystox WFA	Catomance
	Preventol ON	Bayer
3-methyl-4-chlorophenol	Preventol CMK	Bayer
p-Chloro-*m*-cresol	Nipacide	BTP Cocker
Di-(5-chloro-2-hydroxylphenyl) methane	Dichlorphen	BDH Ltd
2,2′-Methylene bis-(4-chlorophenol)	Dichlorophene	Givaudan
Pentachlorophenol	Mystox G	Catomance
Pentachlorophenol	Preventol P	Bayer
Sodium pentachlorophenate	Mystox D	Catomance
Pentachlorophenyl laurate	Mystox LPL	Catomance
Halogenated amine compounds	Actifresh L-8110	British Sanitized
Halogenated aryl-hydroxy compounds	Actifresh T-506	British Sanitized
Halogenated thio compounds	Actifresh T-304	British Sanitized
Quarternary ammonium compound	Preventol GDB	British Sanitized

TABLE 1. *Continued*

Active ingredient	Trade name	Supplier
Alkyl dimethyl benzyl ammonium chloride	Preventol R50	Bayer
Dodecyl dimethyl dichlorobenzyl ammonium chloride	Preventol RB50	Bayer
Dodecyl dimethyl benzyl ammonium cyclopentane carboxylate salt	Nuodex 100	Durham Chemicals
Bis (2-hydroxy-5-chlorophenyl) sulphide	Fentichlor	BTP Cocker
Sodium N-methyl dithiocarbanate	—	Robinson Bros.
Tetramethyl thiuram monosulphide	—	Robinson Bros.
Dipentamethylene thiuram monosulphide	—	Robinson Bros.
Dipentamethylene thiuram disulphide	—	Robinson Bros.
2-*n*-Octyl-4-isothiazol-in-3-one	Kathon 4200	Rohm and Haas
Sodium salt of 2-mercapto-benzothiazole	Nuodex 84	Durham Chemicals
Arylemthanol-mono (poly)-hemiformal	Preventol D2	Bayer
2-bromo-2-nitropropane-1, 3-diol	Bronopol	Boots
2, 4-dichlorobenzyl alcohol	Myacide TF	Boots
Didecyl dimethyl ammonium chloride	Bardac 22	Lonza
3-(trimethoxysilyl)-propyloctadecyldimethyl ammonium chloride	Dow Corning 5700	Dow Corning
bis 1-hydroxy-2 pyridinethionato-zinc	Zinc Omadine	Olin Chemicals
Pentachlorophenyl laurate + Cetyl pyridinium bromide + *p*-chloro-m-cresol	Mystox QL	Catomance
Zinc dimethyldithiocarbamate + Zinc 2-mercaptobenzothiazole	Vancide 51Z	Vanderbilt
N, N-methylolchloracetamide + isothiazolinones	Parmetol DF12	Sterling Industrial
Benzalkonium chloride + tri-n-butyltin oxide	Mystox QT	Catomance

*Lists of specialized products are given in international trade directories (*e.g.* Textile Chemists & Colorists Buyer's Guide, published by AATCC, see footnote pp 102); in occasional publications (Anon. 1982; Allsopp & Allsopp 1983) and British Standard (BS2087; 1981a). These sources and technical leaflets were used to compile the above list.

Industrial Biocide Application

Biocides may be applied to wool and wool blends during the following stages:

1 In the final rinse bowl of a scouring set, during the scouring of loose stock greasy wool.

2 As a final treatment to fabric in a dyeing winch during the dyeing process (Mebes 1975).

3 Applied directly to yarn as an aqueous conditioning fluid.

4 In the final rinse during piece (fabric) scouring.

5 *Via* an oil emulsion or related additive onto yarn.

6 Applied at the lecking stage before greasy milling.

Detection of Microbial Spoilage on Wool Textiles

Difficulties associated with the recognition of biodeterioration have been discussed repeatedly (Eggins 1967; Seal & Allsopp 1983). The following methods have been found most useful in detecting microbial growth on wool.

Macroscopic examination

Microbial spoilage may be readily apparent due to severe pigmentation and fibre tendering (Fig. 1). A characteristic fusty odour is normally associated with fungal growth (Wehrner 1902; Lewis 1981). Odour production may be due also to bacterial degradation. Detection at this stage normally implies that significant microbial growth has occurred.

Microscopic examination

Sellotape may be used to recover microbial fragments from textile fibres (Brown 1959; Lloyd 1965). Direct microscopic examination may reveal extensive fibre damage (Fig. 2) due to bacterial activity (Burgess 1924; Lewis 1981; Mahall 1982) or fungal growth on the surface or within the fibre (Mandels *et al.* 1948).

Microscopic/staining procedures

The following staining methods have been recommended for detecting microbial activity: cotton blue-lactophenol (Nopitsch 1953; Anon. 1950); Methylene blue G acidified with acetic acid (Mahall 1982); carbol-fuchsin (Burgess & Galloway 1940); safranin (Sankov *et al.* 1972); bromthymol blue (Burgess 1924), and the Pauly stain (Pauly 1904; Burgess & Rimington 1929).

Examination under UV light

Stained areas on wool produced by mildew sometimes fluoresce under UV light (Hirst 1927; Anon. 1950; Garner 1967; Lewis 1981). As various optical brightening auxiliaries also fluoresce, the method is little used.

Additional detection methods

These are reported as seldom used, alternative methods which may include physical (stress-strain) tests (Geiger *et al.* 1941), trypsin digestion (Burgess 1934), biomass recovery achieved by dissolving wool in 10% (w/v) NaOH (Mandels *et al.* 1948),

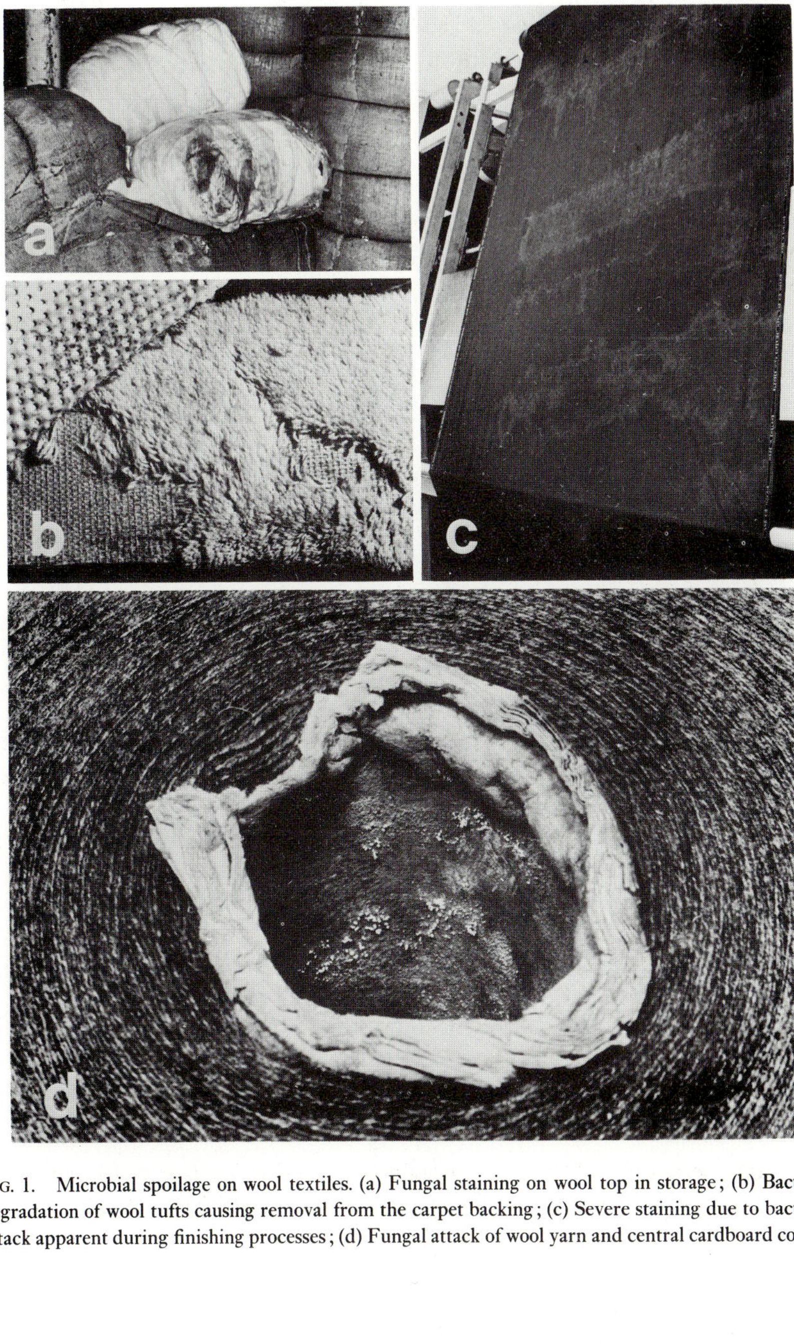

FIG. 1. Microbial spoilage on wool textiles. (a) Fungal staining on wool top in storage; (b) Bacterial degradation of wool tufts causing removal from the carpet backing; (c) Severe staining due to bacterial attack apparent during finishing processes; (d) Fungal attack of wool yarn and central cardboard cone.

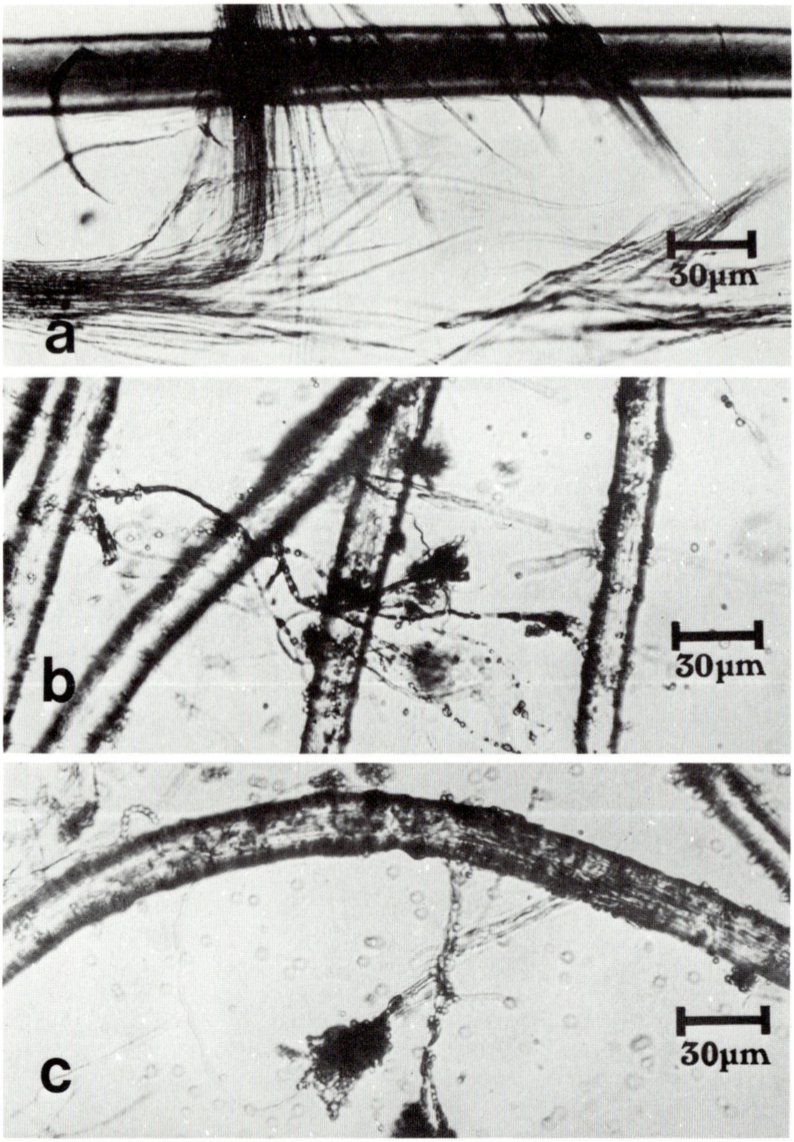

FIG. 2. Microscopic examination of wool fibres. (a) Bacterial attack producing fibre degradation and cortical cell release; (b) and (c) Fungal hyphae growing on wool fibres

recovery of organisms by 'plating' procedures (von Bergen & Mauersberger 1948; Mulcock 1965), and chemical tests for alkali solubility and sulphitolysis (English 1965; Safranek & Goos 1982).

Testing Wool Textile Materials for Resistance to Microbial Attack

Test methods for assessing the susceptibility of wool and woollen materials to microbial degradation, or assessing the efficacy of biocides applied to such materials, are generally based on the 'challenge testing' principle. These involve a method of deliberately infecting the substrate with suitable biodeteriogenic organisms, combined with a method of testing (usually visual assessment or tensile strength testing) the extent of spoilage. The method selected for testing purposes should reflect the nature of the biocide and its expected in-house performance.

The major testing procedures may be classified as follows:

Pure-culture techniques

Samples of material are exposed to attack by a pure culture of a known micro-organism under controlled conditions (Burgess 1954; McQuade 1964; Safranek & Goos 1982). The processes used for sterilization may modify the keratin in wool (McQuade 1964) and produce conflicting results (McQuade 1964; Jain & Agrawal 1980). The results obtained reflect, of course, growth in a man-made environment.

Mixed-culture techniques

Wellman & McCallum (1945) maintain that the advantage of a mixed inoculum is the simultaneous exposure of test samples to a number of organisms which may possess different susceptibilities to biocides and varied metabolic requirements. Toxic metabolites may accumulate in the growth medium (Lewis 1981).

Perfusion techniques

This method allows the continuous replacement of nutrients and removal of waste metabolic products by perfusion during the incubation period (Eggins *et al.* 1968).

Soil-burial methods

This appears to be the most severe and popular test method. Test strips are buried in soil in trays or beds under controlled conditions and assessed for loss of strength (Batson *et al.* 1944; Kempton *et al.* 1963; Raschle 1983). Variations in microbial activity will occur with soils from different locations, with time of year, etc. (Grimm & Kuhne 1969; Gray & Martin 1947; Raschle 1983).

Soil-infection method

Test strips are inoculated by partially coating them with a soil-suspension thickened with kieselguhr, incubated over water and tested after conditioning (Lloyd 1955). An aqueous suspension of horse dung adjusted to a pH of 5.5 to suppress bacterial growth has also been used (Race 1946). Alternatively a manure/soil mixture has been used (Bobkova *et al.* 1975).

Assessment of Microbial Activity on Wool Textile Materials

The term 'bioburden' may be used to refer to the number, types and resistance of viable micro-organisms on or in the test substrate (Carpenter *et al.* 1981). A variety of methods have been devised for assessing microbial contamination (or bioburden) and/or the presence of biocides on textiles:

1 Agar-plate—zone of inhibition test (Ruehle & Brewer 1931).
2 Passage of Nutrient Broth repeatedly through fabric followed by normal plate count procedures (Thomas & van den Ende 1941).
3 Air sampling from the fabric (Lemon 1943; Hughes & Davis 1970).
4 Percussion and air sampling (Puck *et al.* 1946; McQuade & Sutherland 1960).
5 Manometric measurements (Mandels & Siu 1950).
6 Maceration (Church & Loosli, 1953; Wilkoff *et al.* 1969; Andersen 1969; Nichols 1970).
7 Sweep technique—drawing a Petri dish across the fabric surface—(Blowers & Wallace 1955).
8 Determination of ammonia production by *Proteus* spp. (Latlief *et al.* 1951; Sherrill 1956).
9 Overlay of sample with agar (Quinn 1962; Charnley & Eftekhar 1969; Lashen 1971).
10 Agitation (Wilksell *et al.* 1973; Cody *et al.* 1984).

Vigo (1978) has reviewed the methods available for testing antibacterial and antifungal activity on textile fibres. Vigo & Benjaminson (1981) have concluded that:

> 'none of the quantitative or qualitative test methods approved and in use are entirely satisfactory because they do not address the question of (*biocide*) efficacy in use'.

Laboratory Testing Techniques for Substrate Susceptibility and Biocide Efficacy

Methods selected for testing of wool textile materials for susceptibility to microbial attack or to assess biocide performance will be determined by the nature of the information required (e.g. problem solving or testing to contract specification).

Sample preparation and biocide application

Samples of biocide-treated yarn or fabric as received from industrial sources are normally tested 'as received' or after being subjected to various durability procedures. The samples may be treated as follows:

Water leaching

Immerse the textile sample in a large beaker or flask of tap water flowing at a rate which will give approximately two to three volume changes per hour. Use an initial liquor ratio of 1:100 (w/w). Continue the process for 24 h. Air dry the samples at room temperature on a 'washing-line' arrangement composed of two laboratory clamps and connecting string.

Heat ageing

Expose the sample in a dry oven at between 100 and 105 °C for 24 h. Ensure that the oven is well ventilated.

Light exposure

Expose the samples as specified in method B02 of *BS 1006* (Anon. 1978).

Weathering

Expose the samples as specified in method B04 of *British Standard 1006* (Anon. 1978). Durability trials are normally required only for very specialized applications (e.g. woollen felt for use in paper-making).

Application of biocides

For laboratory biocide applications, the following test cloths may be used:

1 100 % worsted flannel (style 526) supplied by Testfabrics Inc., Middlesex, NJ, USA.

2 100 % wool cloth supplied by the Society of Dyers and Colourists (SDC), Bradford, UK.

3 100 % cotton control cloth (Walton & Allsop 1977) to confirm inoculum viability supplied by The Shirley Institute, Manchester, UK.

Biocides may be applied to the test fabrics in an aqueous solution with a Jeffreys Trial Dyeing rig (or comparable equipment). The Jeffreys rig consists of a large water-filled reservoir with a central heating element and temperature controls. Large glass test tubes containing the appropriate biocide concentration are placed in the water bath and allowed to equilibrate. Fabric samples are attached to metal strips which have hooks on one end (for the fabric) and circular hoops at the other for attachment to a central moving assembly. A motor drives the central metal support providing a continuous up and down motion. The fabric samples are drawn through the body of the liquid, thus providing continuous agitation at a predetermined rate. The conditions in the test tubes may be adjusted as appropriate to conform to conditions in an industrial dyeing winch or scouring set. For example, test samples of SDC wool were immersed (60 oscillations per minute) in a 1 % (v/v) solution of the textile biocide, Mystox QL (liquor ratio 30:1), at 55 °C for 30 min. The system of application will be determined by the nature of the biocide, the particular processing conditions (i.e., duration of process, temperature, etc.), the textile substrate and the degree of protection required. For comparing biocide efficacy, a uniform system (determined by the individual constraints of the industrial problem) is required. Biocides may also be applied using full scale processing machinery viz.:

1 application to yarn packages using a Morel conditioning machine;
2 direct application to yarn *via* a lick roller on conventional spinning machines;
3 incorporation into an oil emulsion applied to loose stock before carding and spinning, and
4 direct application to dyeing machines and scouring sets.

Biocides may be applied also during industrial processing, under controlled conditions, with samples taken as required. This method provides the most useful information, i.e. an assessment of biocide efficacy following application in the mill environment.

Standard test methods

The most widely used and accepted testing standards employed by government and industry alike in determining antimicrobial effectiveness in textiles are those set forth by the American Association of Textile Chemists and Colourists (AATCC)* and by the British Standards Institution (BSI). A wide variety of national and industrial test methods are under review by the ISO/TC 38 Textiles Committee of the International Organisation for Standardisation (ISO). A draft International Standard is expected some time in 1985/6. The following discussion of the major

*Technical Manuals are published annually by the American Association of Textile Chemists and Colorists P.O. Box 12215, Research Triangle Park, NC 27709, USA.

standard methods aims to provide a practical workbench summary of the techniques involved. The reader should refer to the standard texts as appropriate.

The foreword to *British Standard 6085* (Anon. 1981b) states that the general purpose of these methods is to determine susceptibility of textiles to micro-organisms and to evaluate the resistance of treated textiles and textile-containing products to attack by micro-organisms. The standard specifies three areas of testing whereby specimens of the material under test are subjected to microbial attack by soil burial, mixed inoculation on an agar plate or in a saturated atmosphere. Biocide efficacy is observed as a significant reduction in strength loss compared with similar material without biocide addition, or prevention of visual surface discolouration.

Soil burial

This is a very severe test for most wool products. Fabric or yarn specimens (150 × 25 or 600 mm) are buried to a depth of 30–50 mm in John Innes No. 1 potting compost in plastic or metal trays. A loose fitting lid is required to maintain the moisture content of the soil between 20–30 %. Small, weighed samples of the compost, taken at random, and dried to constant weight at 105 °C may be used to calculate the moisture content and then discarded. Glass plates are used to separate neighbouring fabric specimens in the trays. Ten specimens of the material under test are required and five of a cotton, control cloth (which should show a strength loss of at least 80 %) for each experiment. Place the prepared soil trays in an incubator at 28 ± 1 °C with a relative humidity of at least 95 % r.h. Remove only the cotton control cloths after 7 days incubation; remove particles of soil by gentle agitation and brushing before rinsing with water. Sterilize the specimens or fragments by soaking in 70 % (v/v) ethanol for 4 h at ambient temperature, allow to condition (at 20 ± 2 °C and 65 ± 2 % r.h.) and determine residual tensile strength (compared with unexposed specimens) in accordance with BS 2576 (fabric; Anon. 1977) or BS 1932: Part 1 (Anon. 1965). If a strength loss of > 80 % has been achieved, incubate the wool specimens for a full 28 d. Prepare the specimens for testing as above and record as a percentage the observed strength loss between the exposed material and the unexposed controls.

Mixed-inoculum agar plate test

The organisms specified in the standard (Table 2) may be obtained from the Commonwealth Mycological Institute, Kew, U.K.

The cultures (Fig. 3) may be used only when they are between 14–28 d old. Cultures less than 21 d old may be stored, securely stoppered, in a refrigerator at 4 °C for up to 3 months prior to use. Prepare a mineral salts solution as follows (g/l of distilled water): KH_2PO_4, 0.7; K_2HPO_4, 0.3; $MgSO_4.7H_2O$, 0.5; KCl, 0.5; $FeSO_4$, 0.01; $NaNO_3$, 2.0.

All the chemicals must be of 'Analar' grade as supplied by BDH. Add 15 g of agar (Bacto: Difco) per l of mineral salts solution, or 0.05 % (w/v) dioctyl sodium

TABLE 2. *Cultures specified for testing in accordance with BS 6085 : 1981*

Name	Strain	Culture collection number
Aspergillus niger	V. Teighem	ATCC 6275
Aspergillus terreus	Thom	PQMD 82j
Aureobasidium pullulans	(De Barry) Arnaud	ATCC 9348
Paecilomyces variotii	Bainier	IAM 5001
Penicillium funiculosum	Thom	IAM 7013
Penicillium ochro-chloron	Biourge	ATCC 9112
Scopulariopsis brevicaulis	(Sacc.) Bain Var Glabra Thom	IAM 5146
Trichoderma viride	Pers. Ex. Fr.	IAM 5061
Chaetomium globosum		CMI 45550

sulphosuccinate (BDH) to form a mineral salts/wetting agent solution. Add 30 g/l sucrose to the agar, if necessary, to encourage fungal growth. Gently add 10 ml of mineral salts/wetting agent to each of the nine culture slopes. With a sterile wire loop, scrape the surface of the slope to liberate spores without unduly detaching mycelial fragments. Decant the nine suspensions into a 500 ml sterile flask, containing a number of glass beads, and shake vigorously. Filter through a glass filter funnel containing sterile cotton wool to remove large clumps. If a large number

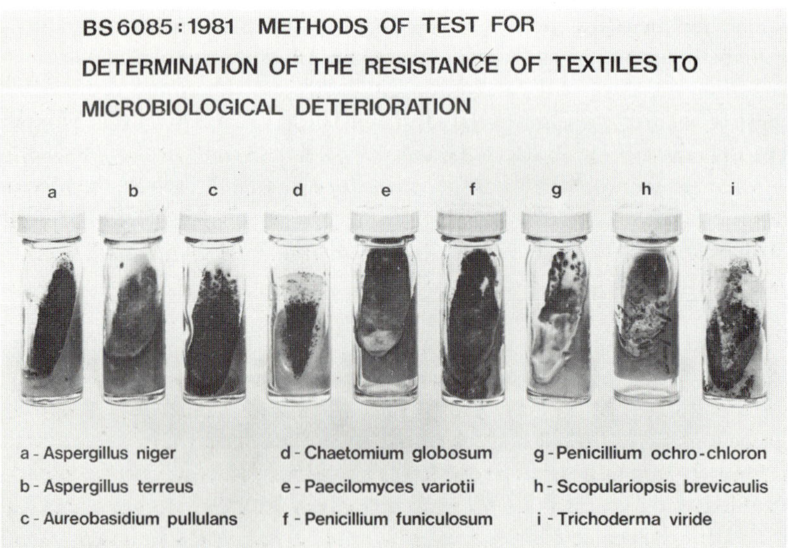

FIG. 3. Cultures specified in British Standard 6085 : 1981

of specimens are to be inoculated, pool the suspensions prepared from a number of fungal sets. Centrifuge the mixed spore suspension three times at 3500 **g**, discarding the supernatent and re-suspending the spores in sterile mineral salts solution. Wet out the test specimens (2.5 × 2.5 mm for visual assessment—four specimens per sample) and controls by immersing the fabrics in water containing 0.05 % (w/v) dioctyl sodium sulphosuccinate for several minutes. Place each fabric piece on individual mineral salt agar in Petri dishes and add 0.5 ml of the mixed inoculum to each side of the material. Work in a microbiological safety cabinet when manipulating the cultures. Autoclave and discard unused spore suspensions at the end of the day. Assess the cotton controls after 14 d and the wool specimens after 28 d. Strength loss may be assessed as above. Alternatively assess specimens visually in accordance with Table 3.

TABLE 3. *Assessment of visible growth on test specimens*

Observed growth	Grading
None	0
Growth only apparent under the microscope	1
Traces of growth (< 10%) visible to the naked eye	2
Light growth (10 to 30 %)	3
Medium growth (30 to 60 %)	4
Heavy growth (> 60%)	5

Examples of visual surface discolouration on both cotton control and wool specimens are shown in Fig. 4. The top left hand corner of the Figure shows two untreated (Grade 0) and two exposed (Grade 5) cotton specimens. Wool samples treated with sufficient biocide to proof the material should show a grading of 0 after 28 d incubation (see bottom right in Fig. 4).

Saturated atmosphere test

This may also be used for assessing visual only (V) the efficacy of biocides. The procedure is intended for textiles that will not come into contact with fungi in the vegetative form but may be contaminated with airborne spores before storage in humid conditions. Inoculate 1 ml of the mixed spore suspension (prepared as above) onto the cloth specimens and controls. Suspend the fabrics above free water (> 90 % r.h.) in sealed glass Kilner jars. Incubate for 14 d and assess the cotton controls. If the controls are graded four or above, continue the trial and assess the wool specimens after 28 d.

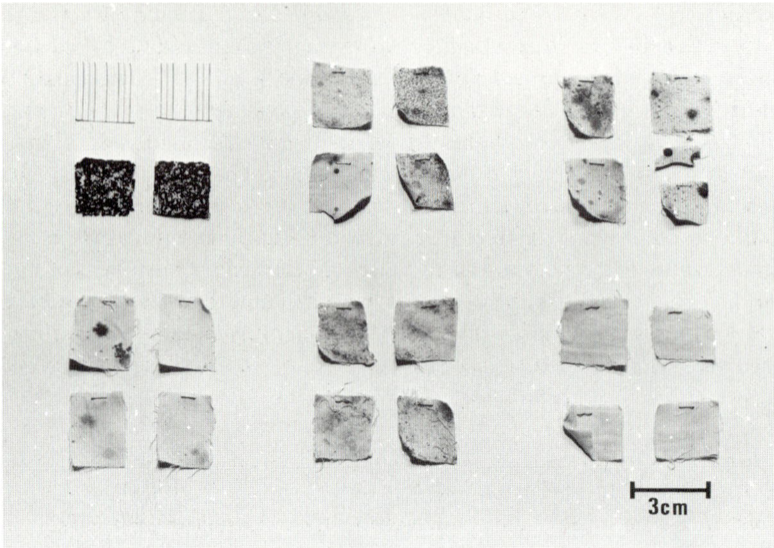

FIG. 4. Biocide treated and untreated specimens tested in accordance with British Standard 6085:1981

AATCC Method 147–1982 'Antibacterial Activity of Fabrics, Detection of: Parallel Streak Method' (see p. 90)

The method (Anon. 1982) is adaptable for biocide-treated wool yarns, loose stock, and most carpet samples as well as fabric (Arnold 1984). The defined objective is the detection of bacteriostatic activity exhibited by textile fabrics. The bacterial species cited in the standard are given in Table 4. Details of maintenance of culture of the test organisms are presented in the Standard.

TABLE 4. *Cultures specified in AATCC 147-1982**

Test bacteria	Culture collection number
Staphylococcus aureus	ATCC 6538
Klebsiella pneumoniae	ATCC 4352
Salmonella cholerasuis	ATCC 10708
Pseudomonas aeruginosa	ATCC 13388

*Prepare AATCC bacteriostasis broth (g/l distilled water): peptone (Difco Laboratories) 10; beef extract (Difco Laboratories) 5,
and sodium chloride (BDH, Ltd).

Prepare AATCC bacteriostasis broth (g/l distilled water): peptone (Difco) 10; beef extract (Difco) 5, and NaCl (BDH, Ltd) 5. Adjust the medium to pH 6.8 with NaOH. Add 15 g of Bacto (Difco) agar/1, adjust pH 7.0 to 7.2 with NaOH to prepare AATCC Bacteriostasis agar. Prepare the inoculum by transfering 1 ml of a 24-h broth culture (37 °C) of a test organism to 9 ml sterile distilled water control. After adequate mixing with a 4 mm inoculating loop, transfer one loopful of diluted inoculum to AATCC agar in a Petri dish; make five approximately 7.5 cm long parallel streaks 1 cm apart without refilling the loop. Care must be taken not to disrupt the surface of the agar. Such streaks would reduce the area of contact between the specimen and the seeded agar layer. Locate the textile specimen (normally 1 to 4 cm diameter) transversely across the inoculum streaks and ensure intimate contact. Small sterile glass weights may be used to prevent the material curling away from the agar. Incubate all samples at 37 °C for 24 h.

Results may be evaluated in two ways:

1 growth free zone—assess the width (in mm) of the growth free zone surrounding the test specimens, and

2 contact inhibition—lift the specimens from the agar surface and, using a low power binocular microscope, assess the percentage growth—free contact area.

The latter method may be used to provide a rough estimate of biocide efficacy in that the count of the inoculum will decrease from initial contact on the nutrient agar to the end of the final streak, resulting in increasing degrees of sensitivity. *Staphylococcus aureus* and *Pseudomonas aeruginosa* may be selected as representative organisms.

AATCC 30–1981 'Fungicides, Evaluation of Textiles: Mildew and Rot Resistance of Textiles'

This standard (Anon. 1981*c*) is arranged in three sections. Test 1 presents a soil-burial method which may be used as an alternative to BS 6085:1981. The agar plate—pure culture—sterile specimen method, which forms that basis of Test 2 is for cellulosic materials only.

Test 3 is recommended for evaluation of textile specimens where surface growing organisms are important. The method involves the growth of *Aspergillus niger* (ATCC 6275) on glucose mineral salts agar. Prepare a spore suspension from 10–14 day-old cultures of the test organism by scraping the fungal mycelia growing on agar slopes into a sterile flask containing 50 ml of sterile water and glass beads. Shake the flask thoroughly to prepare a spore suspension. Prepare the following culture medium (g/l of distilled water): $(NH_4)\ NO_3$, 3.0; KH_2PO_4, 2.5; $K_2H\ PO_4$, 2.0; $MgSO_4.7H_2O$, 0.2; $FeSO_4.7H_2O$, 0.1; glucose, 30.0; agar (Difco), 20.0.

Autoclave at 121 °C for 15 min; wet the specimens (3 × 3 cm) in water containing 0.05 % (w/v) dioctyl sodium sulphosuccinate (BDH); position on the poured medium in a Petri dish and inoculate evenly the surface with 1 ml of the spore suspension using a sterile dispenser. Incubate the inoculated and control specimens at 28 °C for 14 d only. Examine the specimens visually and microscopically for evidence of the growth of *Asp. niger*. The initial specimens may be subjected to various pre-treatments before testing (e.g. leaching, weathering, exposure to dry heat). Test 1 may be used to demonstrate long term biocide proofing of material, and test 3 may be used to assess biocides applied during processing to provide temporary antifungal protection during manufacture.

AATCC 100–1981 'Antibacterial Finishes on Fabrics, Evaluation of'

This method (Anon. 1981*d*) presents four qualitative techniques for demonstrating bacteriostatic activity. Specimens of treated and control fabrics (3 × 3 cm) may be tested qualitatively for antibacterial efficacy by any one of the following methods:

1 AATCC Agar plate method (AATCC Test Method 90: Anon. 1981*e*)
2 Streak test.
3 Majors test (Majors 1959)
4 Quinn test (Quinn 1962).

For example, in the 'Streak test', streak sufficient agar (AATCC Bacteriostasis agar) in Petri dishes with either *Staph. aureus* or *Klebsiella pneumoniae*. Place individual treated or control specimens at right angles to the direction of streak. Examine for clear areas (no growth) either visually or with a low power microscope.

Specimens showing activity are tested quantitatively. Prepare sufficient circular specimens (4.8 cm diameter) of the test (and controls—made of the same material but untreated) fabric to absorb 1 ml of test inoculum (note the number used). Apply 1 ml of an appropriate dilution of a 24-h culture of the test organism (one of the above) so that recovery at time 0 from the control or the sample is 1 to 2 × 10^5. Dilute using AATCC broth. Sterilize the wool samples before the inoculation using ethylene oxide or intermittent sterilization. Place the discrete sets of specimens in 0.237 l (8 oz) sterile glass jars and apply 1 ml of the inoculum evenly to the specimens using a sterile dispenser. Include uninoculated but treated controls and seal the containers. For time 0 testing, add 100 ml of AATCC broth (or neutralizer solution if available) and shake vigorously for 1 min. Make serial dilutions in AATCC broth and plate (in duplicate) on Tryptone Glucose Extract agar (Difco). Incubate the remaining jars at 37 °C for 18–24 h and plate out as above. Incubate all plates at 37 °C for 48 h. Report bacterial counts as colony forming units (cfu) of bacteria per fabric sample (i.e. the number in the jar). Calculate percentage reduction of each test organism for each fabric treatment. Uninoculated samples should be sterile and increasing numbers of organisms should be observed on the inoculated, untreated control.

The method is intended for biocide treatments which may not be detectable by agar plate techniques. It is cumbersome, time-consuming and difficult to reproduce (Arnold 1984). Work is in progress to replace the serial dilution evaluation stage with a membrane filter procedure which, it is hoped, will increase reproducibility and reduce the time necessary to perform the test.

Additional national test standards for testing wool textiles

Australian Standard 1157.1972 Part 2—Resistance of Textiles to Fungal Growth

In section 1 (Anon. 1972*a*) sterile test specimens are inoculated with *Asp. niger* van Tieghem (*Defence Standard Laboratory Culture Collection No. 72*) and then incubated for 14 d. Assessments are then made of surface fungal growth.

South African Bureau of Standards SABS Method 472

This Standard (Anon. 1967) describes tests for the attack by *Asp. niger* SABS Culture No. 70.

South African Bureau of Standards SABS Method 484

This Standard (Anon. 1973) is concerned with resistance to the fungal attack by mixed culture.

Canadian Standard CAN2–4.2–M77 Method 28.2–1972 Method of test for resistance to micro-organisms. Surface-Growing Fungus Test-Pure culture

This method (Anon. 1972*b*) is normally used to assess biocide efficacy. *Aspergillus niger* (USDA 215–4247) is again used as the inoculum with growth assessed after 14 d. The extent of growth of the test organism on the specimens is determined together with any staining or colour changes.

Recent Developments in Laboratory Biocide Testing Techniques

The following method has been suggested for testing biocide-treated materials and may be applied directly to wool textiles.

Dow Corning Corporate Test Method (CTM)-0923 'Antimicrobial activity-Dynamic test of surfaces'

The method (McGee & Gettings 1985) is performed by enumeration of microbes in a flask of phosphate buffered water, both before and after a pre-determined shaking. The challenge inoculum is assessed spectrophotometrically before application and verified by viable counts based on the pour plate procedure. Treated and untreated substrates, along with an inoculum-control flask containing only buffer and inoculum are monitored to demonstrate that antimicrobial activity is a result of the treatment, and not a characteristic of some other material on the substrate or simply a moribund culture. McGee & White (1984) cite an example whereby 0.75 g of treated material was shaken for 1 h in a test system containing 7.5×10^5 to 1.5×10^6 viable cells of *Klebsiella pneumoniae* ATCC 4352 in phosphate buffer at room temperature. They stress that clear differentiation between treated and untreated samples must be obtained. The survival of the challenge organism is monitored with time.

Laboratory bioluminescence techniques applied to textile spoilage and biocide testing

Growth of micro-organisms on wool textile materials may be assessed by the firefly bioluminescent detection and/or assay of adenosine triphosphate (ATP).

Quantitative assessment

Prepare a mixed-spore suspension in accordance with BS 6085 (Anon. 1981*a*). Likewise prepare all fabric specimens (pre-weighed and of uniform size) in accordance with Section Three (agar plate test) of the standard. The initial stages of the test method were adopted to prepare standard contaminated conditions to encourage fungal growth. Incubate all specimens (three replicates per test material per assay time) at 28 °C. At each time interval (normally three hour intervals throughout the working day), place the test specimens in individual sterile bottles containing 15 ml of sterile, buffered rinse solution. Prepare a stock buffer solution by dissolving 34 g of KH_2 PO_4 in 500 ml of distilled water adjusted to pH 7.2, 0.1 with 1M-NaOH, and dilute to 1 litre with distilled water. To prepare a working solution of buffered distilled water, add 1.25 ml of the stock solution to 1 litre of distilled water. Prepare 0.02 % (v/v) solution of Tween 80 (BDH) in buffered distilled water. Sterilize the rinse solution by autoclaving at 121 °C for 15 min. Place the bottles in an ultrasonic tank containing a 0.3% (v/v) solution of Tween 80 and sonicate at 25 kHz for 2 min (Anon. 1980). Determine the bulk ATP concentration by assaying three 100 µl samples of the resulting liquor. Dilute test solution where necessary. A Biocounter M2010 (Lumac B.V) and corresponding Lumac reagents (ATP standards, nucleotide-releasing reagents and luciferin mixtures) may be used (Comber, Stockport, UK) for ATP assays (McCarthy 1983*b*; 1984). Samples may be boiled in trichloroacetic acid to extract ATP from the samples (Lundin & Thore 1975). Test 'time 0' samples immediately following addition of the mixed spore suspension. Inoculate all samples simultaneously and monitor growth of the inoculum at regular intervals over a period of 24–72 h (or longer if necessary). Express results as an 'Activity Index' with units of picograms of ATP/0.1 g of test substrate, derived as a mean result from the three replicates. Figure 5 shows the growth of the mixed inoculum (recorded in terms of increasing ATP concentrations) on the SDC wool fabric. The assay thus provides a rapid non-species-specific test for microbial growth, with significant increases in ATP levels on untreated wool materials. The presence of a biocide applied to fabrics will result in prevention of growth or the rapid killing of the organisms in the inoculum. A

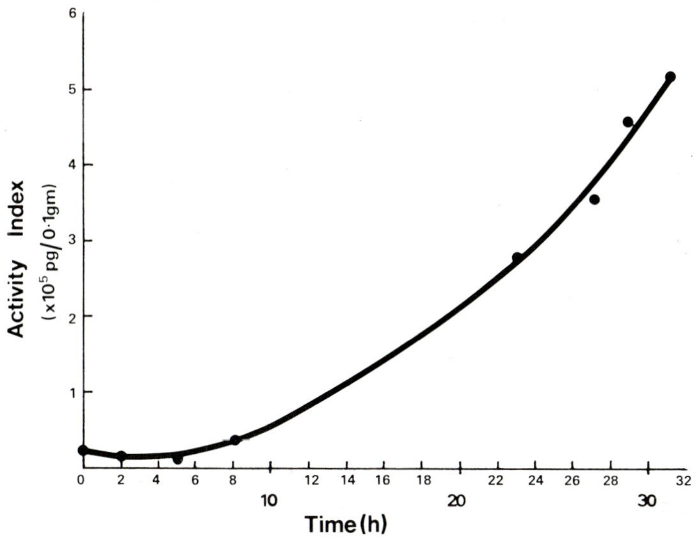

FIG. 5. Growth of mixed fungal inoculum on wool substrate

positive untreated control (with developing biomass) is required to demonstrate inoculum viability. The test may be modified (specifically with regard to testing frequency and test duration) to suit individual requirements.

Photographic detection

Prepare a mixed spore suspension and test specimens in accordance with Section Three of BS 6085 (Anon. 1981*a*). Prepare sufficient samples for regular (e.g. 3 h intervals throughout the working day) testing up to 24–72 h or longer if necessary, or alternatively single specimens for extended incubation duration before photographic testing. Place the specimen in a Perspex sample holder located on top of the shutter of an instant Polaroid Land film assembly (Green *et al.* 1984). Polaroid Land film (type 612, ASA 20 000) may be used to record light production. Add 100 μl of NRB (Nucleotide-releasing agent) followed by 100 μl of Lumit PM (luciferin-luciferase mixture) to the individual fabric specimens. Immediately expose the Polaroid film by opening the shutter for times up to 30 sec. Allow the film to develop for 40 sec and assess the results visually. 'Time 0' samples are tested again immediately following addition of the mixed spore suspension. An incubation period of approx 24–30 h is required for wool samples to develop sufficient biomass (i.e., ATP levels) to be detectable using this technique. This is due to the limited sensitivity of the instant film. If parallel trials are arranged involving biocide-treated and untreated samples, the efficacy of biocides as indicated by a negative photographic image may be assessed by comparison with positive (light producing) controls.

Image-intensifier detection

Prepare specimens as above but observe light emission using image-intensifier light amplification devices.

The above methods are presented as laboratory techniques.

Conclusions

Biocides may be used, where applicable, to protect wool and woollen materials during processing as a temporary treatment, to impart a durable anti-microbial, proofing finish, or to provide a surface-active hygienic effect. A variety of methods are available based on extended incubation techniques. Additional work is required to develop rapid and reproducible test methods for detecting and assessing biocide efficacy for in-house quality assurance purposes.

References

ALLSOPP, C. & ALLSOPP, D. 1983 An updated survey of commercial products used to protect materials against biodeterioration. *International Biodeterioration Bulletin* **19**, 99–146.

ANDERSEN, R. L. 1969 Biological evaluation of carpeting. *Applied Microbiology* **18**, 180–187.

ANON. 1950 The mildewing of wool: causes and prevention. *Wool Science Review* **6**, 31–42.

ANON. 1965 *Determination of Breaking Strength and Extension.* British Standard 1932. Part 1. London: British Standards Institution.

ANON. 1967 *Resistance of Fungal Attack by* Aspergillus niger SABS Culture No 70. South African Standard Method 4–72. Pretoria: South African Bureau of Standards.

ANON. 1972*a Resistance of Textiles to Fungal Growth.* Australian Standard 1157, Part 2. Sydney: Standards Association of Australia.

ANON. 1972*b Method of Testing for Resistance to Microorganisms. Surface-Growing Fungus Test—Pure Culture.* Canadian Standard CAN 2-4.2-M77, Ottawa: Canadian Government Specification Board.

ANON. 1973 *Resistance to Fungal Attack by Mixed Cultures.* South African Standard Method 484. Pretoria: South African Bureau of Standards.

ANON. 1977 *Methods for Testing the Strength of Textiles: Woven Fabrics—Determination of Breaking Strength and Elongation (Strip Method).* British Standard 2576. London: British Standards Institution.

ANON. 1978 *Methods of Testing for Colour Fastness of Textiles and Leather.* British Standard 1006. London: British Standards Institution.

ANON. 1980 *Standard Procedures for the Examination of Space Hardware.* NASA NHB 5340.18. Washington: National Aeronautics and Space Administration.

ANON. 1981*a Preservative Treatments for Textiles.* Part 1. *Specifications for Treatments.* British Standard 2087. London: British Standards Institution.

ANON. 1981*b Methods of Testing for Determination of the Resistance of Textiles to Microbiological Deterioration.* British Standard 6085. London: British Standards Institution.

ANON. 1981*c* AATCC Test Method 30–1981. *Fungicides. Evaluation of Textiles: Mildew and Rot Resistance of Textiles.* AATCC Technical Manual. Triangle Park NC: American Association of Textile Chemists and Colorists.

ANON. 1981*d* AATCC Test Method 100–1981. *Evaluation of Antibacterial Finishes on Fabrics.* AATCC Technical Manual. Triangle Park NC: American Association of Textile Chemists and Colorists.

ANON. 1981*e* AATCC Test Method 90–1981. *Antibacterial Activity of Fabrics, Detector of: Agar Plate Method.* AATCC Technical Manual. Triangle Park NC: American Association of Textile Chemists and Colorists.

ANON. 1982 AATCC Method 147–1982. *Detection of Antibacterial Activity of Fabrics: Parallel Streak Method.* AATCC Technical Manual. Triangle Park NC: American Association of Textile Chemists and Colorists.

ARNOLD, L. B. 1984 Antimicrobial activity on carpet. *Carpet and Rug Industry* **12**, 22–27.

BATSON, D. M., TENNISSON, D. J. & PORGES, N. 1944 Study of a soil burial method for determining rot-resistance of fabrics. *American Dyestuff Reporter* **33**, 423–454.

BLOWERS, R. & WALLACE, K. R. 1955 The sterilization of blankets with cetyl trimethylamine bromide. *Lancet* **i**, 1250.

BOBKOVA, T. S., ZLOCHEVSKAYA, I. V., CHEKUNOVA, L. N., KIRKINA, L.I. & MONAKHOVA, R. I. 1975 Use of a rapid method of soil testing for evaluation of the biodeterioration resistance of textile materials. *Vestnik Moskovskoyo Universiteta Biologia, Pochvovedenie* **30**, 55–59.

BROWN, J. C. 1959 The determination of damage to wool fibres. *Journal of the Society of Dyers and Colourists* **75**, 11–21.

BURGESS, R. 1924 Studies on the bacteriology of wool. *Journal of the Textile Institute* **15**, T575–583.

BURGESS, R. 1934 The use of trypsin for the determination of the resistance of wool fibres to bacterial disintegration. *Journal of the Textile Institute* **25**, T289–294.

BURGESS, R. 1954 Micro-organisms and textiles. *Journal of Applied Bacteriology* **17**, 230–245.

BURGESS, R. & GALLOWAY, L. D. 1940 *Applied Mycology and Bacteriology* 2nd ed. London, Leonard Hill.

BURGESS, R. & RIMINGTON, C. 1929 A technique for the microscopical examination of wool fibres. *Journal of the Royal Microscopic Society* **49**, 341–347.

CARPENTER, D., COLMAN, M., HANSEN, T., HAAS, D., LESHNIOWSKY, W., MELLO, J., PHILLIPS, C., REICH, R. & SHEIDER, P. 1981 Methods of bioburden evaluation. *Developments in Industrial Microbiology* **22**, 323–328.

CHARNLEY, J. & EFTEKHAR, N. 1969 Penetration of gown material by organisms from the surgeon's body. *Lancet* **i**, 172–174.

CHURCH, B. D. & LOOSLI, C. G. 1953 The role of the laundry in the recontamination of washed bedding. *Journal of Infectious Diseases.* **93**, 65–74.

CODY, H. J., SMITH, P. F., BLASER, M. J., LAFORCE, F. M. & WANG, W. L. 1984 Comparison of methods for recovery of *Escherichia coli* and *Staphylococcus aureus* from seeded laundry fabrics. *Applied and Environmental Microbiology* **47**, 965–970.

EGGINS, H. O. W. 1967 The economics of biodeterioration. *Environmental Engineering* **29**, 15–16.

EGGINS, H. O. W. 1983 Biodeterioration past, present and future. In *Biodeterioration* 5. ed OXLEY, T. A. & BARRY, S. pp. 1–9. London: JOHN WILEY.

EGGINS, H. O. W., MALIK, K. A. & SHARP, R. F. 1968 Some techniques to investigate the colonisation of cellulosic and wood substrates. In *Proceedings of First International Biodeterioration Symposium.* eds. WALTERS, A. H. & ELPHICK, J. J. pp. 120–131. London: Elsevier.

EGGINS, H. O. W. & OXLEY, T. A. 1980 Biodeterioration and biodegradation. *International Biodeterioration Bulletin* **16**, 53–56.

ENGLISH, M. P. 1965 The saprotrophic growth of nonkeratinophilic fungi on keratinized substrata and a comparison with keratinophilic fungi. *Transactions of the British Mycology Society* **48**, 219–235.

FRASER, R. D. B. & GILLESPIE, J. 1976 Wool structure and biosynthesis. *Nature,* **261**, 650–654.

GARNER, W. 1967 *Textile Laboratory Manual.* 3rd edn. Vol 6 p. 153. London: Heywood.

GEIGER, W. B., PATTERSON, W. I., MIZELL, L. R. & HARRIS, M. 1941 The nature of the resistance of wool to digestion by enzymes. *Journal of Research of the National Bureau of Standards* **27**, 459–468.

GEORGIEWICS, G. 1924 *Lehrbuch Der Chemisten Technologie Der Gespinstfasern Deuticke.* Leipzig and Vienna.

GRAY, W. D. & MARTIN, G. W. 1947 Improvements on the soil-burial testing method. *Mycologia* **39**, 358–369.

GREEN, K., KRICKA, L. J., THORPE, G. H. G. & WHITEHEAD, T. P. 1984 Rapid assays based on immobilized bioluminescent enzymes and photographic detection of light emission. *Talanta* **31**, 173–176.

GRIMM, H. & KUHNE, C. 1969 The standardization of earth rotting tests of the testing of rot-proofing finishes to cellulosic fibres. *Deutscher Textiltechnie* **19**, 46–49.

HIRST, H. R. 1927 Ultra-violet radiation as an aid to textile analysis. *Journal of the Textile Institute* **18**, 369–375.

HUGHES, W. H. & DAVIES, R. R. 1970 Bacteria and cadmium-treated fabrics. *British Medical Journal* **1**, 430.

ILYICHEV, V. D. 1979 Rtitzy-istochnik bioprovrezhdenii. *Problemy zashchity ot bioprovrezhdenii.* pp. 47–53. Moscow: Znanie.

JAIN, P. C. & AGRAWAL, S. C. 1980 A note on the keratin decomposing capability of some fungi. *Transactions of the Mycology Society of Japan* **21**, 513–517.

KEMPTON, A. G., MAISEL, H & KAPLAN, A. M. 1963 Study of the deterioration of fungicide-treated fabrics in soil burial. *Textile Research Journal* **33**, 87–93.

LASHEN. E. S. 1971 New method for evaluating antimicrobial activity directly on fabric. *Applied Microbiology* **21**, 771–773.

LATLIEF, M. A., GOLDSMITH, M. T., FRIEDL, J. L. & STUART, L. S. 1951 Bacteriostatic, germicidal and sanitising action of quaternary ammonium compounds on textiles. I. Prevention of ammonia formation from urea by *Proteus mirabilis. Journal of Paediatrics* **39**, 730–737.

LEMON, H. M. 1943 A method for collection of bacteria from air and textiles. *Proceedings of the Society of Experimental Biology and Medicine* **54**, 298–301.

LEWIS, J. 1981 Microbial biodeterioration. In *Economic Microbiology* ed. Rose, A. H. pp. 81–130. London: Academic Press.

LLOYD, A. D. 1955 A soil-infection method for the testing of textiles for resistance to microbiological attack. *Journal of the Textile Institute* **46**, 653–661.

LLOYD, A. D. 1965 An adhesive tape technique for the microscopial examination of surfaces supporting or suspected of supporting mould growth. *International Biodeterioration Bulletin* **1**, 10–13.

LUNDIN, A. & THORE, A. 1975 Comparison of methods for extraction of bacterial adenine nucleotides determined by firefly assay. *Applied Microbiology* **30**, 713–721.

MAHALL, K. 1982 Biodegradation of textile fibres as seen under the microscope—actual case histories. *International Textile Bulletin* **4**, 280–292.

MAJORS, P. 1959 Evaluation of the effectiveness of antibacterial finishes for cloth. *American Dyestuff Reporter* **48**, 91–93.

MANDELS, G. R. & SIU, R. G. 1950 Rapid assay for growth: microbial susceptibility and fungistatic activity. *Journal of Bacteriology* **60**, 249–262.

MANDELS, G. R., STAHL, W. & LEVINSON, H. 1948 Structural changes to wool degraded by ringworm fungus. (*Microsporum gypseum*). *Textile Research Journal* **18**, 224–231.

MCCARTHY, B. J. 1983*a* Biodeterioration in wool textile processing. *Biodeterioration 5.* ed. Oxley, T. A. & Barry, S. pp. 519–527. London: John Wiley.

MCCARTHY, B. J. 1983*b* Bioluminescent assay of microbial contaminants on textile materials. *International Biodeterioration Bulletin* **19**, 53–57.

MCCARTHY, B. J. 1984 Bioluminescent determination of microbial activity on textiles; In *Analytical applications of bioluminescence and chemiluminescence* ed. Kricka, L. J., Stanley, P. E., Thorpe, G. H. G. & Whitehead, T. P. pp. 46–49. London: Academic Press.

McGee, J. B. & Gettings, R. L. 1985 Method for the evaluation of immobilized antimicrobial agents. In: *Book of Papers,* 1985 International Conference of the American Association of Textile Chemists and Colorists.

McGee, J. & White, W. C. 1984 Evaluation of surface bonded cationic antimicrobials. *Carpet and Rug Industry* **12**, 24–41.

McQuade, A. B. & Sutherland, W. J. A. 1960 An improved device for sampling bacterial populations on blankets. *Journal of Hygiene,* **58**, 157–158.

McQuade, A. B. 1964 Microbiological degradation of wool. *Dermatologia* **128**, 249–266.

Mebes, B. 1975 Hygiene finishing in the dyebath. *International Textile Bulletin* **2**, 122.

Mulcock, A. P. 1965 *Peyronellaea glomerata*—a fungus growing with the fibres of the unshorn fleece. *Australian Journal of Agricultural Research* **16**, 691–697.

Mulcock, A. P. 1965 The fleece as a habitat for micro-organisms. *New Zealand Veterinary Journal* **13**, 87–93.

Nichols, P. S. 1970 Bacteria on laundered fabrics. *American Journal of Public Health* **60**, 2175–2180.

Nopitsch, M. 1953 Micro-organic attack on texiles and leather. *CIBA Review* No. 100, pp. 3582–3610.

Osborn, J. G. B. 1912 Moulds and mildews: their relation to the damaging of grey cloth and prints. *Journal of the Society of Dryers and Colourists* **28**, 204–208.

Pauly, H. 1904 Ueber die konstitution des histidins (I) *Zeitschrift fur physiologische chemie* **42**, 508–518.

Puck, T. T., Robinson, O. H., Wise, H., Loosli, C. G. & Lemon, H. M. 1946 The oil treatment of bedclothes for the control of dust-borne infection. 1. Principles underlying the development and use of a satisfactory oil-in-water emulsion. *American Journal of Hygiene* **43**, 91–104.

Quinn, H. 1962 A method for the determination of the anti-microbial properties of treated fabrics. *Applied Microbiology* **10**, 74–78.

Race, E. 1946 Problems in the microbiology of protein fibres. *Journal of the Society of Dyers and Colourists* **62**, 67–85.

Raschle, P. 1983 Contribution to the examination of the rot resistance of textiles. *International Biodeterioration Bulletin* **19**, 13–17.

Ruehle, G. & Brewer, C. 1931 Unites States Food and Drug Administration *Methods of Testing Antiseptics and Disinfectants.* Circular No. 198. Washington: US Department of Agriculture.

Safranek, W. W. & Goos, R. D. 1982 Degradation of wool by saprophytic fungi. *Canadian Journal of Microbiology* **28**, 137–140.

Sankov, E. A., Suchkova, G. C. & Andreeva, K. I. 1972 Investigation of biological wool fibre damage by the dyeing method. *Technical Textiles Industry U.S.S.R.*, **4**, 152–153.

Seal, K. & Allsopp, D. 1983 Investigative Biodeterioration. In *Biodeterioration 5* ed. Oxley, T. & Barry, S. pp. 528–534. Chichester: John Wiley.

Sherrill, J. C. 1956 The evaluation of bacteriostatic reagents and methods of application to textile fabrics. *Textile Research Journal* **26**, 342–350.

Thomas, J. C. & van den Ende, M. 1941 The reduction of dust-borne bacteria in the air of hospital wards by liquid paraffin treatment of bedclothes. *British Medical Journal* **1**, 953.

Vigo, T. L. 1978 Antibacterial fibres. In *Modified cellulosics.* Symposium on cellulose, paper and textiles. pp. 259–284 Division of the American Chemical Society.

Vigo, T. L. & Benjaminson, M. A. 1981 Antibacterial fibre treatments and disinfection. *Textile Research Journal* **51**, 454.

von Bergen, W. & Mauersberger, H. R. 1948 *American Wool Handbook.* Textile Book Publishers Incorporation.

Walton, D. W. H. & Allsopp, D. 1977 A new test cloth for soil burial trials and other studies on cellulose decomposition. *International Biodeterioration Bulletin* **13**, 112–115.

Wehrner, C. 1902 Stains on textiles caused by *Aspergillus fumigatus. Journal of the Society of Dyers and Colourists* **18**, 112.

Wellman, R. H. & McCallan, S.E.A. 1945 *Office of Strategic Research and Development Report 5683* Washington: US National Defense Research Communication.

Wiksell, J. C., Pickett, M. A. & Hartmann, P. A. 1973 Survival of micro-organisms in laundered polyester cotton sheeting. *Applied Bacteriology* **25**, 431–435.

Wilkoff, L. J., Westbrook, L. & Dixon, G. L. 1969 Factors affecting the persistance of *Staphylococcus aureus* on fabrics. *Applied Microbiology*, **17**, 268—274.

Preservation Treatments of Textile Materials and their Assessment

D. S. WALES, P. F. HAMLYN, J. E. TODD AND B. F. SAGAR

Shirley Institute, Didsbury, Manchester, UK

Microbiological Deterioration of Textile Materials

Under suitable conditions micro-organisms which normally inhabit soil, water or air can become established and proliferate on textile materials. Species of microfungi, bacteria, actinomycetes and algae can cause biodeterioration of textiles.

Biodeterioration of textile materials may take several forms. One of the most striking effects is the serious loss of strength of cellulosic materials. This may be complete in a few days if conditions are conducive to the rapid growth of contaminating, cellulolytic micro-organisms. Although the vast majority of micro-organisms are not capable of breaking down the polymer structure of cellulose, some can utilize many of the organic substances which may be deliberately or adventitiously present on textile materials. The deleterious effects of the resultant mildew growth on the appearance of the textile are equally serious. The presence of starch-based size and filling agents, certain types of finish, coatings or residual lubricants applied during processing will encourage the growth of micro-organisms by providing an easily assimilable carbon source. Mildew attack of this kind is not confined to cellulosic textiles but may occur on synthetic fabrics such as those derived from polyamides, polyester, polypropylene or polyacrylonitrile. These polymers are intrinsically resistant to attack by micro-organisms but unsightly mildew growth can readily occur on such fibres if metabolizable organic compounds are present. Many micro-organisms produce pigments which can diffuse into the polymer material. Indeed they may become irreversibly fixed. Microbial staining of cellulosic and synthetic textiles involves no loss of mechanical properties. It is objectionable on aesthetic grounds.

Microbial activity can be minimized by keeping susceptible materials dry, as surface growth will occur only when the relative humidity is high. Chemical protection with a biocide is normally only required therefore with textiles likely to be used in conditions in which they remain wet or damp for long periods of time.

Preservatives in the Food,
Pharmaceutical and Environmental
Industries

0–632–01727–9

Natural fibres

Plant fibres such as cotton, flax, jute and hemp are very susceptible to attack by cellulolytic fungi (Mahall 1982). Indeed, complete degradation of cellulose can be effected by the enzymes produced by various cellulolytic microfungi. These complex extracellular enzyme systems catalyse the hydrolysis of the glycosidic bond in cellulose leading ultimately to its total conversion to glucose (Selby 1968).

The best protection against mildew is to make certain that the fabrics are dry when stored and that they remain so during storage. It needs to be stressed, however, that the total amount of water in a material is less important than how 'tightly' the water is held. The latter is related to the ratio of the vapour pressure of the water in the material to that of pure water at the same temperature and pressure. This ratio is called the water activity, the most valid measure of physiological drought in all types of substrates. It is numerically equivalent to the equilibrium relative humidity. If the water activity in the micro-environment of individual fibres is kept sufficiently low, then micro-organisms will not grow. Indeed cellulosic textiles protected from sunlight in the dry Egyptian desert have withstood the ages.

Cellulosic fabrics which are to be used out of doors (e.g. awnings, beach umbrellas, military uniforms, ropes, sails, tarpaulins, tents, truck and boat covers, shoes and shoe linings) may be treated with a fungicide to protect them from mildew damage and rotting.

Man-made fibres

The susceptibility to microbial deterioration of Viscose fibres is similar to that of the natural cellulose from which they are derived. Fibres made from wholly synthetic polymers (e.g. acrylic, Nylon, polyester and polypropylene fibres) are inherently resistant to attack by micro-organisms. This is probably due to their hydrophobicity, their inaccessibility to extracellular enzymes or to their unique chemical structure. Even though a synthetic fibre *per se* will not support microbial growth, low-molecular weight contaminants (e.g. residual traces of the caprolactam monomer of Nylon 6) and processing aids (e.g., lubricants and spinning oils) may provide sufficient nutrients for the modest (surface) growth of micro-organisms (Hueck 1974). In most cases such growth will not affect the strength of the fabric.

Plastics

Plastics are now important components of certain textile products. They are used, for example, in the manufacture of shower curtains or as coatings on fabrics used for foul-weather garments. Most products made mainly or entirely from polymers such as polyethylene or polypropylene do not support microbial growth (Klausmeier

& Andrews 1981). However, two types of polymer (viz. plasticized polyvinyl chloride (PVC) and polyurethanes) used extensively as coatings for textile materials do so. The propensity of PVC formulations to microbial attack is related to the amount and types of plasticizers, fillers, pigments and stabilizers etc., added during processing. Indeed many of the plasticizers used for PVC are metabolized readily by micro-organisms. The embrittlement that follows colonization by micro-organisms is due to the destruction of the plasticizer; the PVC backbone polymers are not attacked (Hamilton 1983). Because of the similarity of some naturally occurring chemical linkages with those found in polyurethanes, the polymer itself may be capable of supporting microbial growth (Darby & Kaplan 1968). As a practical measure, therefore, biocides are often added to both plasticized PVC and polyurethanes intended for textile applications involving a risk of microbiological attack.

Control of Biodeterioration

A biocide is usually incorporated in materials susceptible to microbiological attack especially if they are for use under conditions conducive to microbial proliferation. The attributes of an 'ideal' biocide (Hamlyn 1983) are easily listed:

1 effective against a wide range of micro-organisms, particularly bacteria and microfungi;
2 active during the life of the product;
3 of low mammalian toxicity and non-toxic to humans at the concentrations used;
4 colourless and odourless;
5 effective at low concentrations;
6 inexpensive and easy to apply;
7 resistant to sunlight and leaching from the fabric;
8 without effect on fabric 'handle' or strength;
9 compatible with water-repellent and flame-retardant agents, dyes and other textile auxiliaries;
10 incapable of sensitizing the fabric to damage by light or other influences.

As no preservative agent with this constellation of attributes has been discovered, the search for the 'ideal' biocide continues and thus the choice of the most suitable product (Table 1) is not easy. Indeed some otherwise excellent biocides have fallen out of favour because of unacceptable attributes. Organo-mercury compounds, for instance, are persistent and cumulative pollutants of the environment. Organo-copper compounds, effective biocides of common use in the textile industry, have the disadvantage of imparting a yellow-green colour to treated materials. Pentachlorophenyl laurate, which is relatively resistant to leaching, is widely used for rot-proofing various cellulosic textiles. It is suspected of behaving as a photodynamic agent in certain circumstances.

Table 1. *Commonly used mildew- and rot-proofing agents**

Group	Chemical name	Trade name	Manufacturer	Comments
Phenolics	Sodium *o*-phenyl phenate	Mystox WFA	Catomance	Application level 1–2% (w/v)
Chlorinated phenols	Pentachlorophenyl laurate	Mystox LPL	Catomance	One of the most important agents in U.K. Application level, 1–3.5% (w/v)
	5,5′-Dichloro, 2,2′-dihydroxy diphenyl methane	Panacide	B.D.H.	Broad spectrum of microbiological activity. Application level, approximately 1% (w/v)
Organo-metallic compounds	Copper-8-hydroxy quinolate	Mystox 8	Catomance	Imparts yellow-green colour. Application level, 1.7–6.7%
	Bis (tri-n-butyltin) oxide	Irgarol B1	Ciba-Geigy	Application level, approximately 0.05% (w/v)
Other halogenated compounds	Antimicrobial product with halogenated cyclic & heterocyclic compounds	Actifresh PL8111	British Sanitized	For use in polyurethanes. Application level, 0.3–1.0% (w/v)
Other aromatic compounds	*N*-(2-Methyl-1-naphthyl) maleimide	Vinyzene T-129	Ventron	For use in plastics. Application level, 0.125–0.5% (w/v)
Ammonium compounds	Benzalkylammonium derivative	Preventol R90	Bayer	For use in textiles and textile processing aids e.g. spin finish. Application level, 0.5–5% (w/v).
	Silicone quaternary amine	Q9-5700	Dow Corning	Broad spectrum of microbiological activity. Application level, 1% (w/v)
Organic compounds containing sulphur	*N*-Trichloromethylthio-4-cyclohexene-1,2-dicarboximide	Vancide 89	R. T. Vanderbilt	For use in polymers and starch. Application level, 0.25–1% (w/v)
Other organic compounds	6-Acetoxy-2,4-dimethyl m-dioxane	Giv-Gard DXN	Givauden	Application level, 0.05–0.2% (w/v)

* For further details, see Allsopp & Allsopp (1983).

Application of biocides to textiles for rot- and mildew-proofing is usually done as part of the final finishing process. The fabric is impregnated with either a solvent (usually white spirit) solution or, more usually, an emulsion of the biocide. It is then mangled and dried using a cylinder dryer or stenter. Biocides are often applied along with other finishing agents. For instance an 'FWR' finish consists of a combination of fire-retardant, water-repellent, and rot-proofing agents. Where protection of a plastic-coating is deemed to be necessary, the biocide is usually incorporated into the polymer mix before coating. To aid their subsequent processing, synthetic fibres usually have to be lubricated. The spin finish that is applied soon after extrusion of the fibre often contains a biocide.

Hygienic Finishes

There is an increasing interest, mainly from a hygienic viewpoint, in the application of finishes to textile products in order to prevent the growth of bacteria. Such hygienic or sanitized finishes are designed to eliminate the cause of unpleasant odours associated with bacterial decomposition of perspiration. They are intended to act in moist warm situations which are conducive to bacterial growth, viz. the insides of shoes, mattresses, pillows, the underarm area of clothing etc. Thus hygienic finishes are applied to a wide range of materials including socks, stockings, linings for footwear, underwear, mattresses, pillows, suit linings, dress shields, shower curtains, towels, hospital blankets, floor coverings and carpets. Some control of the microfungi responsible for athlete's foot (*Epidermophyton flocosum* and *Trichophyton mentagrophytes*) is also claimed for many of these agents. Footwear and floor coverings are treated also in order to prevent the build-up of depots of infection. Methods of assessing hygienic finishes are noted on p. 109.

Testing the Propensity to Microbial Attack and Efficacy of Antimicrobial Treatments

Many methods of microbiological testing of textiles have been proposed. The general aims for any test are to induce and to measure the extent of microbial growth under controlled conditions. There are two reasons for carrying out microbiological testing, firstly, to determine the inherent susceptibility of a textile material to attack; and secondly, to evaluate the effectiveness of a biocide in preventing the attack. Different test conditions may be necessary to satisfy these two requirements. Microfungi are used in the majority of microbiological test methods. Bacteria are essential in those used to evaluate hygienic finishes. It is to be regretted that, in general, bacteria are not specifically chosen for biodeterioration testing. Although the existing methods can be modified (Fermor & Eggins 1981), standard test methods are not available to assess the resistance of textile materials to attack from actinomycetes or algae.

Assessment of Efficacy of Anti-fungal Treatments

Textile resistance to microfungal attack is assessed by soil burial or deliberate inoculation. Single or mixed cultures are used and the material under test is placed on agar in a Petri dish or stored in a water-saturated atmosphere. The procedure will be chosen with the end-use of the textile in mind. Thus materials (e.g. sandbags or tents) which are likely to come into contact with soil should be given the soil burial test. A less severe test procedure may be suitable for textiles intended for other uses.

Soil burial test

This test procedure is generally considered to be the most severe exposure for textile products.

Sample preparation

Fabric specimens suitable for subsequent strength determination are prepared. Yarn specimens are similarly prepared.

Procedure

Soil (garden loam or naturally fertile top soil, free from pesticides) with a moisture content of between 20% and 30% and a pH of 5.5 to 7.0 is placed to a depth of 30 mm in the bottom of a suitable container. Specimens of fabric for test are laid on the soil. The middle portions (60–80 mm) of the specimens are covered with a further layer of soil to a total depth of 80–100 mm. The ends of a specimen are then folded inwards to form a U-shape. Soil is added to the remainder of the container. A control, cotton cloth (scoured only)—available from the Shirley Institute, Manchester—is similarly prepared. Yarn specimens, which are tied as a single loop around a stainless steel or glass frame, are buried—the soil coming approximately half way up the vertical frame. The containers are then covered and incubated at $28 \pm 1\,^{\circ}C$.

After incubation for 7 days the control cloths are removed, cleaned and sterilized by soaking in 70% ethanol for 4 h at room temperature, dried and conditioned in the standard atmosphere for physical testing. To confirm the cellulolytic activity of the soil and therefore the validity of the test, these specimens must have lost at least 80% (Anon. 1981*a*) or 90% (Anon. 1981*b*) of their original tensile strength. The incubation of the test specimens is usually continued for up to 28 days. They are then removed from the soil, and treated in the manner described above for the control specimens.

Interpretation of results

Totally resistant samples should not have lost, after soil burial, more than 5% of their strength, measured on test specimens from the original sample and treated with 70% ethanol in a similar manner to the soil burial specimens.

Agar plate tests

Two types of agar plate tests are commonly used. In one, test samples of fabric are placed on a mineral salts agar in a Petri dish and inoculated with either a spore suspension or one of several species of microfungi. The ability of the specimen to serve as a sole carbon source for the growth of microfungi is assessed. In the other test, a metabolizable carbon source is added to the mineral salts agar. The resistance of the specimen of test fabric to encroachment by already established microfungal colonies is assessed.

Sample preparation

Squares (20 × 20 mm) or discs (20 mm diameter) of fabrics are suitable when visual assessment only is required. The test procedure to be used will dictate the size of samples when strength determinations are used for assessment. Samples of yarn are coiled before being placed on the mineral salts agar. The test specimens may be sterilized but there is an attendant risk of affecting the resistance of the textile material to microfungi.

Preparation of agar plates

For the first type of test a suitable mineral salts agar is used (Anon. 1981*a*; Anon. 1981*b*). Sucrose (Anon. 1981*a*) or glucose (Anon. 1981*b*) can be added to the mineral salts agar, or a ready-prepared, dehydrated culture medium used (Anon. 1976*a*) in the second test procedure.

Test microfungi

The choice of micro-organism(s) used for inoculation of the test specimens on mineral salts agar depends on which National Standard is adopted (see footnote to Table 3).

Preparation of the spore suspension

The method of preparation of the inoculum can vary depending on whether a suspension of the spores of one or several species of microfungi is required. For a mixed spore suspension, approximately 10 ml of mineral salts solution containing a suitable wetting agent (e.g. Triton X-100, 0.5% v/v) is added to agar slants of the different species. The spores are harvested by gently scraping the surface of the culture with a sterile loop. The spore suspensions are decanted into a conical flask containing a few glass beads, and shaken to mix the suspensions and break up clumps of spores. The mixed spore suspension is filtered through a sterile cotton or glass wool plug. The spores are washed two or three times by centrifugation at 3500 g, the supernatant fluid is discarded and the spore pellets resuspended in mineral salts solution on each occasion. Single suspensions of the spores of single species may be prepared using the procedure described above, or by the addition of sterile distilled water to the cultured organism (Anon. 1981*b*). Alternatively, a ready prepared medium may be seeded directly from the mother culture when poured into Petri dishes (Anon. 1976*a*).

Test procedures

The specimens of prepared textile as well as those used as controls are placed on the carbon-less mineral salts medium and the one supplemented with glucose or sucrose and inoculated, by spraying or pipetting, with either a suspension of the spores of one or several species of microfungi. To assess the potential for growth of any adventitious spores present on the sample, additional plates containing test specimens may be left uninoculated. All the plates are incubated at 28 ± 1 °C for up to 28 days. When the samples are placed on seeded agar (Anon. 1976*a*), the incubation times are generally shorter, but are at least 2 d.

Assessment

The assessments may be direct, by visual growth of the microfungi, or indirect, when the change in some physical property (usually tensile strength) of the test specimens is determined. In the latter case, the specimens are removed from the mineral salts agar and cleaned before testing as described in the soil burial test (p. 104). The preferred method (British Standard 6085; Anon. 1981*a*) of visual assessment involves grading the amount of microfungal growth on the sample on a scale of 0 to 5 (Table 2).

An alternative method of assessment is used in the Swiss Standard SNV 195 921 (Anon. 1976*a*) for samples on agar containing a carbon source. It differs from the above method of assessment in that, in addition to the recording of the presence of microbial growth on the sample, zones of inhibition of microbiae growth are also measured.

Interpretation of results

Ideally, when using the first method of visual assessment, growth ratings of 0 should be recorded for all tests involving totally resistant samples. However, samples with growth ratings of 1 on some of the specimens should perform satisfactorily in the field. Growth ratings of 2 or more indicate that the sample may be susceptible

TABLE 2. *Grading scale for visual assessment of microfungal growth*

Observed growth	Rating
No growth	0
Growth only visible under microscope	1
Growth visible to the naked eye: less than 10% area covered	2
Light growth: 0–30% area covered	3
Medium growth: 30–60% area covered	4*
Heavy growth: >60% area covered	5

* Control cloths should have a growth rating of at least 4 after 14 days incubation for the test to be valid.

to mildewing under conditions conducive to the proliferation of microfungi; with the second method of visual assessment, samples showing either a zone of inhibition but no growth on the sample are considered to have satisfactory resistance to microfungal attack. Large zones of inhibition around the sample indicate, however, extensive diffusion of the biocide into the agar suggesting that a fabric will have a poor resistance to microbial attack if conditions favour leaching out of the biocide.

Saturated atmosphere test

This procedure is the least severe exposure for textile products.

Test procedure

Samples of fabric and control cloth, inoculated by spraying with *ca.* 1 ml of a mixed spore suspension, are suspended above water in suitable containers in order to maintain a relative humidity greater than 90%. The containers are incubated at 28 ± 1 °C for a minimum of 14 days. The samples are assessed visually by the 0–5 scale described on p. 106 and the results interpreted as discussed above.

Published standard test methods

Essential details relating to three National Standards, viz: British Standard 6085 (Anon. 1981*a*), American Association of Textile Chemists and Colorists, Test Method 30 (Anon. 1981*b*) and the Association Suisse de Normalisation 195 921 (Anon. 1976*a*) are summarized in Table 3.

TABLE 3. *Test procedures for three National Standards used for evaluating the efficacy of anti-fungal treatments*

National Standard	Test Procedure: Soil burial	Agar plate	Saturated atmosphere
BS 6085 (Anon. 1981*a*)	Sample incubated for 14 days (non-coated cellulosics) or 28 days (non cellulosic or coated cellulosic textiles). Visual assessment and/or strength loss	Samples inoculated with mixed spore suspension* and incubated for 14 or 28 days on a mineral salts agar with and without a carbon source (sucrose). Visual assessment and/or strength loss	Samples sprayed with mixed spore suspension* and incubated for a minimum of 14 days. Visual assessment
AATCC 30 (Anon. 1981*b*)	Samples incubated for 2–6 weeks depending on fabric weight and severity of service requirements. Assessment of tensile strength loss	(i) Sterile specimen (autoclaved) inoculated with a single culture† and incubated for 14 days on a mineral salts agar. Visual assessment and strength loss (ii) Non-sterile specimen inoculated with a single culture‡ and incubated for 28 days on a mineral salts agar without glucose or for 14 days on a mineral salts agar with glucose. Visual assessment	
SNV 195 921 (Anon. 1976*a*)		Sterilized circular specimens (ethylene or propylene oxide) placed on a Dextrose-peptone agar seeded with a single culture§ and incubated until control plates without any test material are fully grown. Visual assessment and measurement of any zone of inhibition.	

* Mixed spore suspension of the following cultures: *Aspergillus niger* (AATCC 6275); *Aspergillus terreus* (PQMD 82j); *Aureobasidium pullulans* (ATCC 9348); *Paecilomyces variotii* (IAM 5001); *Penicillium funiculosum* (IAM 7013); *Penicillium ochro-chloron* (ATCC 9112); *Scopulariopsis brevicaulis* (IAM 5146); *Trichoderma viride* (IAM 5061); *Chaetomium globosum* (CMI 45550). † *Chaetomium globosum* (ATCC 6205). ‡ *Aspergillus niger* (ATCC 6275). § *Trichophyton mentagrophytes* (ATCC 9533) or *Candida albicans* (ATCC 10259).

Assessment of the Efficacy of Anti-bacterial Treatments

Two main methods are used to assess the resistance of textiles to bacterial encroachment—a qualitative agar plate test, and a quantitative challenge test in which the numbers of bacteria recoverable from a textile are determined. The latter is also used in the assessment of hygienic textile finishes.

Agar plate tests

Preparation of samples

Discs *ca.* 25 mm diam. of samples are prepared. They may be sterilized by autoclaving or exposure in an atmosphere of ethylene oxide or propylene oxide, the choice of method depending on the nature of the anti-bacterial agent and the intended use of the material.

Procedure

Bacterial cultures (24 h; the bacteria used in these tests are given in the footnote to Table 4; in suitable media are described in Anon. 1976*b* and Anon. 1977) are used to seed agar (1 ml of culture per 150 ml of agar, maintained at 45 °C) which is poured into Petri dishes. The test specimens are carefully pressed onto the centres of the plates to ensure good contact between the specimen and the agar surface, and the plates are incubated at 37 °C for 18–24 h. After incubation, any zone of inhibition around the specimen is measured, and the presence or absence of growth on or underneath the specimen recorded.

Interpretation of results

A clear zone around the specimen indicates anti-bacterial activity. As with the analogous anti-fungal agar plate test, however, a zone of inhibition is also a measure of the diffusibility of the preservative agent through the agar. Therefore growth up to the edge of the specimen, but not on or underneath it, may be a more satisfactory indication of anti-bacterial activity especially where resistance to leaching of the preservative agent is a desirable feature of a fabric.

Bacterial challenge tests

Preparation of samples

Circular swatches (*ca.* 50 mm diameter) are cut from the test fabric. The number of swatches used in each test container is the number that absorbs 1 ml of inoculum without leaving any free liquid in the container. The swatches are sterilized either by autoclaving, or by exposure to ethylene oxide vapour. A control fabric without the anti-bacterial agent should also be prepared.

Test procedure

Swatches of test and control fabrics are placed in sterile 8 oz wide-mouthed glass bottles (with screw tops). The swatches are inoculated with 1 ml of an appropriate dilution of a 24 h bacterial culture (suitable media and culture conditions are described in Anon. 1981*c*) to give $\times 10^4$ to 20 colony forming units (cfu/ml). Immediately after inoculation, 100 ml of a suitable neutralizer solution (e.g. Letheen broth, Difco; see p. 285) are added to half the control and test fabric swatches, and the jars shaken vigorously for 1 min. Serial dilutions of the neutralizer solution are then plated out on a suitable nutrient agar (Anon. 1981*c*). Hence the cfu recoverable after zero contact time can be established. The remaining jars are incubated at 37 °C for 18–24 h, and the cfu determined as described above.

Calculation of results

The cfu recovered from the control fabric after incubation should be higher than the cfu recovered from the control fabric at zero contact time for the results of the test to be valid. Provided this condition is established, the percentage reduction in cfu of the test fabric is given by

$$\left[1-\frac{2A}{(B+C)}\right]100$$

where A is the cfu recovered from the test fabric after incubation; B is the cfu recovered from the test fabric at zero contact time, and C is the cfu recovered from the control fabric at zero contact time.

Interpretation of results

If there is an increase in the cfu recovered from the test specimens after incubation, compared with the cfu recovered from the zero contact time specimens, the anti-bacterial agent is ineffective. No significant difference between these values may indicate that the agent is acting as a bacteriostat. A significant reduction in cfu indicates bacteriocidal activity.

Published standard test methods

Details concerning the two American Association of Textile Chemists and Colorists procedures (AATCC 90, Anon. 1977; AATCC 100, Anon. 1981*c*) and the Swiss method (SNV 195 920, Anon. 1976*b*) are summarized in Table 4.

Durability of Anti-microbial Treatments

Whilst a full-scale field exposure evaluation is undoubtedly the crucial test of effectiveness of a biocide treatment, there are usually difficulties involved in carrying

TABLE 4. *Test procedures for three National Standards used for evaluating the anti-bacterial activity of textiles*

National Standard	Test Procedure	
	Agar plate	Bacterial challenge
AATCC 90 (Anon. 1977)	Sterile circular test specimens placed on a nutritive agar seeded with a single culture* and incubated for 18–24 h at 37 °C. Measure any zone of inhibition	
AATCC 100 (Anon. 1981*c*)		Sterile† circular swatches inoculated with 1 ml culture broth* and incubated at 37 °C. At various time periods up to 18–24 h 100 ml neutralizer solution added (to neutralize the anti-bacterial treatment) and serial dilution plated onto tryptone glucose extract agar. Assessment by bacterial counts and calculating % reduction of bacteria by fabric treatment.
SNV 195 920 (Anon. 1976*b*)	Circular specimens (not sterilized) placed on a nutritive agar seeded with a single culture‡ and incubated for 18–24 h at 37 °C. Assess extent of inhibition of growth of bacteria in contact zone under specimen and measure any zone of inhibition.	

* *Staphylococcus aureus* (ATCC 6538) or *Klebsiella pneumoniae* (ATCC 4352).
† Cotton, acetate and man-made fibres can be sterilized in the autoclave. Wool can be sterilized by ethylene oxide or by intermittent sterilization in flowing steam.
‡ *Staphylococcus aureus* (ATCC 6538) or *Escherichia coli* (ATCC 11229).

out such studies and experimental conditions are rarely reproducible under the uncontrolled climatic conditions occurring in nature. It is necessary, therefore, to simulate in the laboratory the possible loss of anti-microbial compound by leaching with rain or by actinic decomposition when carrying out evaluations of biocide efficacy. An assessment of the likely durability of the anti-microbial treatment is obtained by re-testing, using an appropriate microbiological test, after one or more of the following treatments.

Resistance to leaching

Samples of an appropriate size are immersed in water for 24 h in 2.5 l beakers supplied with a constant flow (6 l/h) of tap water. After leaching, the samples are air dried before microbiological testing.

Thermal resistance of the anti-microbial treatment

Samples are exposed for 24 h in dry heat at 100 °C to 105 °C in a well ventilated oven, or to cycles of 134 °C for 3 min in an autoclave, depending on the textile end-use.

Susceptibility to actinic degradation

Samples are exposed to light from a carbon arc or xenon lamp, for a suitable period of time. The intensity of exposure is assessed by reference to wool fading standards.

Weathering

Samples may be exposed in a weathering chamber. A suitable instrument is the Accelerated Weathering Tester (The Q-Panel Co., Cleveland, Ohio, U.S.A.)

References

ALLSOPP, C. & ALLSOPP, D. 1983 An updated survey of commercial products used to protect materials against biodeterioration. *International Biodeterioration Bulletin* **19**, 99–146.

ANON. 1976*a Examination of the antimycotic effect of impregnated textiles by the agar diffusion test.* SNV 195 921. St. Gallen, Switzerland: Association Suisse de Normalisation.

ANON. 1976*b Examination of the antibacterial effect of impregnated textiles by the agar diffusion test.* SNV 195 920. St. Gallen, Switzerland: Association Suisse de Normalisation.

ANON. 1977 *Antibacterial activity of fabrics, detection of: agar plate method.* AATCC Test Method 90. American Association of Textile Chemists and Colorists Technical Manual **53**, 300–301.

ANON. 1981*a Determination of the resistance of textiles to microbiological deterioration.* British Standard 6085. London: British Standards Institution.

ANON. 1981*b Fungicides, evaluation on textiles; mildew and rot resistance of textiles.* AATCC Test Method 30. American Association of Textile Chemists and Colorists Technical Manual **57**, 309–311.

ANON. 1981*c Antibacterial finishes on fabrics, evaluation of.* AATCC Test Method 100. American Association of Textile Chemists and Colorists Technical Manual **57**, 304–306.

DARBY, R. T. & KAPLAN, A. M. 1968 Fungal susceptibility of polyurethanes. *Applied Microbiology* **16**, 900–905.

FERMOR, T. R. & EGGINS, H. O. W. 1981. The use of actinomycetes in textile biocide testing. *International Biodeterioration Bulletin* **17**, 15–17.

HAMILTON, N. F. 1983 Biodeterioration of flexible polyvinyl chloride films by fungal organisms. In *Biodeterioration 5*, ed. Oxley, T. A. & Barry, S. pp. 663–678. Chichester: John Wiley & Sons.

HAMLYN, P. F. 1983 Microbiological deterioration of textiles. *Textiles* **12**, 73–76.

HUECK, H. J. 1974 Criteria for the assessment of the biodegradability of polymers. *International Biodeterioration Bulletin* **10**, 87–90.

KLAUSMEIER, R. E. & ANDREWS, C. C. 1981 Plastics. In *Economic Microbiology* **6**, ed. Rose, A. H. pp. 431–474. London: Academic Press.

MAHALL, D. K. 1982 Biodegradation of textile fibres as seen under the microscope—actual case histories. *International Textile Bulletin* **4/82**, 280–292.

SELBY, K. 1968 Mechanism of biodegradation of cellulose. In *Biodeterioration of Materials*. ed. Walters, A. H. & Elphick, J. J. pp. 62–78. Amsterdam: Elsevier Publishing Co.

Wood Preservation—Fungal Testing

E. A. HILDITCH AND F. MENDES

Cuprinol Limited, Adderwell, Frome, Somerset

Wood is often regarded and used as a permanent material in buildings etc., but it is a plant product and under certain conditions is rapidly degraded. With the exception of fire, the main destructive agencies are biological, namely, terrestrial and aquatic fungi, bacteria, insects, molluscs and crustaceans. Conventional wood preservatives are designed to protect wood against fungi and insects. Testing resistance to these is essential in wood preservative research, development and evaluation.

In this chapter we consider only the testing of fungicides. Wood preservatives differ from other fungicides by the period over which they are expected to remain effective (25 years or more). Predicting permanence is therefore of major importance. Usually this is done by carrying out fungal tests of the kind described below with material that has, or has not been subjected to artificial weathering or ageing processes.

Fungal Degradation of Wood

Many species of fungi of all classes and bacteria are able to grow in wood and cause deterioration or destruction. Throughout its life wood is subject to repeated, random infection by airborne spores and sometimes hyphal spread. The specific development of each species will be affected by all environmental factors—moisture, temperature, nutrient availability etc. These partly depend on species of wood and how it has been modified by previous microbial growth, and also on the presence of other fungal or bacterial species stimulatory or antagonistic. Development of decay fungi is competitive. Some species grow rapidly, others slowly, and they do not always start growing at the same time.

Experience over the last 25 years has shown that the events leading up to the ultimate destruction of wood are complex and differ in different situations (Findlay & Savory 1954; Savory 1954*a*, *b*; Levy 1965; Carey 1980, 1982; Levy & Dickinson

Preservatives in the Food,
Pharmaceutical and Environmental
Industries

0–632–01727–9

1980, 1982; Smith 1980). Current views (which are likely to be modified as further knowledge is gained) suggest that the following principal events lead to decay:

1 Initial infection is by airborne Basidiomycete spores that germinate and grow into the wood (assuming adequate moisture and temperature). This is followed by mycelial spread to adjacent timbers and is typical of much of the decay in internal buildings where a single species is often responsible for decay over a wide area.

2 A complex pattern, partly of random invasion by a wide spectrum of micro-organisms, with succession of development, is typical in wood that is in contact with soil. Early bacterial development causes very limited destruction of wood but opens the way to fungal invasion and growth. Bacteria are then overtaken by the more destructive micro-fungi. These include the soft rot fungi which may be responsible for the ultimate destruction of the wood, although they may in turn be succeeded by Basidiomycetes. These usually destroy wood very rapidly.

3 Outdoor timbers that are not in contact with the soil appear to occupy an intermediate position. Initial random infection is by bacteria and the spores of micro-fungi and Basidiomycetes. The bacteria increase the permeability of the timber to water. The development of blue-stain fungi and the sporulation of moulds causes visual deterioration by showing through or damaging surface finishes. Final destruction is commonly by Basidiomycetes and two or more species may develop together in the same piece of wood.

4 Fungal decay is mostly due to softrot organisms under very wet conditions, e.g., cooling tower infill timbers, timbers in rivers, lakes or the sea.

These represent a very superficial assessment, and many exceptions are known.

Principles of Current Test Methods

The earliest and many of the current laboratory tests involve the exposure of treated wood blocks to individual Basidiomycetes in pure culture. Such culture tests were for a long time considered to be 'theoretically realistic' although many anomalies were apparent between the conclusions drawn from such tests, field trials and practical experience. Tests with mixed species of Basidiomycetes have not been developed. Resistance of a preservative to current laboratory tests are unable to reflect this varied and complex pattern, except for the simplest situation of single species infection and development. Nonetheless, laboratory tests are most valuable and are widely used in wood preservation development. Such tests alone give little assurance that a chemical will perform well in practice, but one that does not behave satisfactorily in the laboratory is most unlikely to perform well in real use. The cost and scale of field tests makes it essential that materials selected for tests have a good chance of success as far as present technology predicts. Additionally, laboratory tests are most valuable for clarifying points of detail in the performance of individual wood preservatives. There are five main groups of tests used at present for testing the resistance of fungi to wood preservatives:

1 Simple tests where the chemicals are dissolved or dispersed in agar culture media or applied to some convenient base such as filter paper and directly inoculated with fungal spores or mycelium. This is usually done in Petri dishes. This method is suitable only for preliminary screening purposes.

2 Standard or semi-standard tests in which a small block of wood is impregnated with preservative and exposed to a single species of fungus growing on an agar medium or other nutrient base.

3 Non-standard tests using small wood blocks or stakes which are exposed under laboratory conditions to natural (non-sterile) soil.

4 Field tests in which large wooden stakes that have been treated are placed in the ground in an outdoor test plot. Such tests commonly last more than 25 years.

5 Simulated trials in which wood is exposed to the natural environment in test assemblies intended to reflect different use-situations.

Laboratory Screening Tests

Agar medium tests

This general technique is widely used in mycology. It appears to have been applied specifically to wood preservatives by Malenkovic (1904) and was very popular with early American workers (Humphrey & Fleming 1915; Schmitz *et al.* 1930). Candidate chemicals are dissolved or dispersed in melted nutrient agar medium. Water soluble chemicals are added directly at various concentrations. Oil-soluble preservatives are first emulsified. When the agar has solidified, the plates are inoculated with the test fungi. Recommended fungi include *Heterobasidium annosum*, *Gloephyllum trabeum*, *G. sepiarium*, *Lentinus lepideus*, *Poria incrassata*, *Coniophora puteana*, *Polyporus vaporarius*, *Coriolus veriscolor*, *C. hirsutus* and *Antrodia serialis.*

The inoculated plates are incubated at 28 °C and the radial growth measured on alternate days. If there is no signs of radial growth, the original inoculum is transferred to agar slants; if no growth occurs within 14 d, the fungus is considered to be killed. Growth rates are plotted graphically (log scale) against concentration and extrapolated to give the concentration corresponding to zero growth (total inhibition point). With oil-soluble preservatives differences arise with various emulsification techniques and can result from different emulsifying properties of different preservatives (Kneale & Tamblyn 1950; Hilditch 1974).

Filter paper tests

Filter paper squares or discs are impregnated with dilutions of test preservatives. They are placed in Petri dishes on nutrient agar or supported on inverted watch glasses above a liquid medium which has been inoculated with spores of a single or mixed fungal culture. Both water and oil based preservatives can be tested without much difficulty on a cellulose substrate (black filter paper) which is favoured by the cellulolytic organisms inhabiting wood.

In a variation (Dickinson 1974) filter paper is impregnated with test preservatives, inoculated with a test fungus and incubated at 22 °C for 2 weeks inside polythene bags to prevent drying. Average growth is calculated from three measured diameters at 120 °. Toxic concentration is expressed between the percentage concentration of active ingredient that prevents growth and the next highest which just permits it. When agar plates are used, a zone of inhibition may be seen around the treated paper. A large zone suggests diffusion of the preservative and possible limited permanence.

Many varieties of these tests have been used in preliminary studies on wood preservatives. In an attempt to bridge the gap between this type of test which is easy, simple and quick but whose results are a little remote, and wood block tests which involve considerable problems and take a long time, wood shavings (Richardson 1968) and similar materials have been tried but they are not widely used.

Wood block tests

Agar block

Small blocks of wood (*ca.* 50 × 25 × 10 mm) are treated with preservative, placed on fungal mycelium growing on nutrient agar and incubated, usually for 3 months. Such tests were developed in the early 1930s (Liese *et al.* 1935) and were favoured by Europeans. They formed the basis of several national standards (Anon. 1960*a*, *b*; 1961*a*, *b*; 1968). There are major differences between these standards and recently, through the *Comité Européan de Coordination de Norme*, a common technical basis for national standards has been agreed (Anon. 1982*a*). These tests are most useful for testing resistance of Basidiomycetes. They can also be used with soft rot organisms.

The European Standard (Anon. 1982*a*) specifies a method where 50 × 25 × 10 mm blocks of sapwood of *Pinus sylvestris* (Scots Pine) and *Fagus sylvatica* (beech) are impregnated with several dilutions of wood preservative. After the required conditioning, these test blocks are exposed for 16 weeks on glass supports over mycelium on an agar medium in suitable culture vessels. Several fungal species are used separately and on separate sets of blocks (monoculture). Testing against *Coniophora puteana* on soft woods and *Coriolus versicolor* on hardwoods is mandatory; two or more additional species are used at choice. The extent of decay is judged by measuring weight loss. The result is expressed as a toxic value, which is the interval between the lowest concentration that prevented any decay and the highest concentration that allowed some decay.

The standard does not specify any particular weathering treatments for testing the permanence of the test preservative. If these are used they have to be recorded in the test report. The European Standard (Anon. 1982*a*) specifies only tests against Basidiomycetes. The British Standard (Anon. 1961*b*) included the soft rot fungus *Chaetomium globosum*, but it is generally being accepted that this fungus imperfectly represents the complex spectrum of Ascomycetes and Fungi Imperfecti that causes soft rot. Various soft rot test methods have been reviewed by Savory & Bravery (1971), but there is widespread debate on techniques. A suitable method has yet to be developed.

Agar wood block tests have been modified and tested by workers for their individual requirements. Bravery (1970) used 1 cm cubes and exposed them to fungi growing on agar in Petri dishes. The results are reported to be comparable with those obtained by other means (Anon. 1961*a*; Baker *et al.* 1972). Variations on the agar wood block test have been developed to examine specific aspects of wood preservative performance. In an important variant, Becker & Starfinger (1971) treated large blocks by surface application techniques and then cut slices 1, 2 and 4 cm thick from the surface and exposed them to a well developed culture of a decay fungus to determine the effectiveness of the preservative treatment (Anon. 1961*b*). Sutter (1978*a*) developed a thin-slice method to assess the depth of penetration of biologically active components. Using a modification of this technique, Mendes (1982) cut sequentially transverse slices from double vacuum-treated joinery. Assessment of the growth pattern of decay fungi through the thin slice showed the penetration and distribution pattern of the preservative. This was compared with that obtained with reagents applied by spray.

Sutter (1978*b*) proposed another thin slice technique to determine toxic limits by the degree of fungal deterioration. In a further modification Mendes (unpublished) used the time for hyphae to penetrate through cross-cut slices to determine the resistance of a wide range of preservatives and mixtures to leaching and evaporation. Other modifications using thin slices on agar media have been described by Ingleby (1982), Jensen & Imsgard (1982) and Sutter *et al.* (1983). Most of the thin slice methods need refinement, but they appear to be useful in specific investigations.

The techniques described so far either depend on weight loss or on a qualitative assessment of deterioration. Test methods have also been developed to determine strength losses during decay (Cartwright *et al.* 1936; Schaffer 1936; Hopkins & Coldwell 1944; Mateus 1954, 1957; Armstrong & Savory 1959; Kennedy & Ifju 1962; Wilson & Ifju 1965; Richardson 1968; Sharps & Eggins 1968; Bravery & Grant 1971; Bravery & Lavers 1971; Smith 1971). Such methods require very careful sample preparation, large numbers of replicates to avoid widespread variation and very careful examination. They often require rather sophisticated instruments for assessments and are always time consuming. Thus the weight loss methods tend to be favoured by most research laboratories dealing with wood preservative testing.

Among standard methods the major alternative to the agar block is the so-called soil block test. This was developed in the United States (Leutritz 1939, 1946) and now forms the basis of the American Society for Testing and Materials Standard (Anon. 1981).

Wood blocks (19 mm) are treated and conditioned in a manner similar to but not the same as that described in British Standard (Anon. 1982*a*) and are then exposed to attack by the mycelium of single species of Basidiomycetes. The difference lies in the growth medium for the test fungi. Natural soil is sterilized, then inoculated with the fungus and a small feeder block of untreated wood is placed on top of the soil, with the treated test block on top of the feeder block. Betts (1982) described a modified version that is used for investigating the residual effectiveness of creosote-treated poles. Small sample blocks (20 × 5 × 5 mm) were cut at three radii of each pole at ground level, and at 3 and 6 m. The replicate blocks were exposed to *Coniophora puteana* using the same method as in Anon. (1982*b*), but incubated for a shorter period. With similar sized blocks (but an increased number of samples per culture vessel) it was possible to show that difference in resistance to decay exists in the sapwood, juvenile heartwood and mature heartwood region of *Pinus sylvestris* (Scots Pine or European Redwood), but this differential resistance to decay is lost in a biologically active environment (Carey *et al.* 1984).

It is claimed that toxic values in soil/wood block tests approach more closely those found in the field (Duncan 1953, 1958). Test assemblies take more time to prepare, however, and variation in soil sources influences results. Agar/wood block and soil/wood block tests are said to give comparable toxic values (Bravery 1975) but differences have also been reported (Smith 1965; Becker *et al.* 1966). The basic test is suitable for any wood, although pine sapwood is the most commonly used. Pines are the main building timbers in Europe and North America. The sapwood is readily attacked by fungi but is also readily impregnated with preservatives. European tests mostly use European Redwood (*Pinus sylvestris*) or beech (*Fagus sylvatica*) when a hardwood is required.

Selection of blocks for these tests is crucial; there are many variations to contend with when cutting and matching test blocks. There are certain obvious steps to take, such as avoiding knots or areas of abnormal growth, or wood already damaged by bluestain, or anything else. There are major differences, particularly in growth rates and possibly in mineral content etc., in trees of the same species but grown in different places. There are smaller differences between wood taken at different heights from the same tree and between the inner and outer sapwood.

For obvious practical reasons a minimum number of replicate blocks are used, and all reasonable steps are taken to ensure that all the blocks used in a single test run are as nearly identical as possible. It is usual to have between four and six replicate blocks for each treatment and, for each preservative under test, to treat at four to six concentrations, i.e., 24 to 36 blocks per test chemical. Reliable results can be reported only where there is a cut-off concentration above which no decay occurs and below which all blocks are affected. Anomalous results are common.

Laboratory soil tests

Test blocks are buried in unsterilized soil for soft rot tests (Kneale & Tamblyn 1950; Bravery 1968; Bravery & Savory 1970). The Nordic Wood Preservation Council (Anon. 1971*a*) specified a test method, the Mullade method, using unsterile soil. Wood test blocks are usually buried beneath the surface of soil although in some tests many are proportioned as small stakes and half buried. They are exposed to infection and attack by the whole range of soil micro-organisms. Well-composted garden soil is preferred. Culture vessels range from small glass or plastic boxes or bottles each containing two or more blocks, to large beds in which the whole test batch is buried. Incubation is between 3 months and 1 year during which the moisture content of the soil must be maintained at an optimum level for decay. As in monoculture tests replica blocks are usually treated with the same concentration of fungicide and the decay occurring on each block observed.

Both softwood and hardwoods are used. Soil is variable and complex and its source will affect the degree of infection (Hilditch 1978). Defining a standard soil for an acceptable test method, therefore, poses problems. Savory & Bravery (1971) published results of tests carried out at various Institutes using unsterile soil (differing in organic matter, acidity and water-holding capacity). Wide variation in the toxic values was observed. It has not yet been possible to establish an acceptable standard. Soil is one of the most hazardous environments for wood and, despite the shortcomings of tests employing it, should be considered for biodeterioration studies. Results using monocultures and unsterile soil are found to differ significantly (Hilditch 1974; Hilditch & Hambling 1971).

Fungus cellar

In the fungus cellar an attempt is made to simulate conditions existing in a building with severe decay in floor or similar timbers. The development and application of fungus cellars have been described by several authors (Gersonde & Becker 1958; Deppe 1967, 1974; Gersonde & Deppe 1968; Deppe & Gersonde 1969). The original test method involved 50 cm-long boards treated and untreated alternatively arranged on untreated under-floor timbers to a pattern floor of 85 × 50 cm. These pattern floors were erected on a soil foundation that provided a moist underfloor space. The surface was covered by a glass plate, corresponding to an airtight floor cover. The tests were carried out in a large fungus cellar with constant climatic conditions. The floor was inoculated with pure cultures of test fungus (*Serpula lacrymans*, *Coniphora puteana* and *Fibroporia vaillantii*). The boards were assessed visually for decay and thresholds for fungicidal effectiveness were determined. It has been suggested that fungus cellar results can easily be related to practical use, as effectiveness of preservatives are assessed under the conditions similar to those in buildings.

A smaller modified version termed a 'mini fungus cellar' was described by Hansen (1975). Samples were smaller than the original fungus cellar tests of Gersonde & Becker (1958) but large enough to allow the preservatives to be amplified as recommended by manufacturers and to fit within laboratory facilities. The treated pieces were placed over sterile soil in smaller boxes and exposed to attack by monocultures of wood destroying fungi in constant climatic conditions. The results were evaluated by weight loss and visually for decay.

Simulated Field Tests

Large scale but still laboratory-based soil-bed systems have recently been developed in Australia and New Zealand (Anon. 1978; Butcher 1981; Hedley 1980, 1983; Johnson *et al.* 1982, 1983). Soil-bed studies have been carried out on several levels, varied in design and size of bed and based on available facilities and sophistication of control systems (temperature, air and water). The principle is a controlled environment room in which troughs are filled with natural soil—sometimes with other fillings—under the soil-bed to allow for air movement and drainage (Johnson *et al.* 1982, 1983; Murphy *et al.* 1982; Vinden *et al.* 1982). Temperature and moisture content are monitored and maintained. The Australian Commonwealth Scientific and Industrial Research Organisation uses a tropical environment (28 °C air temperature, 85% relative humidity) and the New Zealand Forest Research Institute (Anon. 1978) uses a somewhat cooler climate (21–23 °C, 95% relative humidity). Treated and conditioned stakes are pushed into the soil and examined at intervals for decay. In most cases assessment criteria are based on the usual numerical ratings for stake assessment in field studies. The soil-bed test approach provides some of the realism of field conditions in predicting preservative performance but it does not involve natural climatic variations which influence the action of microflora to a great extent.

The assessment of decay in all tests so far described is by weight loss. Decreases in the oven-dry weight greater than 3% are usually taken to indicate decay. The occasional use of strength methods has been mentioned and for preliminary screening tests visual inspection, aided by thumb nail or small metal probe, is often adequate. Other methods have been considered and have been used in specific studies.

The evolution of carbon dioxide and utilization of oxygen as a measure of the activity of fungi colonizing wood have been described by several workers (Klingstrom 1965; Smith 1967; Good & Darrah 1967; Halabisky & Ifju 1968). A typical method using carbon dioxide evolution as an indication of activity was compared with weight loss test results and a similarity in toxic limit was observed (Smith 1969). Wood blocks were exposed to fungal attack using the US Standard soil/block method (Anon. 1981). Blocks were removed at weekly intervals, placed inside a specially designed flask under asceptic conditions and CO_2-free compressed air was passed through a needle valve flow controller, through a Drechsel wash bottle of water to humidify the air, then through a rotameter type of flow-meter into the respiration flask and out into the gas sampling valve of the gas chromatograph. After each respiration measurement blocks were returned to their original soil jars for continuation of decay. After the final carbon dioxide measurement at 12 weeks, blocks were weighed, oven-dried and the weight loss recorded. Oxygen utilization as an indicator of fungal activity can lead to serious error as non-respiratory oxygen uptake in test wood samples has been recorded (Good & Darrah 1967). Such tests show progressive fungal activity from an early stage in incubation but require complex instruments and involve a lot of work. They are, therefore, unsuitable for routine studies and are unlikely to be used in most laboratories.

Some consideration has been given to biochemical methods that use enzymatic activity to indicate fungal colonization of wood and to assess the effectiveness of wood preservatives (Braid & Line 1980; Line 1982, 1983).

Catalase enzyme assay has been related to microbial activity in wood. As catalase is a constitutive intracellular enzyme produced by all aerobic organisms, its activity in wood reflects the microbial biomass. There are various methods for detecting catalase (Line 1982, 1983). Line (1982) used a spectrophotometric method with a redox indicator (2, 6-dichloroindophenol). Core samples of sound and decayed wood were partially dried at 26 °C for 4 h and powdered in a hammer-mill. Weighed quantities of sieved powder were placed in bottles (30 ml) containing 0.1 M (pH 7.0) phosphate buffer (de-gassed by boiling before cooling) and evacuated in a vacuum desiccator for 20 min. The buffer was decanted and the bottles filled to capacity with freshly prepared redox indicator solution and placed in a water bath at 30 °C for 30 min. The solution was assayed spectrophotometrically at 6.2 nm for catalase.

These enzyme assay techniques are likely to gain popularity as more and more microelectronic instruments for detecting enzymes become available. This offers a considerable potential for reducing test time.

Field Tests

Laboratory tests so far outlined give no more than in indication of likely preservative performance. They have not been developed to a stage where they can serve as a reliable indicator of true performance in service. Field tests are, therefore, essential.

Stake tests

The stake test is pre-eminent and predominant among field tests. In principle it is very simple. Stakes of wood are treated and are then buried for part of their length in the ground in a totally natural environment. They are observed over a long period (10–15 years) during which they are subject to the natural soil, chemical and microbiological environment and to all weathering and similar influences. Many variations are used. The usual stake is 50 × 50 mm cross-section by about 600 mm long, but sizes range from below 20 × 20 mm sections to full-sized telegraph posts. Timbers used and methods of treatment cover both the full range of commercial practice and many experimental varieties. Plywood and similar composite materials have been used.

The methods are described in standards, American Society for Testing and Materials (Anon. 1980; Nordic Wood Preservation Council Standard Anon. 1971*b*), and accounts of many variations have been published (Hilditch & Hambling 1971; McQuire 1971; Gambetta & Orlandi 1973; Gersonde 1973; Cockroft 1974; Hilditch 1974, 1984; Purslow 1975; Perkins & Hunt 1977; Beesley 1978; Sparks 1978; Clubbe & Levy 1982; Hilditch *et al.* 1983).

Major differences occur (as would be expected) between test sites in widely different geographical locations. By far the most experienced and longest established laboratory carrying out field trials of this type is the United States Department of Agriculture Forest Products Laboratory at Madison, Wisconsin, which implants stakes at several sites (Gjovik & Davidson 1975, 1979).

A variation occasionally used in the field trial is laying the treated wood on top of, but not actually buried in, the ground. This has been used particularly by the author's company (unpublished) where it is wished to retain some element of the certainty of infection and natural environment of stake tests but without the full and complex interchange with the soil.

Tests not involving Ground Contacts

The most important economic use for wood is probably in buildings. Mostly this is not in contact with the ground, but decay is common. Performance of preservatives in this situation is not reliably indicated by soil tests since practically all elements of the immediate environment of the timber are different.

Because of the wide variety of situations in which wood is used and the difficulty in re-creating these environments experimentally in such a way that decay of untreated wood is certain, the development of test methods relating immediately to this use lags considerably behind the need. Some methods which have been published are given below. These mostly relate to joinery since it is in this area that there has been the most recent interest. Several test methods have been tried (Morgan 1971; Purslow & Williams 1978; Sparks 1978; Fougerousse 1981; Carey 1982; Mendes 1982; Hilditch *et al.* 1983). In most of these a joint of the type used in joinery is exposed to the weather in such a way that water can get in, and the natural microbiological infection can occur.

Exposure trials at Princes Risborough Laboratory used T- or L-joints (Morgan 1971; Purslow & Williams 1978). The T-joints were two pieces with sawn surfaces 45 × 45 mm in cross-section, each arm 600 mm long, and assembled using a stub tenon fitted into a stoppered mortise without glue. A hole was drilled at one end of the mortise to fix it to the racks. Exposure was at an angle of 45°, pointing south. The L-joint, of two pieces with planed surface (51 × 51 mm in cross-section and each arm 455 mm long) was assembled with a through tenon into an open mortise. The joint was neither glued nor primed. Each specimen was supported on a ledge in a water-trap frame with a backing board and exposed to the weather on racks, in a near vertical position facing south. Results of exposure tests were summarized by Purslow & Williams (1978). Comments were made on the two different joint designs and the L-joints were found most suitable. The T-joints did not remain wet for long because they were exposed on all sides to the drying action of the wind. With the L-joint, air movement was restricted by the presence of the supporting framework and remained wet for very long periods. Untreated, unpainted pine sapwood decayed in 7 years (L-trials) and 10 years (T-trials) indicating more severity in test conditions reflected in L-joints. L-joints were easier to make and inspect. Purslow & Williams also commented that similar performance could be achieved by reducing the size of the specimen.

As the result of these observations Carey (1982) exposed L-joints of similar design (Purslow & Williams 1978) but of reduced dimensions and with modification in exposure design and inspection criteria. Each L-joint consisted of two Scots pine sapwood components, measuring 203 × 38 × 38 mm, joined by a tongue and groove bridle joint, with the horizontal member providing the tenon. The arrises of the horizontal members were rounded to reduce the loss of paint. The end grain, remote from the joint of both members was sealed using an epoxy/coal tar pitch mixture. The L-joints were exposed on south facing plywood racks, *ca.* 900 mm above the ground, sloping at 10° to provide a water trap in the joint area. The upright position of the L-joint was positioned by spacing blocks fixed to the rack between vertical members. Two 5 mm square stickers were placed under the L-joints to limit excess wetting from the water trap.

These tests which were designed to give information about the performance of preservatives and treatment techniques appropriate to building timbers, have been found to provide an ideal basis for studies of the progressive colonization and physical changes that occur in wood during exposure (Carey 1982*a*, 1983). Similar studies have been carried out on stakes (Levy 1962). L-joints are well suited for studying the effect of paint coats, exterior woodstains and the like on the total durability of the system. Fungal colonization, physical changes in wood and moisture retention under these circumstances have also been studied (Mendes 1982).

A variety using a Y-joint in which a short piece is joined at an upward angle to a longer piece to form a Y has been used in France and Canada (Sedziak *et al.* 1970;

Fourgerousse 1981). This assembly tends to collect rain and force it into the joinery. This Canadian study lasted 14 years.

The small dimension L-joint design is gradually being accepted as the most suitable system for conducting out-of-ground-contact tests, as reflected by the setting up of cooperative trials in many countries under the auspices of the International Research Group on Wood Preservation (Bravery *et al.* 1983).

Conclusions

We have sought to give an outline of test methods that are used to determine the fungal resistance of preservative-treated wood. There are many kinds of tests and many variations of each kind. Variety is inevitable when in a certain type of test there is, say, no clear reason why a block 50 × 25 × 10 mm should be preferred to a block 10 × 10 × 10 mm or any other size, yet a direct consequence of this variety is that results obtained cannot be directly compared. Thus there is a case for standardization to reduce the number of varieties which have no clear and separate features, but it is most difficult to arrive at an agreement on what is, and is not, useful.

The available laboratory tests do not enable the practical performance of a wood preservative to be predicted reliably in the absence of confirmatory long-term field tests. Laboratory tests that can subject a test piece of preservative-treated wood to the chemical, physical and biological assault that it will face in nature have not yet been developed. This is a major barrier to wood preservative development and is the main need in wood preservative testing.

References

ANON. 1960*a Methode d'essasi rapide des produits fongicides utilises pour la protection des bois par impregnation superficielle.* NFX pp. 41–510 Paris: Association Francaise De Normalisation (AFNOR).

ANON. 1960*b Procedure for testing anti-rot wood preservatives. Swedish Standard K60–1 : 1960.* Stockholm: The National Swedish Institute for Materials Testing.

ANON. 1961*a Methods of Testing for Toxicity of Wood Preservatives.* British Standard 838. London: British Standards Institution.

ANON. 1961*b Prufung von Holz-schuttzmitte—Mykologische Kurzprufung Klotzchen—Verfahren.* DIN 52.176. Berlin: Deutscher Normenausschuss.

ANON. 1968 *Method for Laboratory Testing of Wood Preservatives against Fungi.* Indian Standard 4873. New Delhi: Indian Standards Institute.

ANON. 1971*a Testing of Wood Preservatives. Mycological Test—Mullade Method—a Soil Block Test in Unsterile Soil.* Standard 1.4.1.2/70. Helsinki: Nordic Wood Preservation Council.

ANON. 1971*b Standard for Testing Wood Preservatives. Mycological Test—A Field Test with Stakes.* Standard 1.4.2.1/71. Helsinki: Nordic Wood Preservation Council.

ANON. 1978 A new method for testing wood preservatives. In *What's new in Forest Research* No. 65. September. New Zealand Forest Service. Rotorua: Forest Reserach Institute.

ANON. 1980 *Standard Method for Evaluating Wood Preservatives by Field Tests with Stakes.* Standard D1758–74. Philadelphia: American Society for Testing and Materials.

ANON. 1981 *Standard Method of Testing Wood Preservatives by Laboratory Soil Block Culture.* Standard D1413–76. Philadelphia: American Society for Testing and Materials.

ANON. 1982*a* EN113: *Wood preservatives. Determination of the toxic values against wood destroying Basidiomycetes cultured on an agar medium.* Bruxelles: Comite European de Normalisation.

ANON. 1982*b* *Wood Preservatives. Determination of the Toxic Values against Wood-destroying Basidiomycetes Cultured on an Agar Medium.* British Standard 6009. London: British Standards Institution. (European Standard EN113: EEC, Brussels).

ARMSTRONG, F. H. & SAVORY, J. G. (1959) The influence of fungal decay on the properties of timber. Effect of progressive decay by the soft rot fungus. *Chaetomium globosum* on the strength of beech. *Holzforschung* **13**, 84–89.

BAKER, J. M. MORGAN, J. W. W. & SAVORY, J. G. 1972 *Wood protection research at the Forest Products Research Laboratory.* Record of the Annual Convention of the British Wood Preserving Association. Cambridge 1972. pp. 183–199. London: British Wood Preserving Association.

BECKER, G. HOF, T., JACQUIOT, C., LOHWAG, K. RENNERFELT, E. & WALCHLI, O. 1966 Comparative investigations on the laboratory testing of the effectiveness of wood preservative to fungi. *Holz als Roh-und Werkstoff* **24**, 53–58.

BECKER, G. & STARFINGER, K. 1971 Uber die Reichweite der fungiziden und insektiziden Wirksamkeit Pigment-haltiger Obigen Schutmittel in Holz. *Holz als Roh-und Werkstoff* **29**, 344–348.

BEESLEY, J. 1978 An Australian test of wood preservatives. Part 1. Preservative Principles and Practices. *Material und Organismen* **13**, 31–50.

BETTS, W. D. 1982 *Creosote: A century of service as a wood preservative. Record of the Annual Convention of the British Wood Preserving Association.* Cambridge 1982. pp. 10–27. London: British Wood Preserving Association.

BRAID, G. H. & LINE, M. A. 1980 Biological techniques for determining relative soft-rot attack in hardwoods. *Holsforschung* **34**, 1–4.

BRAVERY, A. F. 1968 Determining the tolerance of soft rot fungi to wood preservatives. *Material und Organismen* **3**, 213–227.

BRAVERY, A. F. 1970 Preliminary observation on some effects of wood cell wall penetration by organic solvent type wood preservatives. *International Biodeterioration Bulletin* **6**, 145–147.

BRAVERY, A. F. 1975 *Microbiological assay of chemicals for the protection of wood.* Current Paper CP 92/75. Princess Risborough: Building Research Establishment.

BRAVERY, A. F. & GRANT, C. 1971 Preliminary investigation into the use of thin strip tensile strength tests for rapid evaluation of wood preservatives against Basidiomycetes. *International Biodeterioration Bulletin* **7**, 169–173.

BRAVERY, A. F. & LAVERS, G. M. 1971 Strength properties of decayed softwood measured on miniature test beams. *International Biodeterioration Bulletin* **7**, 129–141.

BRAVERY, A. F. & SAVORY, J. G. 1970 Collaborative experiments in testing the toxicity of wood preservatives to soft rot fungi. *Material und Organismen* **5**, 59–80.

BRAVERY, A. F., DICKINSON, D. J. & FOUGEROUSSE, M. 1983 *Progress report on co-operative research project on L-joint testing.* Document IRG/WP/2192. International Research Group on Wood Preservation. Brisbane.

BUTCHER, J. A. 1981 *Questionnaire; facility for accelerated stake tests in unsterile soil.* Document IRG/WP/2166. International Research Group on Wood Preservation. Sarajevo.

CAREY, J. K. 1980 *The mechanisms of infection and decay of window joinery.* Ph.D. Thesis, London University.

CAREY, J. K. 1982 Assessing the performance of preservative treatments for window joinery. *Holz als Roh-und Werkstoff* **40**, 269–274.

CAREY, J. K. 1983 Colonization of wood joinery. In *Biodeterioration 5*. p. 13–25. ed. Oxley T. A. and Barry S. Chichester: John Wiley.

CAREY, J. K. SAVORY, J. G., MENDES, F. & BRAVERY, A. F. 1984 Loss of resistance to decay of Scots pine heartwood in a biologically active environment. *Applied Biochemistry and Biotechnology. Abstract.* **9**, 343–344.

CARTWRIGHT, K. St. G., CAMPBELL, W. G. & ARMSTRONG, F. H. 1936 Influence of fungal decay on the properties of timber. I. The effect of progressive decay by *Polyporous hispidus* on the strength of English ash. *Fraxinus excelsior* L. *Proceedings of Royal Society* B120, 76–85.

CLUBBE, C. P. & Levy, J. F. 1982 Microbial ecology of CCA treated stakes. *Material und Organismen* **17**, 21–34.

COCKROFT, R. 1974 *The performance of pentachlorophenol in a stake test in the United Kingdom.* Current Paper CP21/74. Princess Risborough: Building Research Establishment.

DEPPE, H. J. 1967 Untersuchungen Zum Schulz von Holzspan-platten gegan Pilzbefall. *Holz-Zentrablatt* **93**, 2271–2274.

DEPPE, H. J. 1974 Zum Stand der verwendung von Holzfaser-platten als Baustoff in Sinne der deutschen bauafichtelcken Bestimmungen. *Holz-Zentrablatt* **100**, 43–46.

DEPPE, H. J. & GERSONDE, M. 1969 Hertellung und Prufung von geschutzten Holz-spanplatten. *Material und Organismen* **4**, 123–136.

DICKINSON, D. J. 1974 A new technique for screening fungicides for wood preservation. *International Biodeterioration Bulletin* **10**, 49–51.

DUNCAN, C. G. 1953 *Soil block and agar block techniques for evaluating of oil-type wood preservatives: creosote, copper naphthenate and pentachlorophenol.* Forest Pathology, Special Release No 37, January 1953. Madison: Department of Agriculture.

DUNCAN, C. G. 1958 *Studies on the methodology of soil block testing.* Report No 2114. Madison: Forest Products Laboratory.

FINDLAY, W. P. K. & SAVORY, J. G. 1954 Decomposition of wood by lower fungi. *Holz als Roh-und Werkstoff* **12**, 293–296.

FOUGEROUSSE, M. 1981 *Field test out of ground contact in France. Definition of the test procedure and preliminary result after 18 months.* Document IRG/WP/2161. Sarajevo: International Research Group on Wood Preservation.

GAMBETTA, A. & ORLANDI, E. 1973 The preservation of wood put to use in the open. *Contributi Scientifico—practici per una Migliore Conoscenza ed utilizzione de Legno* **17**, 37–60.

GERSONDE, M. 1973 Field tests with wood preservatives in wooden stakes. *Holz als Roh-und Werkstoff* **31**, 89–96.

GERSONDE, M. & BECKER, G. 1958 Prufung von Holzshutz mitteln fur den Hochbau auf Wirksammkeit gegan Pilze an praxisgemassen Holzproben 'Schwamm keller' = 'Versuche'. *Holz als Roh-und Werkstoff* **16**, 346–357.

GERSONDE, M. & DEPPE, H. J. 1968 Moghelchkeiten und verfahrens—technische Varaussetzungen einer Schutzbehandlung von Spanplatten gegan holzzerstorende Pilze. *Holz-Zentralblatt* **94**, 876–877.

GJOVIK, L. R. & DAVIDSON, H. L. 1975 *Comparison of wood preservation in Mississippi Post studies.* USDA Forest Service Research Note FPL-01. Madison: United States Department of Agriculture.

GJOVIK, L. R. & DAVIDSON, H. L. 1979 *Comparison of wood preservation in stake tests* (1979 Progress Report). USDA Forest Service Research Note FPL–02. Madison: United States Department of Agriculture.

GOOD, H. M. & DARRAH, J. A. 1967 Rates of decay in wood measured by carbon-dioxide production. *Annals of Applied Biology.* **59**, 463–472.

HALABISKY, D. D. & IFJU, G. 1968 Use of respirometry for fast and accurate evaluation of wood preservatives. *Proceedings of American Wood Preservers Association* **64**, 215–223.

HANSEN, J. 1973 The 'Mini-Fungus Cellar'—A mycological test method for wood protection. *International Biodeterioration Bulletin* **9**, 82–84.

HEDLEY, M. E. 1980 *Comparison of decay rates of preservative treated stakes in field and fungus cellar tests.* IRG/WP/2135. Paper presented to 11th Annual Meeting Raleigh U.S.A. 1980. Stockholm: International Research Group on Wood Preservation.

HEDLEY, M. E. 1983 *Comparison of decay rates of preservative treated stakes in field and fungus cellar tests—results after 40 months cellar experience.* IRG/WP/2200. Paper presented to 14th Annual Meeting Brisbane, Australia. 1983. Stockholm: International Research Group on Wood Preservation.

HILDITCH, E. A. 1974 *Wood preservative tests.* Record of the Annual Convention of the British Wood Preserving Association. Cambridge 1974. pp. 95–104. London: British Wood Preserving Association.

HILDITCH, E. A. 1978 *Soil chemistry and wood decay.* IRG/WP/2109. Paper presented to the 10th Annual Meeting Peebles, Scotland 1978. Stockholm: International Research Group on Wood Preservation.

HILDITCH, E. A. 1984 A rapid wood preservative test over thirteen years. *Journal of the Institute of Wood Science* **10**, 24–28.

HILDITCH, E. A. & HAMBLING, R. E. 1971 *Wood preservative tests and their assessments.* Record of the Annual Convention of the British Wood Preserving Association. Cambridge 1971. pp. 95–113. London: British Wood Preserving Association.

HILDITCH, E. A., SPARKS, C. R. & WORRINGHAM, J. H. M. 1983 *Further development in metallic soap based wood preservatives.* Record of the Annual Convention of the British Wood Preserving Association, Cambridge 1983. pp. 61–70. London: British Wood Preserving Association.

HOPKINS, C. Y. & COLWELL, B. B. 1944 *Surface coatings for rot proofing wood.* Canadian Chemistry and Process Industries. National Research Council Report No. 1256. Ottawa: National Research Council of Canada.

HUMPHREY, C. J. & FLEMMING, R. M. 1915 *The toxicity to fungi of various oils and salts, particularly those used in wood preservation.* Bulletin no. 227. Washington: United States Department of Agriculture.

INGLEBY, A. 1982 *New techniques for the evaluation of remedial treatment formulations in the laboratory.* Record of the Annual Convention of the British Wood Preserving Association, Cambridge 1982. pp. 39–43. London: British Wood Preserving Association.

JENSEN, B. & IMSGARD, F. 1982 *The importance of axial penetration in relation to double vacuum treatment of window joinery.* Record of the Annual Convention of the British Wood Preserving Association, Cambridge 1982. pp. 55–62. London: British Wood Preserving Association.

JOHNSON, G. C., THORNTON, J. D. & GREAVES, H. 1982 *The accelerated field simulator (=fungal cellar).* IRG/WP/2170. Paper presented to 13th Annual Meeting. Cesme, Turkey. 1982. Stockholm: International Research Group on Wood Preservation.

JOHNSON, G. C., THORNTON, J. D., CREFFIELD, J. W. & HOWICK, C. D. 1963 *Natural durability studies in an accelerated field simulator—a novel approach.* IRG/WP/2197. Paper presented to 14th Annual Meeting. Brisbane, Australia 1983. Stockholm: International Research Group on Wood Preservation.

KENNEDY, R. W. & IFJU, G. 1962 Applications of micro-tensile testing in thin wood sections. (TAPPI). *Technical Association of the Pulp and Paper Industry* **45**, 725–733.

KLINGSTROM, A. 1965 *CO_2 production as a measure of decay activity in wood blocks.* Stockholm: Studia Forestali Suecica No. 26.

KNEALE, R. E. & TAMBLYN, N. 1950 *Development of a laboratory technique for testing the toxicity of copper naphthenate.* Progress Report No. 2. Melbourne: Commonwealth Scientific and Industrial Research Organisation. Australian Division of Forest Products.

LEUTRITZ, J. 1939 Acceleration of toximetric tests of wood preservatives by the use of soil as a medium. *Phytopathology* **29**, 901–903.

LEUTRITZ, J. 1946 A wood soil contact culture technique for laboratory study of wood destroying fungi, wood decay and wood preservation. *Bell System Technical Journal* **25**, 102–135.

LEVY, J. F. 1962 *Studies on fungi attacking underground mining timbers and fence posts.* Record of the Annual Convention of the British Wood Preserving Association Cambridge 1962. pp. 3–13. London: British Wood Preserving Association.

LEVY, J. F. 1965 The soft rot fungi, their mode of action and significance in the degradation of wood. In *Advances in Botanical Research.* Preston, R. D. pp. 323–357. London: Academic Press.

LEVY, J. F. & DICKINSON, D. J. 1980 *Wood protection research at Imperial College.* Record of the Annual Convention of the British Wood Preserving Association. Cambridge 1980. pp. 67–73. London: British Wood Preserving Association.

LEVY, J. F. & DICKINSON, D. J. 1982 The role of research in relation to wood preservation. Record of the Annual Convention of the British Wood Preserving Association, Cambridge 1982. pp. 72–79. London: British Wood Preserving Association.

LIESE, J., NOWAK, A., PETERS, F. & RABANUS, A. 1935 Toximetrische Bestimmung von Holzkonservier ungsmitteln. *Angewandte Chemie* **48**, 21–23.

LINE, M. A. 1982 *Towards a colour assay of wood degradation.* Stockholm: International Research Group on Wood Preservation Document IRG/WP/2180.

LINE, M. A. 1983 Catalase activity as an indication of microbial colonization of wood. In *Biodeterioration,* 5. pp. 38–43, ed. Oxley, T. A. and Barry, S., Chichester: John Wiley.

MALENKOVIC, B. 1904 Zur Lehre und Anwendung der Holzkonsvierung in Hochbaue. *Mitteilung Geganstand Artillerie und Geniew (Vienna)* **35**, 311–333.

MATEUS, T. J. E. 1954 *Evaluation of the effectiveness of wood preservatives by a new method based on the measurement of deflection.* Publication No. 48. Portugal Ministry of Public Works. National Laboratory of Civil Engineering. Lisbon.

MATEUS, T. J. E. 1957 *A mechanical test for studying wood preservatives.* Record of the Annual Convention of the British Wood Preserving Association. Cambridge 1957. pp. 137–157. London: British Wood Preserving Association.

MCQUIRE, A. G. 1971 *Results of fifteen years of exposure testing with three-quarter inch stakes.* Technical Paper No. 56. Wellington: New Zealand Forest Service.

MENDES, F. 1982 *Microflora associated with treated window joinery.* Ph.D. Thesis, London University.

MORGAN, J. W. W. 1971 Field tests on preservatives out of contact with the ground. Timberlab News. *Newsletter of the Forest Products Research Laboratory* **8**, 2–3.

MURPHY, R. J., SCHASCHING, K. & DALLEY, R. (1982). A simple soil-bed design for durability testing of wood. *Journal of the Institute of Wood Science* **9**, 155–160.

PERKINS, R. H. & HUNT, M. O. 1977 *Field service tests of cold soak preservative treated products.* Research Bulletin 939. Purdue Agricultural Experimental Station, Purdue University, West Lafayette.

PURSLOW, D. F. 1965 *Results of stake tests on wood preservatives* (Progress Report to 1974). Current Paper CP86/75. Princes Risborough: Building Research Establishment.

PURSLOW, D. F. & WILLIAMS, N. A. 1978 *Field trails on preserved timber out of ground contact.* Current Paper CP78/78. Princes Risborough: Building Research Establishment.

RICHARDSON, B. A. 1968 A new technique for the comparative evaluation of some organometallic wood preservatives. *International Pest Control.* **10**, 14–19.

SAVORY, J. G. 1954a Breakdown of timber by Asomycetes and Fungi Imperfecti. *Annals of Applied Biology* **41**, 336–347.

SAVORY, J. G. 1954*b* Damage to wood caused by micro-organisms. *Journal of Applied Bacteriology* **17**, 213–218.

SAVORY, J. G. & BRAVERY, A. F. 1971 Observations on methods of determining the effectiveness of wood preservatives against soft rot fungi. *Mitteilungen Deutschen Gesselschaft für Holzforschung* **57**, 12–17.

SCHAEFFER, T. C. 1936 *Progressive effects on* Polyporus versicolar *on the physical and chemical properties of red gum sapwood.* Technical Bulletin, 527. Madison; United States Department of Agriculture.

SCHMITZ, H., BATEMAN, E. E., COLLEY, R. H., CHURCH, S. R., HARTLEY, C., WATERMAN, R. E., KAMMERER, A. L., SCHRENK, H. von. FULKS, E. B., HUBERT, E. E., REEVE, C. S., SNELL, W. H., RICHARDS, C. A., LINDER, D. H. & BURNES, J. D. 1930 A suggested toximetric method for wood preservatives. *Industrial and Engineering Chemistry, Analytical Edition* **2**, 361–363.

SEDZIAK, H. D., SHIELDS, J. K. & KRZYZEWSKI, I. 1970 Effectiveness of brush and dip preservation treatments for above ground exterior exposures of wood. *International Biodeterioration Bulletin* **6**, 149–155.

SHARP, R. F. & EGGINS, H. O. W. 1968 A rapid strength method for determining the biodeterioration of wood. *International Biodeterioration Bulletin* **4**, 63–66.

SMITH, D. N. R. 1965 *The evaluation of wood preservatives.* Record of the Annual Convention of the British Wood Preserving Association. pp. 123–149.

SMITH, D. N. R. 1971 A possible method for the rapid evaluation of wood preservatives. *Mitteilungen Deutschen Gesselschaft für Holzforschung* **57**, 18–22.

SMITH, D. N. R. 1980 Study of decay of preservative treated wood in soil. *Journal of the Institute of Wood Science* **8**, 194–200.

SMITH, R. S. 1967 Carbon dioxide evolution as a measure of attack of wood by fungi and its application to testing wood preservatives and sapstain preservatives. *Annals of Applied Biology* **59**, 473–479.

SMITH, R. S. 1969 Wood preservative toxicity evaluation using wood weight loss and fungal respiration methods. *Wood Science* **2**, 44–53.

SPARKS, C. R. 1978 *A new brand of metallic chemicals for use in wood preservatives.* Record of the Annual Convention of the British Wood Preserving Association. Cambridge 1978. pp. 48–51. London: British Wood Preserving Association.

SUTTER, H. P. 1978*a Methods to determine the depth of penetration of the biologically active components of wood preservatives.* IRG/WP/2108. Paper presented to 10th Annual Meeting. Peebles, Scotland 1978. Stockholm: International Research Group on Wood Preservation.

SUTTER, H. P. 1978*b* A new technique for screening fungicides for wood preservatives. *International Biodeterioration Bulletin* **14**, 95–99.

SUTTER, H. P., GARETH JONES, E. B. & WALCHLI, O. (1983) The mechanims of copper tolerance in *Poria placenta* (Fr.) Cke. and *Poria vaillantii* (Pers.) Fr. *Material and Organisemen* **18**, 241–262.

VINDEN, P., SAVORY, J. G. DICKINSON, D. J. & LEVY, J. F. 1982 *Soil-bed studies.* IRG/WP/2181. Paper presented to 13th Annual Meeting, Cesme, Turkey. 1982. Stockholm: International Research Group on Wood Preservation.

WILSON, J. W. & IFJU, G. 1965 *Measuring density and strength properties of minute wood specimens.* Wood characteristics. VI. Technical Report 423. Montreal: Pulp and Paper Research Institute of Canada.

Evaluation of Biocides for Use on Building Materials

C. GRANT AND A. F. BRAVERY

Building Research Establishment, Princes Risborough Laboratory, Aylesbury, Buckinghamshire, UK

Building materials such as stone masonry, concrete, bricks, asbestos-cement sheeting and painted surfaces are frequently subject to disfigurement and attack by the growth of organisms. In exterior situations algae, mosses and lichens most commonly occur, while indoors the growth of moulds gives cause for concern (Anon. 1981, 1982; Richardson 1973).

Many chemical formulations are currently available for the eradication and control of organisms on building surfaces (Bravery 1981; Anon 1982; Allsopp & Allsopp 1983) and realistic and reliable laboratory test methods are necessary to enable screening and evaluation of candidate compounds without early recourse to site trials. Indeed, the latter might only prove necessary for final selection from a small number of potential treatments. An assessment of several test methods by Grant & Bravery (1981*a*) led to the development of a laboratory technique employing actual building materials and permitting full substrate/organism/biocide interaction during testing. Use of the so-called moist vermiculite-bed technique to evaluate the efficacy of a quaternary ammonium biocide against algae on a range of natural building stones and on mortar has been reported (Grant & Bravery 1981*b*). Since then many more biocides and substrates have been tested using the technique which has undergone further development to include testing of algicidal and fungicidal paints. It may also prove suitable for evaluation of toxic washes for treatment of mould growths on painted surfaces.

Test Procedure—Materials and Methods

Substrates

Ideally the chosen test substrate should correspond as closely as possible to that used in the practical situation since, for example, it has been found that a biocide

Preservatives in the Food,
Pharmaceutical and Environmental
Industries

0–632–01727–9

can vary in effectiveness on different types of natural stone (Grant & Bravery 1981*b*). Therefore if a biocidal treatment is required for a building constructed largely of Portland limestone, testing should be carried out on samples of this material. In this way a realistic assessment of biocidal efficacy is gained and an indication given of possible adverse interactions between the treatment and the substrate. Examples of the latter are brown or purple staining from a reaction between phenolic biocides and the iron content in some stone or the formation of soluble salts with subsequent crystallization and damage after treatment with sodium pentachlorophenoxide or related compounds (Richardson 1980).

Manufactured sheet materials such as asbestos-cement and the recently introduced calcium silicate boards (Masterboard and Masterclad, Cape Boards and Panels Ltd. Uxbridge) together with plasterboard can be useful model substrates for evaluation of biocides and surface finishes intended for application to cementitious surfaces. Alternatively, required substrates can be cast in moulds to produce samples of concrete with specific cement : aggregate : water ratios (British Standard (BS) 4550:Part 3:1978 (Anon. 1978*b*), plaster (Bravery *et al.* 1983) or mortar (BS 4550:Part 5:1978 (Anon. 1978*c*)).

Conditioning of test samples

Various types of conditioning, including both artificial and natural weathering, can be done on samples prior to laboratory testing. Fresh concrete and mortar surfaces are usually about pH 12; freshly-prepared plaster has a pH between 10 and 11. Thus with freshly-prepared samples of such materials some form of conditioning will be necessary to reduce pH, at least at the surface, to a level that will permit growth of test organisms—generally at $<$pH 8-9. Recently produced sheets of asbestos-cement and other types of cement-based building boards may also have a high alkalinity and should be checked before use. It may also be desirable to 'age' treated or painted samples to assess the degree of loss of protection. A number of methods of achieving this are available. Some currently used conditioning techniques are described below.

Artificial carbonation

Atmospheric carbon dioxide in the presence of moisture reacts with hydrated cement minerals. In the process, the rate of which is determined by the concentration of CO_2, the moisture content of the sample and the relative humidity of the surroundings, $Ca(OH)_2$ is converted to $CaCO_3$ and other cement compounds are decomposed with an accompanying fall in pH (Neville 1975). Artificial carbonation can therefore be used to accelerate reduction of pH to levels conducive to the growth of organisms.

Carbonation can be conveniently done in a desiccator containing a saturated solution of sodium bromide. The attained relative humidity of 58% is within the range favouring rapid carbonation (Vénuat & Alexandre 1969). Carbon dioxide gas is admitted and displaced air vented until the desiccator is full of CO_2. This can be checked by bubbling the vented gas through a small reservoir of lime-water ($Ca(OH)_2$). During the course of carbonation the vent is left open and fresh CO_2 admitted daily. Any moisture condensing inside the desiccator is removed periodically. The progress of carbonation towards neutrality can be monitored by narrow-range pH indicator papers. They are placed momentarily on the dampened surface of test samples. Alternatively a phenolphthalein indicator solution (1 g phenolphthalein dissolved in 50 ml ethanol diluted to 100 ml with water; Anon. 1978*a*) may be used.

Artificial weathering

Exposure for a specified number of hours to radiation from a continuous carbon arc and a cycled water-spray is one way of achieving artificial weathering. An apparatus such as the Marr 'Weatherometer' (J. B. Marr Ltd., Kingston-on-Thames, Surrey) or equipment operated according to BS 3900:Part F3:1971 (Anon. 1971) is used. Exposure to condensation conditions (BS 3900:Part F2:1973 (Anon. 1973)) and natural weathering (BS 3900:Part F6:1976, Anon. 1976) may also be used for the ageing of samples.

Post-conditioning treatment

The handling of samples following conditioning depends upon the nature of any pre-treatments or subsequent treatment operations. Where it is required to apply treatments after conditioning, samples will need to be dried under laboratory conditions until equilibrium has been achieved. This is especially important if they have been subjected to some form of artificial weathering.

Samples which have already been treated with biocide or finished with biocidal coating can be placed directly into incubation chambers. They are inoculated as appropriate and assessed for resistance to either algal or mould infection. Those which are to be infected with algae to assess eradicant treatments can be similarly positioned, although it may be advantageous to soak air-dry samples in water or in nutrient solution to expedite colonization by test organisms.

Culture vessels

A variety of vessels, fitted with transparent lids for tests with algae, have been used. In our experience transparent polystyrene boxes (220 × 110 × 80 mm) with close-fitting lids (C. E. Payne & Sons, Slough, Berks) have proved well-suited for this purpose. Their size permits the use of three replicate test samples (60 × 80 mm) per treatment. These can be readily accommodated in each box (Fig. 1). Small plant propagators have also been used.

FIG. 1. Transparent polystyrene box used for incubation of three replicate test samples.

Test samples are bedded horizontally onto a moisture-retaining medium such as vermiculite (grade DSF, Dupré Vermiculite Ltd., Hertford) moistened to *ca.* 400% moisture content—equivalent to approx. 130% of its water-holding capacity (Carey & Grant 1975). Other grades of vermiculite can be used but their moisture retaining properties must be checked. The polystyrene boxes (Fig. 1) are charged with 100 g air-dry vermiculite which is lightly tamped down to give a flat surface. It is then moistened to give the required moisture content.

Test organisms

Algae

The algae chosen for laboratory test work should be representative of those encountered in practice; they should ideally be easy to culture and sufficiently robust to withstand the inoculation procedures. The algae responsible for fouling of terrestrial substrates have been reviewed and those occurring most frequently listed by Grant (1982). Selection of test organisms can be made to include representatives from the Chlorophyceae and the Cyanophyceae. Those routinely used at PRL are: *Gloeocapsa alpicola*, CCAP 1430/1; *Nostoc commune*, CCAP 1453/29; *Pleurococcus* spp., CCAP 464/1; *Stichococcus bacillaris*, CCAP 379/1a; *Trentepohlia aurea*, CCAP 483/1.

The species are cultured separately in liquid mineral salts media. Modified Knop's solution (Wright 1975) has proved suitable for the majority of species. The compositions of other nutrient media are given by Stein (1973). The very slow-growing *Tr. aurea*, on the other hand, is better cultured in Bold's basal medium (Nichols & Bold 1965) modified after Ho *et al.* (1983). Adams & Dobbs (1984) have proposed a further modification of Bold's basal medium in which molybdenum trioxide is replaced by the more soluble ammonium molybdate.

The same requirements governing selection of suitable test organisms apply to mould species. Bravery *et al.* (1983) gives a comprehensive list of those commonly encountered on paint films. The moulds routinely used at PRL for test on emulsion-painted surfaces are: *Alternaria alternata*, PRL No. S1051; *Aspergillus versicolor*, PRL No. S1064; *Cladosporium cladosporioides*, PRL No. S1065; *Cladosporium sphaerospermum*, PRL No. S1045; *Penicillium corylophilum*, PRL No. S1028; *Phoma nebulosa*, PRL No. S1048; *Ulocladium atrum*, PRL No. S1043. The importance of certain species of pink yeasts as colonizers of interior paint surfaces has now been recognized. Current PRL tests include *Rhodotorula glutinis* var. *dairiensis* (PRL No. S1056) and *Sporobolomyces roseus* (NCYC No. 717) in the inoculum.

All of the organisms noted above are cultured routinely on slants of 2% (w/v) malt extract agar in test tubes.

Inoculum preparation

Algae

Individual cultures of algae are combined to produce a mixed inoculum. For non-filamentous and non-colonial species, where cell counts can be done with a haemocytometer, inoculum density should exceed 10^6 cells/ml. Alternatively, inocula can be prepared to a standardized optical density at a specific wavelength of light (Sorokin 1973). Filamentous types such as *Trentepohlia* and those, such as *Nostoc*, which form clumps of cells embedded in mucilage require some maceration. Short periods (up to 10 s) in a partially enclosed blade macerator (type × 1020, Intern Laborat App, GmBH, Döttingen, W. Germany) produce viable inocula. These can be applied *in situ* either from a wide bore pipette or with a soft artist's brush. Inocula of single celled organisms can also be applied by these methods or by a laboratory spray gun (Shandon Southern Products Ltd., Runcorn, Cheshire). A uniform deposit of algae over the entire, exposed face of test samples must be a feature of any method of inoculation.

Where testing is designed to assess the eradicant activity of biocidal treatments, a heavy overgrowth of algae on test samples must be established. Inoculation at weekly intervals may be done until the necessary density of growth is achieved. For assessment of the protective effectiveness of biocides or algicidal paint films, a continued challenge to the treatment is required. Repeated inoculation at regular intervals throughout the exposure period achieves this objective.

Mould fungi

Details of the preparation of spore suspensions and the method of inoculation of test samples are given in Bravery *et al.* (1983). Aerosol OT (BDH, Poole, UK) is used as a wetting agent because it does not adversely influence spore viability, a feature that has been noted, for example, with Tween 80. Samples can be spray- or swab-inoculated *in situ* in the culture vessels. One inoculation is generally sufficient provided uniform coverage is achieved.

Illumination and incubation

Fluorescent strip lights are used in tests with algae. It needs to be stressed, however, that light intensity is not uniform along the length of such strips, there being a marked diminution towards the ends. Furthermore, as light output changes with the age of a tube, checks of light intensity must be made in tests of long duration. Although it is unlikely that the choice of the type of fluorescent light is critical in work of this nature, the Northlight type tube has good spectral output characteristics at about 660 nm which corresponds to a peak in the light absorption for many algal cells. In practice the lights can be mounted in a variety of ways to produce adequate uniformity and intensity of lighting; one example is shown in Fig. 2. For routine work with most algae a light intensity of 2500 lx at the surface of the test sample, with a daily 16 h-photoperiod, has given good results. Light intensity can be measured by a light meter such as the Megatron Type EA2 (Megatron Ltd., London). With new 1500 mm long 65-80W Northlight tubes the light level noted above is achieved with the tube positioned approximately 225 mm from the sample. A light intensity of <1000 lx (Ho *et al.* 1983) is needed to achieve optimum growth in tests involving *Trentepohlia.*

Illumination of mould cultures is unnecessary. Incubation temperatures of between 18 to 22 °C are suitable for both algae and moulds.

The duration of the exposure period can be varied as appropriate. With mould tests, results are normally assessed after 28 days. The test may be extended when attempting to reveal the ultimate failure of apparently durable fungicidal paints and eradicant treatments. With algal tests, it has been found necessary to extend incubation for up to 12 weeks following application of eradicant treatments or exposure of pre-treated test samples, particularly where the treatment has been challenged by repeated re-inoculation. Occasional moistening of test samples with

FIG. 2. Illumination of test samples.

a spray of water or nutrient solution may be necessary when paint or water-repellent treatments are being evaluated.

Until further results from field trials are available there is at present insufficient evidence to correlate performance over a given period of laboratory testing with likely duration of performance under conditions in service.

Biocide treatment

Protective treatments

The rates of application by spraying, brushing or dipping of paints, water-repellents and other protective treatments to samples must be equivalent to those used in practice. Moreover any recommended or specified preparative procedures for priming, undercoating *etc.*, the samples should be observed. Care should be taken to ensure full drying and curing of the treatment. If this is not done, the vapours, especially those of solvents, in culture vessels may adversely affect the test organisms.

Eradicant treatments

As treatments for the assessment of eradicant action are applied to fully colonized samples, care should be taken to prevent physical damage to the test organisms. For algae, application by a spray or soft artists' brush is suitable; swabbing is also acceptable for moulds. One needs to ensure, where possible, equivalent application rates whatever method is employed. In all tests, suitably infected, control samples should be set up and treated with either water or the appropriate solvent used in dilution or preparation of the biocide.

Assessment of results

Results of tests are usually assessed by the naked eye and by use of a microscope at a magnification of × 25–30. Incident u.v. illumination can be most useful for microscopic observation of algal cells which fluoresce red due to the presence of chlorophyll pigments.

Assessment of growth using a subjective 0–5 scale (Springle 1975) is useful for reporting results. It can give results that are consistent between different operators and even between different laboratories (Bravery *et al.* 1978, 1984). Removal of subjectivity can be attempted also by use or adaption of surface reflectance measurement techniques (Grant 1972).

Practical Test Results

Results representative of those obtained with the vermiculite-bed algal test using the subjective growth rating scale are given to illustrate the test in operation (Fig. 3). Test samples of asbestos-cement were allowed to become heavily colonized by a culture of mixed algae and treated with aqueous solutions of biocides as follows:

1 Dichlorophen (5,5′-dichloro-2,2′-dihydroxydiphenyl methane)—1% v/v
2 Sodium pentachlorophenoxide (NaPCP)—2% w/v
3 Quaternary ammonium salt (quat) (alkyl-aryl-trimethyl ammonium chloride)—3% v/v
4 Commercial mixture of tri-n-butyl tin oxide and a quaternary ammonium salt (tnbto/quat)—5% v/v

Control samples were treated with water alone. During the course of exposure (12 wks) all samples were re-inoculated with a suspension of mixed algae 4 and 8 wks after treatment.

With all treatments there was a reduction in the extent of overgrowth on the test samples; the controls remained heavily colonized throughout the test period. Initially the dichlorophen, NaPCP and the tnbto/quat mixture eliminated algal growths; the quat alone permitted a trace of viable growth to remain on the surface.

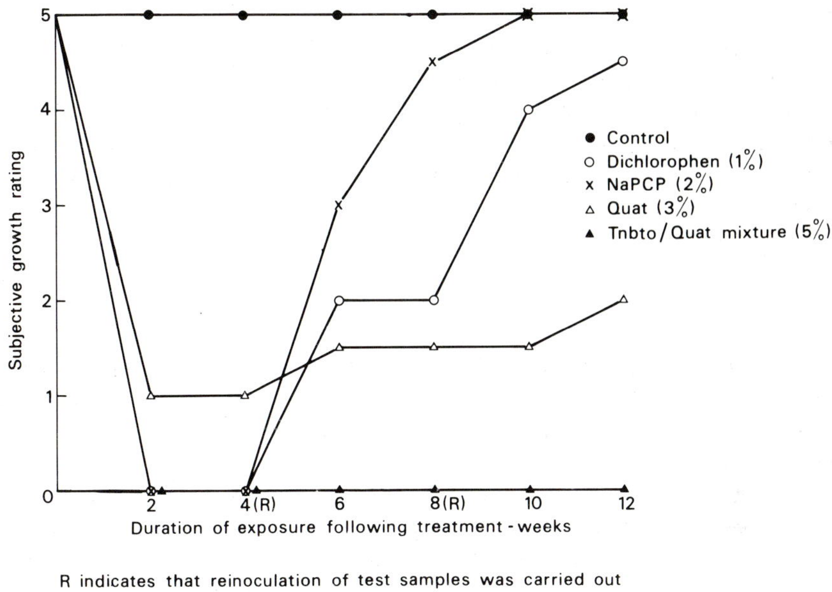

FIG. 3. Eradicant performance of biocides against algae in moist vermiculite-bed test.

After the test samples were re-inoculated at 4 and 8 weeks after treatment, there was extensive re-colonization of dichlorophen and NaPCP-treated samples. Those treated with the quat were only partially re-colonized. Treatment with the tnbto/quat mixture completely prevented any re-growth of algae; indeed the samples remained clean and bright in appearance with no trace of algal debris. The appearance of typical laboratory test samples is shown in Figs. 4 and 5.

Although these results do not as yet allow precise prediction of the likely service life, they do permit ranking and thus selection of appropriate treatments. Furthermore these and other results obtained in a comprehensive programme of laboratory tests of biocides for masonry treatment are in good agreement with those currently accumulating from a number of in-service trials.

Conclusions

The moist vermiculite-bed technique, a simple and straightforward operation, has proved suitable for the rapid evaluation of algicidal and fungicidal paints, water-repellent treatments and biocides intended for use as toxic washes against organisms on the surfaces of buildings. Because the technique uses actual building substrates, such as natural stone and asbestos-cement, full substrate/organism/biocide

FIG. 4. Moist vermiculite-bed test of algicidal biocides three weeks after application of treatments. A control sample is at top left. Note incomplete eradication by some treatments.

FIG. 5. Moist vermiculite-bed test of aligicidal biocides 12 weeks after application of treatments. A control sample is at top left. Note extensive regrowth of algae on certain treated panels. Only those treated with tnbto/quat mixture remain free of algae.

interactions are permitted. In this way both beneficial, and possibly detrimental effects of treatments, can be evaluated and differences in biocide performance on different substrates identified.

Until further results from field trials are available the technique cannot at present be used to predict actual service life of treatments. However, results from laboratory tests give good general indications of the performance being achieved in practice for those treatments where comparison is possible. The method thus permits selection and ranking of candidate compounds and finishes.

References

ADAMS, N. & DOBBS, A. J. 1984 A comparison of results from two test methods for assessing the toxicity of aminotriazole to *Selenastrum capricornutum. Chemosphere*, **13**, 965–971.

ALLSOPP, C. & ALLSOPP, D. 1983 An updated survey of commercial products used to protect materials against biodeterioration. *International Biodeterioration Bulletin*, **19**, 99–145.

ANON. 1971 *Resistance to artificial weathering (enclosed carbon arc)*. British Standard BS 3900 : Part F3 : 1971 (1979) London : British Standards Institution.

ANON. 1973 *Determination of resistance to humidity under condensation conditions.* British Standard BS 3900 : Part F2 : 1973 (1983) London : British Standards Institution.

ANON. 1976 *Notes for guidance on the conduct of natural weathering tests.* British Standard BS 3900 : Part F6 : 1976. London : British Standards Institution.

ANON. 1978*a Carbonation of concrete made with dense natural aggregates.* Building Research Establishment Information Sheet IS 14/78. Garston : Building Research Establishment.

ANON. 1978*b Strength test.* British Standard BS 4550 : Part 3 :Section 3,4 : 1978. London : British Standards Institution.

ANON. 1978*c Standard sand for concrete cubes.* British Standard BS 4550 : Part 5 : 1978. London : British Standards Institution.

ANON. 1981 *Mould growth in buildings.* Proceedings of a seminar on mould growth in buildings 11 and 12 June 1980. Building Research Establishment, Princes Risborough Laboratory, Princes Risborough, Buckinghamshire.

ANON. 1982 *Control of lichens, moulds and similar growths.* Building Research Establishment Digest 139. Garston : Building Research Establishment.

BRAVERY, A. F. 1981 Preservation in the Construction Industry. In *Principles and Practice of Disinfection.* ed Russell, A. D., Hugo, W. B. & Ayliffe, G. A. J. pp. 379–402 Oxford: Blackwell Scientific Publications.

BRAVERY, A. F., BARRY, S. & COLEMAN, L. J. 1978 Collaborative experiments on testing the mould resistance of paint films. *International Biodeterioration Bulletin*, **14**, 1–10.

BRAVERY, A. F., BARRY, S. & WORLEY, W. 1983 An alternative method for testing the mould resistance of paint films. *Journal of the Oil and Colour Chemists' Association* **66**, 39–43.

BRAVERY, A. F., BARRY, S., PANTKE, M. & WORLEY, W. 1984 Further collaborative experiments on testing the mould resistance of paint films. *Journal of the Oil and Colour Chemists' Association* **67**, 2–8.

CAREY, J. K. & GRANT, C. 1975 Moisture control in laboratory tests with wood-rotting fungi. *International Biodeterioration Bulletin* **11**, 101–105.

GRANT, C. 1972 The use of a reflectance method for estimating surface mould growth on chipboard. *International Biodeterioration Bulletin* **8**, 139–140.

GRANT, C. 1982 Fouling of terrestrial substrates by algae and implications for control—a review. *International Biodeterioration Bulletin* **18**, 57–65.

GRANT, C. & BRAVERY, A. F. 1981*a* Laboratory evaluation of algicidal biocides for use on constructional materials. 1. An assessment of some current test methods. *International Biodeterioration Bulletin* **17**, 113–123.

GRANT, C. & BRAVERY, A. F. 1981*b* Laboratory evaluation of algicidal biocides for use on constructional materials. 2. Use of the vermiculite bed technique to evaluate a quaternary ammonium biocide. *International Biodeterioration Bulletin* **17**, 125–131.

HO, K. K., TAN, K. H. & WEE, Y. C. 1983 Growth conditions of *Trentepohlia odorata* (Chlorophyta, Ulotrichales). *Phycologia* **22**, 303–308.

NEVILLE, A. M. 1975 *Properties of concrete.* pp. 338–341 London: Pitman Publishing Ltd.

NICHOLS, H. W. & BOLD, H. C. 1965 *Trichosarcina polymorpha* gen. et sp. nov. *Journal of Phycology* **1**, 34–38.

RICHARDSON, B. A. 1973 Algae, lichens, mosses: what they are and what they do. *Building Maintenance* **7**, 29–31.

RICHARDSON, B. A. 1980 *Remedial treatment of buildings.* pp. 118–123 Lancaster: The Construction Press Ltd.

SOROKIN, C. 1973 Dry weight, packed cell volume and optical density. In *Handbook of Phycological Methods.* Culture Methods and Growth Measurements. ed Stein, J. R. pp. 321–343. Cambridge: Cambridge University Press.

SPRINGLE, W. R. 1975 Testing biocidal paints. In *Some Methods for Microbiological Assay* ed Board, R. G. & Lovelock, D. W. pp. 191–201 London: Academic Press.

STEIN, J. R. 1973 Culture Methods and Growth Measurements. In *Handbook of Phycological Methods.* pp. 7–24 Cambridge: Cambridge University Press.

VÉNUAT, M. & ALEXANDRE, J. 1969 *De la carbonatation du béton.* Centre d'Etudes et de Recherches de l'Industrie des Liants Hydrauliques. Publication No. 195. Paris.

WRIGHT, S. J. L. 1975 Use of micro-algae for the assay of herbicides. In *Some Methods for Microbiological Assay.* pp. 257–269 ed Board, R. G. & Lovelock, D. W. London: Academic Press.

Test Procedures for Marine Antifouling Paints

S. E. J. Furtado

Transocean Marine Paint, PO Box 25107, 3001 HC The Netherlands

and R. L. Fletcher

Portsmouth Polytechnic, Hayling Island, Hampshire

All marine organisms that possess a sessile stage as part of their life history are capable of causing a fouling nuisance. In practice, however, only a relatively small number of genera are implicated in ship and offshore structure fouling but these nevertheless impose serious penalties for operations in the marine sphere. The penalties include speed reduction (or extra fuel consumption to maintain speed), blockage of ducts or inlets, increase of the loading on offshore structures, increase in corrosion under decaying fouling and loss of effectiveness of operation (e.g. of sonar domes).

In response to the fouling phenomenon man has developed antifouling coatings which embody one of two principles:

1 the repulsion or poisoning of the settling stages at or just before settlement on the surface or
2 the creation of surfaces to which the organisms cannot adhere.

Most of the commercial antifoulings in use today rely on the first principle and therefore these coatings contain a relatively high loading of biocide which is released when the coating is immersed—the antifouling chemist strives for steady state, constant emission of biocide throughout the life of the coating. There is concern, however, over the environmental impact of biocides and products are now emerging which rely on non-toxic principles to prevent fouling. This paper considers largely the testing of biocides and biocidal formulations. Some of the field tests are relevant to non-toxic antifoulings however, and other pertinent techniques are discussed in the ultimate section of this chapter.

Preservatives in the Food,
Pharmaceutical and Environmental
Industries

0–632–01727–9

Biocides

As a potential user of biocides, the antifouling industry ranks after agriculture, domestic (and hospital) use and wood preservation (pp. 115). Most biocides are therefore conceived for use other than in antifouling paints. The biocides for marine antifouling use have to satisfy a number of demanding criteria, many of them conflicting:

1 high toxicity to a wide range of species (including bacteria, diatoms, algae, barnacles, hydroids, bryozoans, mussels etc.);
2 low mammalian toxicity—to permit manufacture and application of such paints in relative safety;
3 relatively low sea-water solubility in order to prevent rapid loss from the submerged film and a consequent rapid reduction in service life;
4 relatively non-toxic to other marine life—they should not accumulate in the biomass nor persist in the environment;
5 compatibility with resins used in antifouling paints, and
6 suitable price/performance characteristics.

Of course no biocides satisfy all these criteria and the number that come closest to these ideals and have found widespread use in antifouling paints is very limited—cuprous oxide, cuprous thiocyanate, tributyl tin compounds (principally oxide and fluoride), triphenyl tin compounds (especially the hydroxide and flouride) and certain organic compounds (e.g. tetra methylthiuram disulphide, TMT). Much more hazardous compounds have been used in antifouling paints in the recent past including compounds of mercury, arsenic and lead. Their use has been discontinued because of hazards to operators and the environment even though some of these compounds were very effective biocides (e.g. 10,10′-oxybisphenoxarsine). Indeed, concern about the environmental impact of currently used biocides continues to mount. In particular the effects of very low concentrations of the tributyltins on marine organisms in enclosed waters have been observed with increasing alarm (Alzieu *et al.* 1982; Beaumont & Budd 1984).

When presented with a biocide from a reputable manufacturer and with known toxicological and environmental impact data, an investigator will seek to ascertain points 1, 3 and 5 above. He may well be restricted in his choice of media (from a commercial or availability viewpoint) and therefore resin compatibility trials will take the form of extended stability (preferably at 30 ° or 40 °C to accelerate any deterioration) and field trials which will be described later. Nevertheless, he may rapidly avail himself of the sea-water solubility characteristics and effectiveness of the biocide toward the major fouling organisms.

Sea-water solubility

Generally the solubility of compounds being sought for antifouling use will be in the order of 50 ppm or less. At these levels, detection of candidate compounds will require sophisticated analytical facilities—atomic absorption spectrophotometry, gas-liquid chromatography or u.v.—visible spectrophotometry—the choice depending on the manufacturers' recommendations. For example, the solubility of metal-containing compounds can be determined by equilibriating a saturated solution of the compound in sea-water and then using atomic absorption spectrophotometry to analyse for the metal in the filtered supernatant fluid.

In certain instances the presence of sea-water may interfere with chemical detection in which case a simple physical method may be used to estimate solubility:

1 Make a solution of the biocide in a water-miscible solvent (e.g. 10 000 ppm in acetone).
2 Add incrementally small (known) volumes of the solution to a large volume of filtered sea-water on a stir-plate at 20 °C—1 ml additions of a 10 000 ppm stock solution in acetone to a 5 l beaker of sea-water will give 2 ppm increments in the biocide content in sea-water.
3 As the limit of the compound's solubility in sea-water is passed, a turbid solution will result. This technique can be monitored visually or a nephelometer used to quantify turbidity.

This method is a variation on that proposed by Brooker & Ellison (1974).

In vitro *biocide tests*

The relative efficacy of biocides may be gauged from an *in vitro* assessment of their toxicity to target organisms under laboratory conditions. The organisms used should be those present routinely on fouled vessels, e.g. representatives of the green 'grass' (*Enteromorpha* spp.) brown mats (*Ectocarpus* spp.) and barnacles (*Balanus* spp.). It is important also that the toxicity tests involve the settling stages of these organisms, viz. asexual zoospores of *Enteromorpha* and *Ectocarpus* and cyprid larvae of barnacles. As the availability of these test organisms and their respective settling stages cannot be assured at all laboratories, toxicity tests may be based on the brine shrimp *Artemia salina* (Mawatari 1973) as a substitute for barnacle larva and the unicellular, flagellate green alga *Chlamydomonas* (Rivett 1965) as a substitute for *Enteromorpha* spores.

The use of barnacle cyprids for bioassay work, involving as it does the collection of the naupliar larvae and their copious feeding on axenically-cultured planktonic diatoms through development to the settling cyprids, is outside the scope of this paper. The reader is referred to Freiberger & Cologer (1966) for details.

The isolation and use of *Enteromorpha* spores for the toxicity tests, however, is relatively easy. Both geographically and locally, *Enteromorpha* spp. are quite abundant. In North Atlantic waters they are quite common shore inhabitants on rocks and in pools at high water mark, usually forming a zone above the Fucoids. They are also common colonizers of floating structures, again usually at and above the water line. Reproductive plants can be obtained throughout the year and only a simple procedure is necessary to obtain spores for the tests. Studies on North Atlantic populations of *Enteromorpha* have revealed a lunar periodicity in spore production and release (Christie & Evans 1962), with maximum release occurring during the spring tide period. After cleaning the surface of epiphytes with cotton wool, the fertile tips can be excised and immersed in a small amount of pasteurized and filtered sea-water to stimulate spore release. Released zoospores (recognisable by their four flagella) can then be concentrated by their negative phototactic response and drops of this spore suspension used for the tests. The spores can either be pre-settled on coverslips for 2 h in the dark or pipetted directly into compartments containing a dilution series of the test compound.

The dilution series of the compounds can conveniently be made in pasteurised and filtered sea-water. Perspex Repli dishes (10 cm^2, 25 compartments) also enable a number of replicates to be done in close proximity. The dilution series can be prepared from a stock solution of either a saturated sea-water solution (of previously determined biocide content) or from a solution in a water-miscible solvent such as acetone. Serial dilution of the biocide can then be made down individual rows of the Repli dishes and portions of the spores or larvae (newly attached or still swimming) may be introduced into each compartment. The Repli dishes are then maintained under cool-lights (algal spores) or in the dark (barnacle larvae) at 20 °C. The settlement and growth of the settling stages can be monitored after a few days using a stereo-zoom microscope. At slightly higher magnifications (×200–400), assessment of percentage germination/filament production of the *Enteromorpha* spores at the different toxic levels can be carried out. The relative survival/performance of the test organisms may then be used to determine LD_{50} values for each biocide. Table 1 shows some results extracted from publications by a number of workers on the effects of selected biocides against marine organisms. Although there is considerable variation due to the use of different techniques, the Table does demonstrate the very low levels at which these compounds are effective.

Field tests

The biocide screen tests described above, although yielding very precise results, can only be applied to a limited range of organisms and they do not always simulate field conditions. For this reason several attempts have been made to devise tests which have recourse to field tests of the biocides.

TABLE 1. *Effects of some biocides on marine test organisms*

Source	Biocide	Organism	Minimum growth-inhibiting concentration	
			ppm	Molar
Rivett (1965)	$[(C_4H_9)_3Sn]_2O$	*Chlamydomonas* sp.	6.0×10^{-3}	1.0×10^{-8}
	$(C_4H_9)_3Sn.F$		5.0×10^{-3}	1.6×10^{-8}
	$(C_6H_5)_3Sn.OOC.CH_3$		1.6×10^{-2}	4.0×10^{-8}
Skinner (1971)	CuCl	*Enteromorpha* zoospores	7.0	7.1×10^{-5}
	CuSCN		7.0	5.8×10^{-5}
	$[(C_4H_9)_3Sn]_2O$		3.5×10^{-1}	5.9×10^{-7}
	$(C_4H_9)_3Sn.F$		3.0×10^{-1}	9.7×10^{-7}
	$(C_6H_5)_3Sn.OOC.Ch_3$		2.0×10^{-1}	4.9×10^{-7}
	$(C_6H_5)_3Sn.Cl$		3.0×10^{-1}	7.8×10^{-7}
	CuCl	*Ectocarpus* zoospores	$>1.0 \times 10$	$>1.0 \times 10^{-4}$
	$[(C_4H_9)_3Sn]_2O$		$>6.0 \times 10^{-1}$	$>1.0 \times 10^{-6}$
	$(C_4H_9)_3Sn.F$		$>6.0 \times 10^{-1}$	$>1.9 \times 10^{-6}$
Christie (1973)	$(C_6H_5)_3Sn.Cl$	*Enteromorpha* zoospores	2.0×10^{-3}	5.2×10^{-9}
Millner & Evans (1980)	$(C_6H_5)_3Sn.Cl$	*Enteromorpha* zoospores*	1.5	4.0×10^{-6}
	$(C_6H_5)_3Sn.Cl$	*Ulothrix* zoospores*	>1.9	$>5.0 \times 10^{-6}$
Callow & Evans (1981)	$(C_6H_5)_3Sn.Cl$	*Achnanthes subsessilis**	8×10^{-2}	2.0×10^{-7}
Davies *et al.* (1984)	$[(C_4H_9)_3Sn]_2O$	*Enteromorpha* zoospores	3.2×10^{-4}	5.4×10^{-10}

*Concentration required to produce 50% inhibition of $H^{14}CO_3$ fixation after 1 h exposure.

Porous pot technique

In this method (first described by Christie & Crisp 1963) the biocide is put into a porous pot in sea-water and the sealed pot placed in the sea. The biocide goes into a saturated sea-water solution which seeps from the porous pot and prevents fouling developing on it. The thickness of the pot wall, however, may retard the diffusion rate of a biocide to the exterior. In a refinement of this technique, a membrane of known porosity is used to cover the mouth of small plastic diffusion cells mounted in a rack for easy assembly. The same quantity of toxicant is placed in each cell. The length of time during which the membrane resists fouling is a function of the solubility of the compound in sea-water and its toxicity to the fouling organisms involved (Lovegrove 1978).

Porous panel technique

Miller (1964) developed a technique in which compounds under test were lodged within the interstices of a permeable material from which they would be released (at the binder/sea-water interface) at a rate consistent with the compound's inherent solubility in sea-water. The chemical is first dissolved in a carrier solvent (50 g/l) and this solution used to impregnate blocks (63 mm square and 5 mm thick) of fine porous carbon which are dried and then exposed in sea water. When the chemical is insoluble in water, the blocks may be impregnated with a dispersion of the compound in water, xylene or a ketone (with or without the aid of a wetting agent). Evaluation of the blocks over a period of time under raft-exposure (p. 151) conditions continues until 50% of the panel faces are covered with fouling. The relationship between time to 50% fouling and biocide content then gives an assessment of the compound's effectiveness.

74% by volume toxicity test

This test derives from the theory that visualizes biocides as discrete spherical particles. To obtain a paint in which the biocide particles are in contact (hexagonal close packing) with each other and ultimately the sea-water, the dry film needs to be 'loaded' with at least 74% by volume of the bio-active material (van Londen 1971). Simple paints are made using the '74% biocide and 26% binder' (volume relationships in the dried film) formula by dissolving the binder in a suitable solvent and then dispersing the biocide in the binder solution by high speed mixing or by milling with a ball mill. The vinyl resin, VYHH (Union Carbide), is suggested by van Londen as particularly suitable.

Johnsen & Rendbaek (1976) suggest a modified formulation to be used with liquid biocides viz. (% by volume): rosin, 34; tricresylphosphate, 10; bio-active material, 26; iron oxide, 4, and vinyl resin, 26.

The 'paints' are applied on suitable test panels at two dry-film thicknesses (50 μm and 100 μm are suggested for convenience) by brushing on evenly weights of wet paint which will give the required dry film thickness. Evaluation of these panels over a period of exposure on a raft trial will reveal differences in fouling settlement. These are a function of the compound's leaching rate from the matrix and its toxicity to the fouling organisms present.

Although these tests are valuable, they suffer because the biocide release-rate cannot be controlled (and the tests are therefore not quantitative). Moreover, the medium (e.g. the vinyl resin, above) used in these tests may not be optimal for the biocide under test.

Nevertheless, the results of the foregoing biocide assessment trials may lead the investigator to make paints based on these biocides. He can now formulate with the biocides, binders, plasticizers, wetting agents, thixotropic agents, solvents etc. at his disposal in order to arrive at paints with good storage, application and dried-film characteristics.

Paint Testing

Before launching a new antifouling product, there are a number of tests to which such paints can be subjected in order to glean sufficient information about their performance and service characteristics.

Leaching rate tests

It may be instructive to assess the rate at which a biocide is being released from the paint film under laboratory conditions. This is most easily accomplished by painting small panels with the trial formulation. The panels may be rotated in a small volume of sea-water which is analysed for whatever biocides are in the formulation. This analysis may be accomplished chemically or by bioassays. The latter may be used to determine approximate leaching rates using organisms which are very sensitive to particular biocides (Rivett 1965). If the bioassay method is used, it must be recognized that the toxicity to the test organism may not be representative of the toxicity to all fouling organisms. Cologer (1984) has adapted the bioassay technique for evaluating new antifouling formulations. Glass coupons painted with the antifouling agent under test are put into Petri dishes containing newly attached barnacles. The response of the animals to the poisons so released is noted.

Some workers (Anon. 1952) have suggested methods which accelerate the leaching of toxicants. A sodium glycinate/sodium chloride solution as a leaching medium has been proposed. These tests accelerate the leaching of selected components only, particularly copper oxide. Although such tests will eliminate many unsatisfactory formulations, they do not permit a prediction of the length of time over which a formulation will maintain an adequate leaching rate.

Raft testing

Because of relative accessibility and controllability, raft testing has been generally adopted for antifouling trials. The location of the site is of prime importance. An ideal location would be an unpolluted harbour containing all the important fouling organisms (especially *Ectocarpus, Enteromorpha, Balanus* spp.), an extended fouling season (if possible) and a good water current to effect water flow over the panels. In practice, the most used sites in the UK have been along the south coast of England and the west coast of Scotland and Wales. The rafts, which are anchored in the direction of the tidal current, are usually located in the main channels of a large harbour or estuary. The twice-daily tides ensure a convenient water flow over the panels.

Raft exposure panels may be of sizes that suit the needs of the investigator. In view of the statistical nature of fouling, it is contended that about 300 cm^2 may be a useful minimum. Anon. (1980) advises a minimum panel size of 465 cm^2. The authors have used routinely mild steel panels of 7.5 × 45.7 cm (3 × 18 inch), six to a

FIG. 1. Frame with panels used in raft testing.

frame. These have given consistent results over a number of years (Fig. 1). The spaces between the panels dissipates the tidal forces which have been known to bend and fracture the holding frames during use. Complete underwater systems (anticorrosive and antifouling paints, each at the appropriate film thickness) should be applied to the mild steel panels, grit blasted to near-white metal finish. To minimize transport costs, Bakelite panels are often used in place of metal ones. These have the paint applied to an abraded, degreased surface.

Panels are scrutinized periodically not only for the species of fouling organisms present but for the condition of the paint film. Signs of physical distress e.g. blistering, cracking, chalking, loss of adhesion to the anticorrosive, etc. are noted. As such features may be especially apparent above the waterline, it is recommended that a part of a panel should project above the water. In this situation it is subjected to occasional immersion, rain, sunlight, etc.

The fouling present on the panels is recorded on appropriately designed forms. Although slime can be of a rather transient nature and of tenuous attachment, its presence is still noted as it may adversely affect ship performance. When a panel becomes irretrievably covered by a solid band of fouling, be it ectocarpoid only or a mixture of algal and animal types, it can be removed from test.

Periodic insertion of non-toxic panels (e.g. Tufnol or Bakelite) allows monitoring of fouling potential during a given exposure period. The thorough worker will also obtain relevant hydrological data (surface water temperature, salinity, pH, current strength and degree of pollution) for consideration in the total evaluation of the antifouling agent under test. The raft should be kept free of large organisms, especially kelp. By 'sweeping' across the surfaces of the panels, colonization may be impeded. A regular change of panel position within the raft is recommended also so that the influence of biocides from neighbouring panels is not constant.

Although the classical test involves the vertical suspension of panels from the raft, some workers have sought a more critical assessment of panel fouling. Sparrow (1966) described a 'Turtle raft'. This attempted to simulate the illumination conditions on the broadsides, turn-of-bilge and flat areas of ships by exposing panels at different angles to incident sunlight. Thus, animal fouling can settle on the panels which have little or no direct light without interference from already formed algal mats.

Finally, manufacturers of antifouling paints should not rely on test data from one site alone, but several from worldwide sources. Indeed, they ought to subject their paints to as diverse fouling communities as ships in service are likely to encounter. Moreover, the data obtained from raft trials must be considered alongside those obtained from ship trials. Taken in isolation, the former test procedure may sometimes yield rather anomalous results.

Rotor testing

The more or less static nature of the exposure is the major disadvantage of raft tests. Rotors were adopted in order to introduce tests of a more dynamic nature into trials. In the early models, rotating cylinders in a continuously changing volume of sea water were fitted with flexible panels which followed the contours of the cylinder. After being spun at a given speed for a known length of time—this corresponded to a given distance—the panels could then be removed from the cylinder and fixed to raft frames for quasi-static exposure. In practice, this regime simulates more closely the conditions experienced by a ship in service, i.e. alternating lay and steaming periods (van Londen 1970).

With the advent of the ablative antifouling paints, rotor testing has assumed a much more important role. The slow hydrolysis of the polymer in flowing sea-water leads to a diminution of film thickness with time. This can be followed subjectively (the time to 50% visual loss of film) or quantitatively (measurement of film thickness). Here too, the rotor can be a spinning disc (Milne & Hails 1974) with radially applied stripes of antifouling material. The latter can be applied with a cube applicator designed to deposit known film thickness on the upper and lower surfaces. These surfaces may well give different results. Fig. 2 shows a typical cylinder rotor with a number of panels attached before immersion.

As with raft testing, it is helpful to include a standard paint—one with known ship-performance characteristics—in such trials so that the results may be related approximately to operating conditions. It is more usual, however, to regard rotor

FIG. 2. Cylinder rotor with attached panels.

testing merely as a screen and any paints having ablation rates which may be commercially useful are selected for ship trials.

The supply of sea-water to the rotor tanks should be plentiful and clean; the rotors should be sited preferably at a point along the coast where industrial effects on the sea-water—especially heat and contaminants—are negligible and where uptake of sea-water can be accomplished throughout the tidal cycle. A settling tank in-line to the rotors can help to sediment sand and larger detritus which in severe cases can scour the surface of the paints under test. Filters too are used, but they require frequent cleaning. Some workers also use sterilizers, viz. u.v., to prevent microbial growth, as well as pH and water temperature controls in the sea-water supply circuit. Consistent results have been obtained, however, with clean unsterilized sea-water at ambient temperature.

The rate at which the volume of sea-water in the rotor tanks is renewed should be controlled so that there is no increase in temperature by friction due to the rotating disc or cylinder and so that the solubility products of the biocides (in the paints under test) are not exceeded. As the rate at which the rotor is spun should approximate ship conditions, speeds of 15 to 25 knots are used routinely. In order to explore the effects of slow-steaming on ablative antifouling paints, speeds of 10 knots and lower are sometimes used.

To ensure fully turbulent flow, baffles are set at intervals around the circumference of the rotor tanks. These prevent the body of sea-water simply swirling around with the spinning disc or cylinder within the tank. The very severe problems of trying to reproduce on a laboratory scale turbulent flow as prevail on ships, are such that in the final analysis one must resort to testing on vessels.

Ship tests

The following tests will need the permission of shipowners/operators and co-operation from the shipyard. They call for advance preparation of samples and, after application, intermittent monitoring of the trials and removal of samples.

Bilge keel panels

There are, apart from the bilge keels of vessels, no other areas of the underwater hull where panels may be attached and detached routinely. The bilge keels of most vessels are of sufficient dimensions to accommodate test panels of a reasonable size. This method of testing permits the antifouling agents on trial to be tested under 'real ship' conditions as it exposes them to fouling and panels can be retrieved in port, by a diver, without the need for the vessel to drydock.

Test panels can be of various sizes. Bohlander & Preiser (1984) described the use of 10 × 12 inch panels for attachment to vessels of the US Navy. For ease of handling and attachment, long steel panels are preferred—10 × 100 cm have been used. A number of different antifoulants under trial can be applied at intervals over the length of the panel with sufficient space being left at the panel ends so that turbulence around the attaching bolts does not interfere unduly with the paints. The test panels should be protected with adequate anti-corrosive coats (at least 250 μm of coal-tar epoxy is recommended on steel, blast-cleaned to a good standard). The antifoulants on trial may be painted on to this coat. Alternatively, the trial paints may be applied to smaller panels (e.g. Formica, 10 × 15 cm, lightly abraded and degreased before application of the trial formulations). These smaller panels are attached with a strong sea-water resistant adhesive to the carrier panel. The latter are held in place by strong bolts (at least $\frac{3}{8}$ inch) with 'locking' nuts and plastic washers at each end. The panel is underlayed with a continuous thick strip of plastic of the same dimensions. This will insulate the panel from cathodic over-protection effects. This may be an important feature as sacrificial anodes are often located on the bilge keel.

Panels should be so marked that the sequence of trial formulations can be easily traced when they are recovered—at least 6 months service is recommended before the first retrieval—and returned to the laboratory with fouling intact. Apart from monitoring the fouling cover, flakes of paint removed from the panels can reveal loss of biocide, especially loss of cuprous oxide, when examined with light microscopy or with X-ray micro-analysis combined with scanning electron microscopy.

The technique outlined above has been of especial use with the advent of ablative antifouling systems. In particular it has allowed a correlation of ship service ablation-rates with those obtained in rotor tests.

Patch tests

Patch tests have been widely used in the industry (Fig. 3). They have allowed paint manufacturers to obtain valuable ship experience with new products and they have also enabled shipowners to evaluate the rival claims of different manufacturers. A number of points need to be made:

1 The test patch should be of a reasonable size—between 100 to 200 m^2—tending towards the larger size if possible. Thus, when part of the patch is damaged (ships are prone to mechanical damage on the broadsides area, particularly from berthing), a substantial part remains to be surveyed.

2 The test patch should run from the deep load line to the bilge keel to encompass areas of continuous immersion as well as those which are periodically exposed when the ship is unladen.

FIG. 3. The side of a ship with test patches (grey rectangles against light background)

3 It is worth preparing the patch area to a standard acceptable to the characteristics of the trial formulation being applied. This may involve grit-blasting the patch area, particularly if the ship has a build-up of systems on its broadsides.
4 Test areas should be designated forard or aft of midships. Aft is preferable because turbulence may be higher, fouling more severe and berthing damage less likely.
5 Test patches should be monitored regularly in service so that the period of fouling initiation is noted. The condition of the surrounding 'standard' antifouling material should be noted also as it may be a well-tried product.
6 Test patches should be placed on a variety of ships operating in different parts of the world. This will ensure that the trial formulations are subjected to as wide a spectrum of environmental and fouling variables as possible. One clean test patch never 'made' a product!

By combining these field results, even before the conclusion of the patch tests, an investigator may feel confident enough to use the formulation in whole-ship trials. When in-service information confirms or surpasses those from laboratory and static tests, a high performance antifouling coating is ready for the market.

In all of the foregoing it should be remembered that although only antifouling paint development was considered, such studies go hand-in-glove with anti-corrosive research for underwater areas. All reputable marine paint manufacturers market underwater systems for fouling resistance *and* corrosion protection and the testing that is undertaken is therefore on the whole system or system possibilities.

Ablative paints

The development of this antifouling concept has revolutionized the industry. Because of the steady diminution in film thickness with time in service, these paints have called for special monitoring techniques. Chief of these is the measurement of the polishing (ablation) rate.

Ablation rate measurements depend on a knowledge of the antifouling film thickness at commencement of the test and at time of sampling. There are two ways of doing this. Both can be applied equally to rotor panels, raft panels and ship tests.

Film thickness measuring devices

A number of well-defined points (using, for example, a template) on the panel or ship are measured for total dry-film thickness. Any one of a number of widely available electronic film thickness gauges can be used. After a period on trial, the same points are sampled again for film thickness. The loss in thickness related to the time of exposure gives the ablation rate. The main difficulty is in sampling consecutively at the same points. The method is however non-destructive and the test may be resumed after the measurements have been taken.

Microscopical measurement

If a 'pip' of a non-ablating paint (e.g. vinyl) is placed on the ablative antifouling coat, the paint below the 'pip' will not be in contact with the sea-water and hence it will not ablate. After exposure, a flake incorporating a non-ablating 'pip' is removed from the panel/ship and mounted in solventless epoxy in the laboratory. After polishing, it is viewed in cross-section under a microscope. The difference in height between the paint under the 'pip' and the paint on either side is the loss in film thickness during the period of exposure. The method is destructive and, as the 'pip' is proud of the surface, it may create local turbulence thereby increasing the ablation of the immediately surrounding area. Within these constraints, the method does provide a ready and accurate method of measuring ablation rate (Fig. 4).

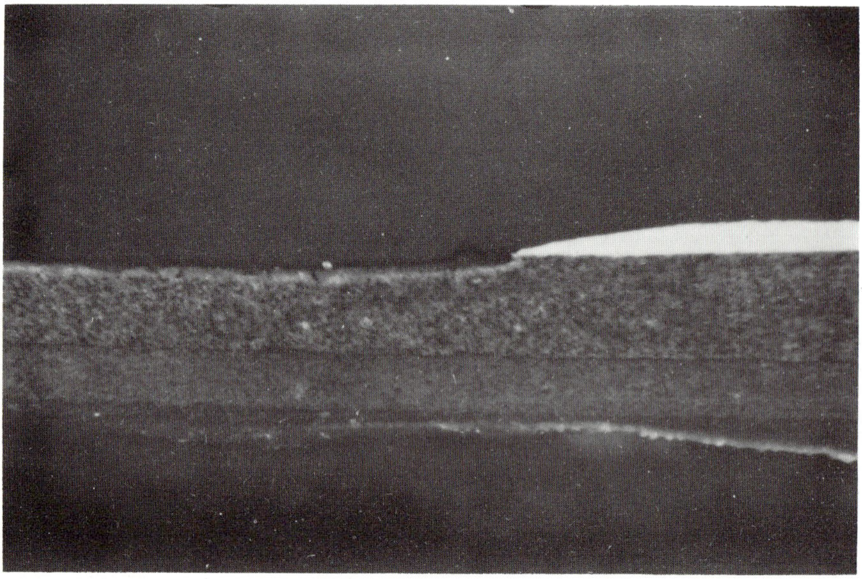

FIG. 4. Cross-section through an ablative paint flake showing original film thickness under the white non-polishing 'pip'. On the left hand side of the pip the film thickness has been polished down during exposure.

Apart from benefits of fouling prevention, ablative paints have brought about the possibility of containing, if not alleviating, hull roughness. Milne & Hails (1974) noted that the rate of ablation of organotin copolymer-based antifouling paints was much greater on asperities of roughness than it was on level or depressed areas. The economics of investment in systems that not only control fouling but also contain roughness are now being carefully scrutinized (Townsin *et al.* 1980; Svensen 1983) and shipowners are now more than ever before aware of the penalties of hull roughness and the rewards to be gained from keeping underwater areas of ships both biologically and physically smooth.

Non-toxic Antifoulings

Two factors have spurred the development of antifoulings which rely on non-toxic principles to keep fouling at bay rather than releasing biocide to deter fouling:

1 All biocidal formulations have finite lifetimes—when the biocide release rate falls below the critical level necessary to deter fouling organisms, the surface will foul.
2 As noted earlier there is concern about the release of biocides into the environment and in this case the precise effects of biocides released into the sea. Recent work has shown that although very small amounts of organotin (TBT) do not cause total oyster mortality, they do reduce growth and cause shell thickening. Biocontration effects can also lead to much higher levels of organotin in oyster tissue than in the surrounding water (Waldock & Thain 1983).

The research in this field has included:

1 'Natural' fouling resistant layers—either surface moieties that repel primary film formers, adsorbed enzymes that effect adhesive secreted by settling organisms or even the encouragement of natural film (either molecular or microbial) formation on surfaces which are then repellent to subsequent fouling.
2 Self cleaning, ablative surfaces—based on non-toxic compounds.
3 Surfaces with low surface free energies—to which fouling may not adhere quite so tenaciously.
4 Hydrophilic polymer surfaces—which swell on immersion in sea water and may then be fouling resistant.

The materials that have attracted the greatest attention in the search for a non-toxic antifouling have been those with a low critical surface tension; of these, fluorinated compounds (Griffith & Bultman, 1978) and silicone elastomers (Milne 1976; Montgomery, 1985) have yielded the best prospects for successful commercial exploitation.

Many techniques have been used to designate the antifouling propensities of these coatings:

1 Intensity of fouling has been an obvious yardstick. Dexter (1976) demonstrated that the rate of bacterial film formation and density of attached bacteria was related to the critical surface tension of a range of materials subjected to static marine exposure. Milne & Callow (1984) were able to demonstrate considerably lower

fouling ratings for silicone elastomers after 9 months raft panel exposure compared to a polyurethane control. In addition they were able to demonstrate that significantly lower numbers of bacteria and diatoms attached to these elastomers than to controls when exposed under laboratory conditions.

2 Low surface energy coatings elicit marked changes in the morphology of rhizoids produced by algae settling on these surfaces (Fletcher *et al.* 1984). These changes influenced the organisms contact with and adhesion to (by secretion of mucilaginous 'glue-like' material) the surface. These changes were all visible at the light microscope level.

3 The forces required to dislodge fouling from surfaces can give a reliable estimate of the adhesion of fouling to that surface. Griffith & Bultman (1978) measured the forces required to dislodge barnacles settled on PTFE and aluminium surfaces (Table 2).

TABLE 2. *Force required to dislodge barnacles settled on PTFE and aluminium surfaces*

	Dislodging force (kg/cm²)	
	Vertical	Shear
PTFE—epoxy	3.2	4.9
Aluminium	5.8	7.7

When they tested a fluorinated polyurethane containing PTFE, the dislodging forces were further reduced to an average of 1.1 kg/cm^2. These measurements represent substantial reductions in adhesive forces of the barnacle cement on the lower energy surfaces.

Young & Crisp (1982) describe the use of a strain gauge to measure the force of adhesion of mussel byssus pads on a range of substrata. They suggest that the mussel attachment forces to natural polar surfaces (e.g. slate) could be reduced almost a hundred-fold by the application of low energy coatings (e.g. PTFE or paraffin wax).

Milne & Callow (1984) used the same technique to demonstrate that the degree of fouling of a range of silicone elastomers (under raft conditions) correlated well with adhesion of the mussel byssus—a conventional paint had highest fouling accumulation and strongest byssus adhesion whilst a variety of silicone elastomers had reduced fouling and intermediate byssus adhesion. A room temperature vulcanizing silicone elastomer modified by the inclusion of a silicone fluid gave lowest fouling settlement and lowest adhesive strength of the byssus.

5 As a variation on the foregoing, the development of the radial flow chamber (RFC) by Fowler & Mackay (1979) has opened up the possibility of studying the attachment forces of the 'primary' fouling species—bacterial and diatom slimes—on a variety of substrates under laboratory conditions. Milne & Callow (1984) have used this technique to record a critical shear stress for detachment of *Amphora* diatom cells from glass of 18 Nm^{-2} while that for the same diatom on a modified silicone elastomer was approximately 2 Nm^{-2}. Correlating these results with raft tests they conclude that a reduction in adhesion of over 90% is necessary for satisfactory antifouling performance of these non-biocidal materials.

Gunn *et al* (1984) have used the RFC to investigate the adhesion of *Enteromorpha* spores to glass and it should be possible to investigate adhesion of other macro algal spores and even animal larvae by this technique.

The prospects for a generally usable physical effect antifouling are still some way off the silicone elastomers have poor mechanical properties (low abrasion and tear resistance) and fluorinated epoxies and polyurethanes foul to an unacceptable degree (although it is claimed that the adhesion is somewhat tenuous and even shell fouling can be dislodged by high pressure washing, Griffith & Bultman, 1980). The commercial use of silicone elastomers is at present limited to offshore oil rigs where service lives the duration of the structure itself are being claimed for it. However, spiralling biocide and paint raw material costs, environmental concern and the trend to longer interdocking service periods will combine to ensure that the search for materials with inherently fouling-free capabilities continue to be sought.

References

ALZIEU, C., HERAL, M., THIBAUD, Y., DARDIGNAC, M. J. & FEUILLET, M. 1982 Influence des peintures antisalissures à base d'organostanniques sur la calcification de la coquille de l'huitre *Crassotrea gigas*. *Revue des Travaux de Institut des Peches Maritimes* **45**, 100–116.

ANON. 1952 *Marine fouling and its prevention*. Annapolis, Md: US Naval Institute.

ANON. 1980 D3623—78a. Standard method for testing antifouling panels in shallow submergence. In *1980 Annual Book of the American Society for Testing and Materials Standards, Part 27. Paint—Tests for formulated products and applied coatings*. pp. 775–783. Philadelphia, Pa. ASTM.

BEAUMONT, A. R. & BUDD, M. D. 1984 High mortality of the larvae of the common mussel at low concentrations of tributyltin. *Marine Pollution Bulletin* **15**, 402–405.

BOHLANDER, G. C. & PREISER, H. S. 1984 New technology antifouling paints: U.S. Government research and assessment. *Naval Engineers' Journal* **96**, 125–131.

BROOKER, P. J. & ELLISON, M. 1974 The determination of the water solubility of organic compounds by a rapid turbidimetric method. *Chemistry & Industry, October* 785–787.

CALLOW, M. E. & EVANS, L. V. 1981 Some effects of triphenyltin chloride on *Achnanthes subsessilis*. *Botanica Marina*, **24**, 201–205.

CHRISTIE, A. O. 1973 Spore settlement in relation to fouling by Enteromorpha. In *Proceedings 3rd International Congress of Marine Corrosion and Fouling*, ed. Acker, R. F., Brown, B. F., DePalma, J. R. & Iverson, W. P., pp. 674–681. Evanston, Ill: Northwestern University Press.

Christie, A. O. & Crisp, D. J. 1963 A diffusion technique for assessing antifouling activity. *Annals of Applied Biology* **51**, 361–366.

Christie, A. O. & Evans, L. V. 1962 Periodicity in the liberation of gametes and zoospores of *Enteromorpha intestinalis* Link. *Nature, London* **193**, 193–194.

Cologer, C. P. 1984 Development of long life antifouling paints and shipboard evaluation. *Naval Engineers' Journal* **96**, 25–32.

Davies, R. J., Fletcher, R. L. & Furtado, S. E. J. 1984 The effects of tributyltin compounds on spore development in the green alga *Enteromorpha intestinalis* (L.) Link. In *Proceedings 6th International Congress of Marine Corrosion and Fouling, Marine Biology*, Athens, Greece *1984*. pp. 557–565.

Dexter, S. C. 1976 Influence of substrate wettability on the formation of bacterial slime films on solid surfaces immersed in natural sea water. In *Proceedings 4th International Congress on Marine Corrosion and Fouling*, Antibes *1976*. pp. 137–144.

Fletcher, R. L., Baier, R. E. & Fornalik, M. S. 1984 The influence of surface energy on spore development in some common marine fouling algae. *In Proceedings 6th International Congress on Marine Corrosion and Fouling, Marine Biology, Athens 1984*, pp. 129–144.

Fowler, H. W. & Mackay, A. J. 1979 The measurement of microbial adhesion. In *Microbial adhesion to surfaces.* Ed. Berkeley, R. C. W., Lynch, J. M., Melling, J., Butler, P. R. & Vincent, B. pp. 143–161. Chichester: Ellis Harwood.

Freiberger, A. & Cologer, C. P. 1966 Rearing acorn barnacles in the laboratory for marine fouling studies. *Naval Engineers' Journal* **78**, 881–890.

Griffith, J. R. & Bultman, J. D. 1978 Fluorinated naval coatings. *Industrial Engineering Chemistry Product Research and Development* **17**, 8–9.

Griffith, J. R. & Bultman, J. D. 1980 Fouling release coating. *Naval Engineers' Journal* **92**, 129–132.

Gunn, N., Fletcher, R. L. & Jones, E. B. G. 1984 Observations on the strength of attachment of spores and germlings of the marine fouling alga Enteromorpha. *In Proceedings 6th International Congress on Marine Corrosion and Fouling, Marine Biology*, Athens *1984*, pp. 81–98.

Johnson, S. & Rendbaek, V. 1976 A screening method for bio-active materials in antifouling paints. In *Proceedings 4th International Congress on Marine Corrosion and Fouling*, Antibes, France, *1976*, pp. 271–279.

Lovegrove, T. 1978 Techniques for the study of mixed fouling populations. In *Techniques for the study of mixed populations.* ed. Lovelock, D. W. & Davies, R. W. Society Applied Bacteriology Technical Series No. 11, pp. 63–69. London: Academic Press.

Mawatari, S. 1973 New methods of screening test of antifouling toxicants and coatings. In *Proceedings 3rd International Congress on Marine Corrosion and Fouling* ed. Acker, R. F., Brown, B. F., DePalma, J. R. & Iverson, W. P., pp. 956–965. Evanston, Ill: Northwestern University Press.

Miller, S. M. 1964 Techniques for evaluation of chemicals as marine antifoulants. *I. & E. C. Production Research & Development* 3, 226–230.

Millner, P. A. & Evans, L. V. 1980 The effects of triphenyltin chloride on respiration and photosynthesis in the green algae *Enteromorpha intestinalis* and *Ulothrix flacca. Plant, Cell and Environment* **3**, 339–348.

Milne, A. 1976 *Coated marine surfaces.* British Patent 1 470 465.

Milne, A. & Callow, M. E. 1984 Non-biocidal antifouling processes. *Transactions of the Institute of Marine Engineers* **97**, 229–233.

Milne, A. & Hails, G. 1974 *Marine Paint.* British Patent 1,457,590.

Montgomery, R. M. 1985 Novel antifouling coatings. *Shell Polymers* **9**, 68–70.

Rivett, P. 1965 Biological method for the assessment of leaching rates of antifouling compositions. *Journal Applied Chemistry* **15**, 469–473.

Skinner, C. E. 1971 Laboratory techniques for screening potential toxicants for antifouling paints. In *Antifouling Symposium*, Marienberg, West Germany: Ciba-Geigy.

SPARROW, B. W. 1966 Essais, sur radeaux, de peintures de carenes de navires. *Peintures Pigments Vernis* **42**, 4–21.

SVENSEN, T. E. 1983 The economics of hull and propellor maintenance examined in the face of uncertainty. *North East Coast Institute of Engineers & Shipbuilders*, **100**, 11–34.

TOWNSIN, R. L., BYRNE, D., MILNE, A. & SVENSEN, T. E. 1980 Speed, power and roughness: the economics of outer bottom maintenance. *Transactions of the Royal Institute of Naval Architects,* **122**, 459–483.

VAN LONDEN, A. M. 1970 Evaluation of testing methods for antifouling paints. *Journal of Paint Technology*, **42**, 511–515.

VAN LONDEN, A. M. 1971 A simple method for the evaluation of bio-active materials as antifouling agents. In *Antifouling Symposium*, Marienberg, West Germany: Ciba-Geigy.

WALDOCK, M. J. & THAIN, J. E. 1983 Shell thickening in *Crassotrea gigas*: organtin antifouling or sediment induced? *Marine Pollution Bulletin* **14**, 411–415.

YOUNG, G. A. & CRISP, D. J. 1982 Marine animals and adhesion. In *Adhesion 6*. ed. Allen, K. W. pp. 19–39. London: Applied Science Publications.

Techniques for the Evaluation of Biocide Activity against Sulphate-reducing Bacteria

A. P. SHARMA, N. S. BATTERSBURY AND D. J. STEWART

Department of Brewing & Biological Sciences, Heriot-Watt University, Chambers Street, Edinburgh, UK

Sulphate-reducing bacteria (SRB) are a physiological-ecological group which generate energy by the respiratory reduction of sulphate to sulphide. They are obligate anaerobes which require a redox potential of *ca.* 100 mV for the initiation of growth. SRB can tolerate exposure to oxygen. Several strains possess the O_2^- protective enzymes, catalase and superoxide dismutase (Hardy & Hamilton 1981). The group is widespread in nature, some species are particularly active in marine waters where sulphate is generally abundant (*ca.* 29 mM).

The adverse effects of SRB activity on the environment and industry are diverse and extensive. The major problems are corrosion and pollution by hydrogen sulphide. The subject has been reviewed by Postgate (1984). In the sulphate-rich environment of the off-shore oil industry, SRB can be active throughout the whole oil production system (Stranger-Johannessen 1983). The maintenance of reservoir pressure by injection of sea water can introduce them into the oil reservoir. In addition, sea-water is often used for the hydrostatic testing of pipelines and as ballast (Crough 1982). The growth and reduction of $SO_4^{2-} \rightarrow S^{2-}$ by the SRB can result in: microbial corrosion by cathodic depolarization or by ferrous sulphide production; the production of toxic hydrogen sulphide gas with concomitant health hazards; souring of oil and gas, and rock-pore blockage by bacterial cells and precipitates of ferrous sulphide. As with the oil and gas industries, the corrosive effects of SRB on the exterior of buried iron water-mains have been known for many years (Postgate 1984).

The prevention of corrosion and pollution entails the curbing of the growth and metabolism of SRB. Numerous means of control have been tried but few have been satisfactory since economic considerations often impose limitations. Aeration is the cheapest and most effective inhibitor of SRB (Postgate 1984), although other problems of corrosion due to aerobic, sulphur oxidizing bacteria such as *Thiobacillus*

Preservatives in the Food, Pharmaceutical and Environmental Industries

0–632–01727–9

spp. may arise. Considerable success has been achieved in closed systems by the use of biocides (disinfectants) and there is a great deal of interest among oil and chemical companies in the development of effective biocides for the control of SRB. Biocides commonly used in the oil industry include quaternary ammonium compounds (QACs) and isothiazolones.

The recommended American Petroleum Institute (API) method for testing the efficacy of biocides against SRB (Anon. 1975) is a determination of the minimal inhibitory concentration (MIC) which just prevents the growth of a test strain. This bacteriostatic test involves the incubation of inoculated bottles for 28 d before the cultures are examined and the MIC for the antibacterial agent noted. The API also recommend 'time-kill' tests. SRB contaminated injection-water is exposed to various concentrations of the biocide under test. Samples are removed at intervals and the number of viable cells estimated. This procedure takes *ca.* 28 d (Anon. 1975). In this chapter we describe a simple technique for evaluating the MIC of a biocide against a test strain of SRB within 24 h. We also present techniques for investigating the effect of SRB surface colonization on the efficacy of biocides and for the demonstration of cross-resistance to biocides.

Rapid Method for Evaluating the Efficacy of Biocides against SRB

Growth tubes

Postgate's (1984) medium C is modified (medium Cm) by the addition of 0.0002% (w/v) resazurin, a redox indicator, and 0.1 g/l sodium ascorbate and 0.1 g/l sodium thioglycollate as reductants. Portions (9.5 ml) of the medium are dispensed into 20 ml reductase tubes autoclaved at 121 °C until (*ca.* 10 min) reduced (resazurin redox indicator, colourless), and cooled under a stream of O_2-free nitrogen purified by passage through a carrier gas purifier (Supelco, Bellefonte, USA). The tubes are sealed with butyl rubber roll-tube stoppers (Astell Hearson, London) and autoclaved (15 min at 121 °C) with the tops clamped down. On cooling, the junction of the tube with the stopper is sealed with a silicone rubber 'sealing compound', such as Silcoset 151 (Ambersil, Basingstoke, UK), to ensure an airtight seal. The tubes are injected with 0.2 ml of a 30 h culture of the test strain (e.g. *Desulfovibrio desulfuricans* NCIB 8400) in modified medium C. The absorbance of the contents of inoculated tubes at 580 nm is read against inoculated ones; the tubes are inserted into a colorimeter with a collar modified for the larger diameter (1.5 cm) reductase tubes. Growth is followed by the increase in absorbance at 580 nm with incubation at 30 °C.

In the early logarithmic phase of growth (*ca.* $A_{580}=0.15$), 0.3 ml of the test biocide diluted in sterile distilled water is injected into four tubes which have the same absorbance, and the contents are mixed. The range of biocide concentrations used varies but Crouch (1983) recommends a range from 50 to 500 μg/ml in 50 μg/ml intervals for biocides with potential off-shore applications. Control tubes are inoculated with sterile distilled water. After injections of biocide or water the tubes are incubated at 30 °C and the absorbance of each set is recorded every 3 h. SRB growth is followed by plotting the arithmetic mean of absorbance against time from biocide addition. At each sampling time, the contents of one tube from each set is used for determinations of residual sulphate (see below). The numbers of viable SRB at the time of biocide addition and after 9 h are determined also (see below).

Sulphate assay

Sulphate is estimated by $BaCl_2$-EDTA titration (Anon. 1973). One ml of 0.5 N HCl is added to 2.5 ml culture in a 50 ml conical flask. Ten ml of 0.04 N $BaCl_2$ is then added and mixed. A total hardness indicator tablet (BDH) is dissolved in 3 ml of sulphate buffer which is prepared by adding 4.125 g of NH_4Cl to 56.5 ml of ammonia (sp. gr. 0.88), adjusting the volume to 500 ml with distilled water followed by the sequential addition of 3.72 g of ethylenediaminetetra-acetic acid (EDTA) disodium salt and 2.03 g of $MgCl_2 . 6H_2O$. The buffer mixture is added to a sample which is then titrated with 0.02 N EDTA solution (disodium salt) until the red-brown colour changes to blue-grey. The residual sulphate in the sample is calculated from the volume of EDTA titrated.

Viable SRB count

The numbers of viable SRB present in the samples before the addition of biocide and 9 h after incubation are determined as described by Postgate (1984). Biocide carry-over is generally minimized by sample dilution (up to 10^{-8}). Biocide inactivators can be added to the enumeration medium. Lubrol WX (Sigma), 3% (w/v) acts as an inactivator of the biguanide, Dodigen 181 (Hugo 1982). Supplementation with 2% (w/v) lecithin (Sigma) and 3% (w/v) Lubrol WX neutralizes residual cetytrimethyl-ammonium bromide (CTAB) and other QACs (Anon. 1960). Inoculated agar shake-tubes are incubated at 30 °C for 7 days. Black colonies are scored as SRB.

Biocide effects

Results typical of those obtained by this method are shown in Fig. 1 (CTAB) and Fig. 2 (glutaraldehyde). In each case SRB growth was inhibited by a concentration of 50 μg/ml. It can be seen from the slight increase and levelling off of A_{580} that both biocides are bacteriostatic at 50 μg/ml. At 100 μg/ml the decrease in A_{580} reflects a bactericidal action with concomitant cell lysis. These observations are confirmed by

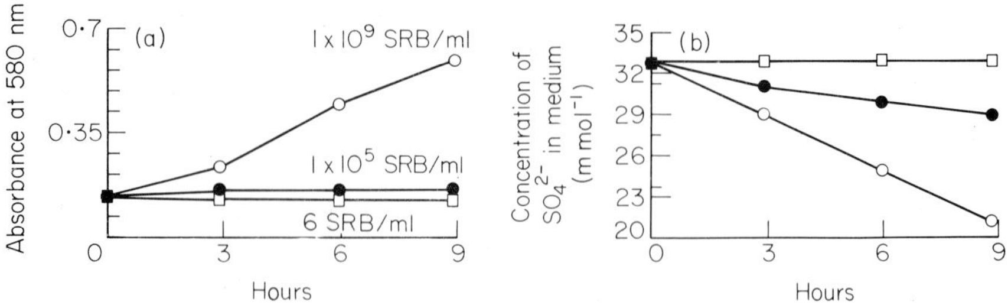

FIG. 1. Action of cetyltrimethyl-ammonium bromide (CTAB) on growth (a) and sulphate reduction (b) by *Desulfobacter desulfuricans* NCIB 8400. Control (○), 50 μg/ml (●) and 100 μg ml⁻¹ (□).

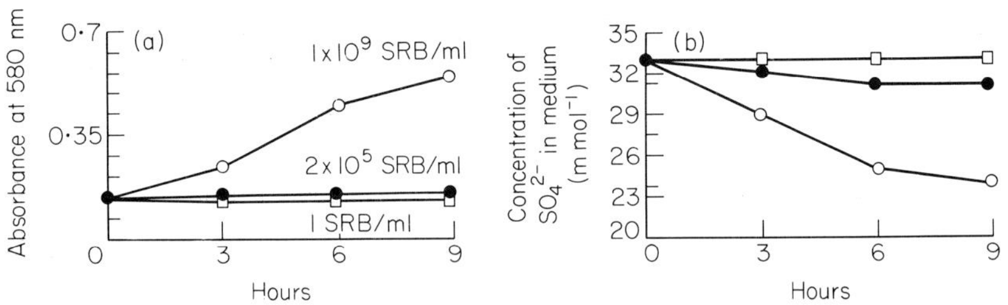

FIG. 2. Action of glutaraldehyde on growth (a) and sulphate reduction (b) by *Desulfobacter desulfuricans* NCIB 8400. Control (○), 50 μg/ml (●) and 100 μg/ml (□).

the viable SRB counts obtained after 9 h incubation. As the reduction of $SO_4^{2-} \rightarrow S^{2-}$ is the most destructive activity of SRB, a knowledge of their respiratory activity in the presence of biocides is important. It can be seen from Figs. 1 and 2 that, while 50 g/ml of CTAB or glutaraldehyde inhibits SRB growth, sulphate reduction still occurs but at a reduced rate. Complete inhibition of sulphide production is obtained only at a higher biocide concentration of 100 μg/ml. Recently radiorespirometric methods have been described for the testing of biocides against SRB (Hardy & Syrett 1983; Norqvist & Roffey 1983). The use of $^{35}SO_4^{2-}$ has attendant safety problems and requires relatively sophisticated and expensive equipment. The method described above is simple, inexpensive and rapid.

Biocide Effectiveness against Surface-Colonizing SRB

The efficacy of biocides against SRB has been tested traditionally against free-floating (planktonic) bacteria. Recent research has shown, however, that adherent bacteria predominate in aquatic systems and they can rapidly colonize submerged structures (Costerton & Gessey 1979). Costerton *et al.* (1978) suggest that the majority of bacteria in these systems occur in a fibrous matrix of polysaccharide. This extracellular material, the glycocalyx, is often recognizable as slime or biofilm. Although the composition of the glycocalyx can show great variation, its chemical structure is essentially that of a highly hydrated, anionic polysaccharide matrix (Costerton *et al.* 1981). The action of aerobic bacteria within the glycocalyx can result in the reduced conditions essential for the growth of SRB at or near the surface of off-shore structures and pipelines in otherwise oxygenated waters (Crouch 1982). SRB can also show sessile growth and biofilm production in pure culture, the slime probably consisting of mucopolymannoside (Postgate 1984). The extensive anionic matrix surrounding SRB within a glycocalyx can act as an ion-exchange barrier between charged biocides and the SRB cells (Costerton *et al.* 1981).

Ruseska and co-workers (1982) have demonstrated recently that cationic biocides had a negligible effect against SRB. This was attributed to the anionic matrix of the biofilm. Here, we present methods to investigate surface-colonizing SRB and the effect of biguanide (Dodigen 181), QAC (CTAB), isothiazolone (Kathon WT) and aldehyde (glutaraldehyde) type biocides on them.

Scanning electron microscopy

The test strain is grown in medium Cm in the presence of iron nails (15 × 1.5 mm) which have been degreased in 95% (v/v) ethanol for 15 min followed by 70% (v/v) ethanol for 10 min. After incubation at 30 °C for 4, 5 and 6 d the nails are removed from the culture, washed twice in 0.8% or 2.5% (w/v) saline and fixed for 2 h at 4 °C in a mixture of 9 parts cacodylate fixative (sodium cacodylate, 14.3 g and sucrose, 57 g/l) and 1 part of 25% (v/v) glutaraldehyde. They are then immersed overnight at 4 °C in cacodylate buffer (sodium cacodylate, 14.53 and sucrose, 102.6 g/l), dehydrated for 15 min in each of the following concentrations of acetone: 25, 50, 75 and 100% (v/v) and stored under 100% (v/v) acetone with critical point drying in liquid CO_2 (Polaron).

The dried nails are coated with *ca.* 50 nm of gold in a sputter-coater and examined with a scanning electron microscope. The electronmicrographs shown in Figs. 3 and 4 are of *D. desulfuricans* NCIB 8400 after incubation for 4 and 6 d respectively. The nail was colonized within 4 d. The extracellular, loose mat of fibres interconnecting the SRB cells is evident. After incubation for 6 d large amounts of polymer had enclosed the SRB within a dense matrix. In SRB strain NCIB 8400 this has been shown to be anionic in character (Sharma 1984). It would clearly protect any sessile cells against the effects of cationic biocides.

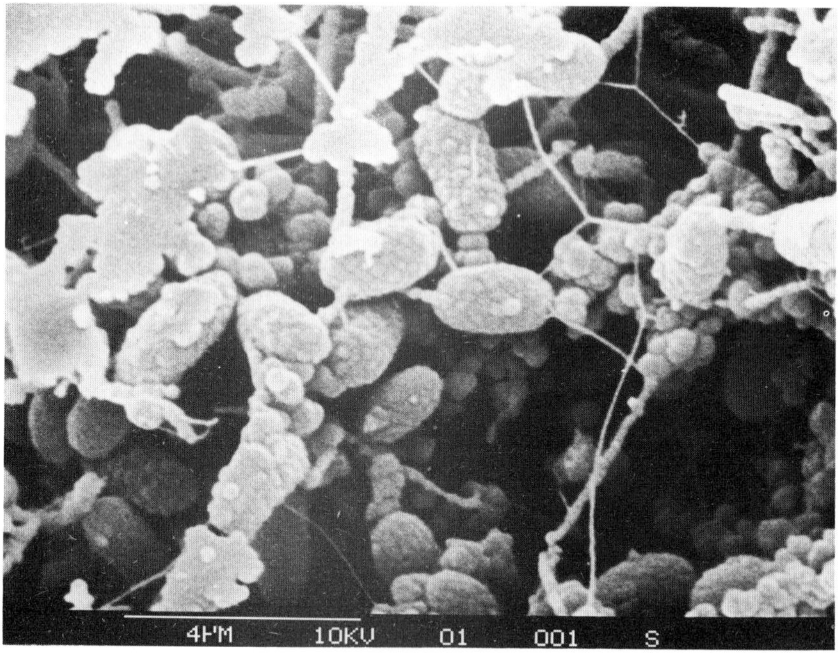

FIG. 3. Scanning electron micrograph of colonization of *Desulfobacter desulfuricans* NCIB 8400 on an iron nail after 4 days incubation.

MICs of biocides against sessile and planktonic SRB

Four-day cultures of the test SRB strain growing on iron nails are prepared as described above. The cells are removed in sterile saline by ultrasonic vibration and counted in a haemocytometer. A colonized nail or an equivalent population of planktonic SRB from a liquid culture are inoculated into three screw-top test tubes containing various concentrations of the biocide in medium B of Postgate (1984). The tubes are incubated at 30 °C for 21 d; growth is indicated by a blackening of the medium due to FeS production. The MIC of the biocide is that concentration which prevents growth in at least two of the tubes. Results obtained for *D. desulfuricans* NCIB 8400 with biguanide, QAC, isothiazolone and aldehyde biocides are shown in Table 1. Cationic surfactants, such as Dodigen 181 and CTAB, showed the greatest loss in activity. This was consistent with the observed binding properties of the anionic glycocalyx. The 5-carbon aldehyde, glutaraldehyde, had a satisfactory inhibitory effect against sessile SRB. Kathon WT, a non-ionic biocide, had its effectiveness diminished by half when evaluated against sessile SRB. This is apparently due to its rapid inactivation by sulphide trapped within the active glycocalyx (Sharma 1984).

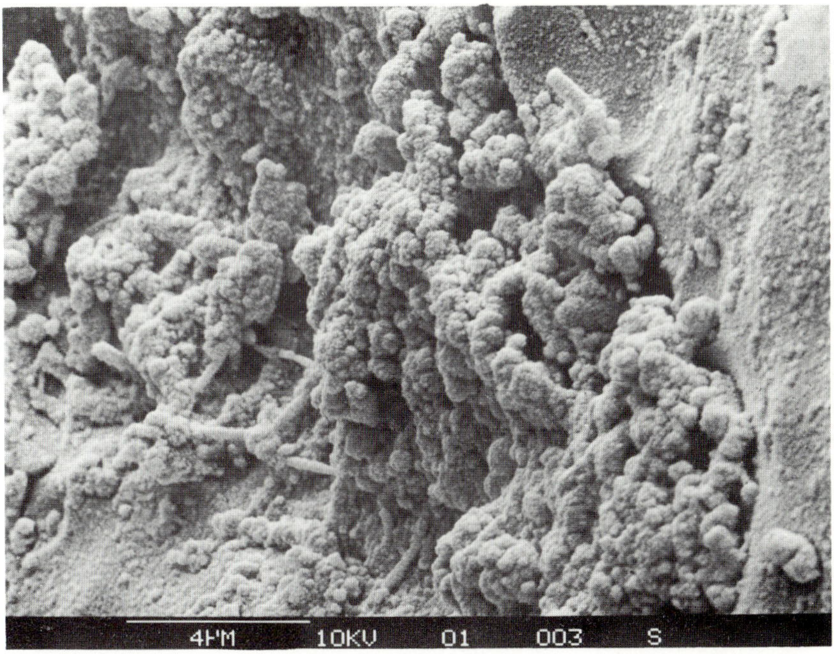

FIG. 4. Scanning electron micrograph of colonization of *Desulfobacter desulfuricans* NCIB 8400 on iron nail after 6 days incubation.

TABLE 1. *MIC of biocides against planktonic and sessile cells of* Desulfobacter desulfuricans *NCIB 8400*

Biocide	MIC planktonic—SRB (μg/ml)	MIC sessile—SRB (μg/ml)
Dodigen 181	30	280
CTAB	95	800
Kathon WT	21	50
Glutaraldehyde	55	65

Adsorption of biocides by the SRB glycocalyx

The loss in effectiveness of individual biocides when in contact with the SRB glycocalyx can be measured biologically. The test strain of SRB is grown at 30 °C for 7 days in medium Cm containing 2 mm ballotini. The beads are washed in 0.1% (w/v) NaCl, then in distilled water and packed into the barrel of a 10 ml plastic syringe. The biocide is passed through the column and filter sterilized. The MIC of the biocide against the test strain of SRB before and after passage through the column is estimated as described above. As many glycocalyces contain alginate-like compounds (Springer & Roth 1973; Sutherland 1983), the adsorption effects of such compounds can be studied with columns produced as described above. Results presented in Table 2 show that both types of column bind the cationic biocides Dodigen 181 and CTAB. Non-ionic biocides, such as Kathon WT and glutaraldehyde, are unaffected.

TABLE 2. *Adsorption of biocides by the glycocalyx of* Desulfobacter desulfuricans *NCIB 8400 and by an alginate column**

Biocide	SRB glycocalyx (% decrease in activity)	Alginate column (% decrease in activity)
Dodigen 181	45	33
CTAB	60	57
Kathon WT	0	0
Glutaraldehyde	0	0

* The glycocalyx and alginate were packed in the barrel of a 10 ml plastic syringe.

The Demonstration of Cross Resistance to Cationic Surfactants by SRB

An important aspect of microbial resistance to a particular biocidal agent is the range of other chemical agents to which a strain has become resistant. Indeed such cross-resistance is an important consideration in the choice of biocides for the control of SRB. Cross-resistance can be determined by measuring the differences in the effectiveness of each member of a selected series of biocides against the genetically modified as well as the parent strains. In the example given below, the resistance of *D. desulfuricans* Loch Eil 19 (a strain isolated from an anoxic sediment from a sea loch in NW Scotland) to the biguanide, Dodigen 181, was determined initially.

Isolation of strains resistance to Dodigen 181 Strain 19 was grown for one week in Medium B of Postgate (1984) containing 15 μg/ml of Dodigen 181 (MIC= 20 μg/ml). This culture was used to inoculate a medium containing twice this concentration of Dodigen 181 and, after incubation, it was subcultured into a medium containing 30 μg/ml of the biocide. After a further four subcultures in a medium containing this concentration, the MIC of this particular strain of SRB designated strain 19T was 80 μg/ml of Dodigen. It was maintained in a medium containing 70 μg/ml of the biocide.

Isolation of single cell-lines resistant to Dodigen 181

A 48-h culture of strain 19T was diluted serially in 2.5% (w/v) NaCl and 1 ml added to 15 ml amounts of molten (45 °C) Postgate's medium E (Postgate 1984) supplemented with 70 μg/ml of Dodigen 181. The agar was poured into Petri dishes, allowed to gel and overlayed with 10 ml of the same medium. After incubation, at 30 °C in an anaerobic cabinet (Forma Scientific), a well isolated SRB colony was inoculated into the maintenance medium. After incubation, this culture (designated strain 19TK) was sub-cultured serially five times in medium B alone. As the MIC of Dodigen 181 against this strain was still 80 μg/ml, the retention of biocide resistance in the absence of biocide was taken as evidence of a permanent change in the genotype of Loch Eil strain 19.

Cross-resistance 19 TK

The MICs of a series of cationic and unrelated biocides to strains 19 (parent) and 19 TK (resistant mutant) were determined as described previously. It is evident

TABLE 3. *Cross-resistance of* Desulfobacter desulfuricans *strain 19 TK and parent strain to biocides*

Biocide	MIC parent strain 19 (μg/ml)	MIC mutant strain 19 TK (μg/ml)
Cationic		
Dodigen 179	40	150
Dodigen 180	30	150
Dodigen 5462	60	150
Dodigen 5594	150	280
Cetyltrimethyl-ammonium bromide (CTAB)	90	20000
Octadecyltrimethyl-ammonium bromide (ODTB)	130	18000
Tetradecyldimethylbenzyl ammonium chloride	60	9500
Dodecyldimethyl-ammonium chloride	90	6500
Dodecyltrimethyl-ammonium chloride	50	400
Armogard 152	20	90
Non-cationic		
Dichoroxylenol (DCMX)	100	100
Kathon WT	13	13
Glutaraldehyde	50	55
Tego 103	2600	2600
Metronidazole	0.8	0.8

from Table 3 that strain TK was resistant to Dodigen and all the pure QAC compounds, especially CTAB and octodecyltrimethyl ammonium bromide. No increase in resistance to DCMX was observed even though this biocide does have a similar mode of action to the cationic QAC biocides. Cross-resistance to other unrelated biocides did not occur.

Our findings indicate clearly that the continuous exposure of SRB to sub-lethal concentrations of a biocide can lead to the selection of resistant mutants. Such mutants may exhibit cross-resistance, sometimes to a marked extent, to a range of chemically-related biocides. Preliminary studies have shown that 19 TK can transfer cationic-biocide resistance to the normally sensitive *D. desulfuricans* NCIB 8400. This occurs at a low frequency of transfer, and may be chromosomally or R-factor mediated, whereas this strain can accept the promiscuous RP_4 plasmid from *Escherichia coli* at a high recombination rate. The broad spectrum in antimicrobial action of the biocides used against SRB and the lowering of their concentration by an anionic glycocalyx could well lead to the presence of a range of resistant SRB and other bacteria in a treated area.

As the glycocalyx enhances the intercellular transfer of genetic material the potential for biocide resistance in the oil and gas industries presents a major problem. This will be resolved only by a detailed knowledge of the biocides available and their activities against SRB.

References

ANON. 1960 *Method for the Laboratory Evaluation of Disinfectant Activity of Quaternary Ammonium Compounds.* British Standard 3286. London: British Standards Institution.

ANON. 1973 *Chemical Methods of Water Testing* 2nd edn. Poole: BDH Chemicals Ltd.

ANON. 1975 *Recommended Practice for Biological Analysis of Subsurface Injection Waters* 3rd edn., API RP 38. Dallas: American Petroleum Institute.

COSTERTON J. W., GEESEY, G. G. & CHENG K-J. 1978 How bacteria stick, *Scientific American* 238, 86–95.

COSTERTON, J. W. & GEESEY, G. G. 1979 Microbial contamination of surfaces. In *Surface Contamination* Vol. 1 ed. Mittal, K.L. pp. 211–221. New York: Plenum Publishing Corporation.

COSTERTON, J. W., IRVIN, R. T. & CHENG, K-J. 1981 The bacterial glycocalyx in nature and disease. *Annual Reviews in Microbiology* 35, 299–324.

CROUCH, B. A. 1982 Biocides in oil field operations. In *Biocides in the Oil Industry*, pp. 29–38. London: The Institute of Petroleum.

CROUCH, B. A. 1983 Evaluation of biocides for use in North Sea oil operations. In *Biodeterioration 5.* ed. Oxley, T. A. & Barry, S. pp. 725–730. Chichester: John Wiley & Sons.

HARDY, J. A. & HAMILTON, W. A. 1981 The oxygen tolerance of sulfate-reducing bacteria isolated from North Sea waters. *Current Microbiology* 6, 259–262.

HARDY, J. A. & SYRETT, K. R. 1983 A radiorespirometric method for evaluating inhibitors of sulphate-reducing bacteria. *European Journal of Applied Microbiology and Biotechnology* 17, 49–52.

HUGO, W. B. 1982 Disinfection Mechanisms. In *Principles and Practice of Disinfection, Preservation and Sterilisation* ed. Russell, A. D., Hugo, W. B. & Ayliffe, G. A. J. pp. 158–185. Oxford: Blackwell Scientific Publications.

NORQVIST, A. & ROFFEY, R. 1983 Alternative method for monitoring the effect of inhibitors on sulfate reduction. *Journal of General and Applied Microbiology* **29**, 335–344.

POSTGATE, J. R. 1984 *The Sulphate-Reducing Bacteria* 2nd edn., Cambridge: Cambridge University Press.

RUSESKA, I., ROBBINS, J., COSTERTON, J. W. & LASHEN, E. S. 1982 Biocide testing against corrosion-causing oil-field bacteria helps control plugging. *Oil and Gas Journal* **10**, 253–258.

SPRINGER, E. L. & ROTH, I. L. 1973 The ultrastructure of the capsules of *Diplococcus pneumoniae* and *Klebsiella pneumoniae*. *Journal of General Microbiology* **74**, 21–31.

STRANGER-JOHANNESSEN, M. 1983 Offshore structures at risk from sulphate-reducing bacteria. *Offshore Engineer* Feb., p. 51.

SHARMA, A. P. 1984 *Biological Aspects of the Control of Sulphate-Reducing Bacteria.* Ph.D. Thesis, Heriot-Watt University.

SUTHERLAND, I. W. 1983 Microbial exopolysaccharides—their role in microbial adhesion in aqueous systems. *Critical Reviews in Microbiology* **10**, 173–201.

Preservatives Available for Use

D. G. CHAPMAN

School of Pharmacy, Robert Gordon's Institute of Technology, Schoolhill, Aberdeen, UK

Micro-organisms survive and grow on the components of pharmaceutical and cosmetic preparations. Many of the modern preparations are more susceptible than older formulations to microbial contamination (Plotkin & Austrian 1958; Burdon & Whitby 1967). This increased contamination risk may be due in part to specific excipients such as non-ionic emulsifying agents which are capable of supporting the growth of moulds and bacteria including pseudomonads (Bean *et al.* 1965; Bryce & Smart 1965). It is thus important that an adequate concentration of preservative is present in a formulation if antimicrobial preservation is essential (Bean 1972).

Several factors, besides formulation components, can markedly influence the availability and the activity of a preservative (Wedderburn 1964; Kurup 1981). The importance of preservative evaluation within the final product has been noted by Lawrence (1955). He showed that while phenylethyl alcohol (0.5% w/v) was more active in a final formulation than on its own, certain antimicrobial agents, notably phenylmercuric nitrate and thiomersal (0.005–0.01% w/v), exhibited the opposite effect. As each preparation requiring preservation has a unique set of problems (Hugo 1978) and as no suitable means of making a theoretical selection of a preservative exists (Croshaw 1977), the antimicrobial agent should be considered a part of and selected early in the development of the formulation (Wedderburn 1958).

Crompton (1962) suggested that ophthalmic solutions should include a bacteriostatic concentration of preservative but he (Crompton 1964) declared that such solutions should contain a rapidly acting bactericide. Tenenbaum (1967) suggested that the activity of a preservative within a product should exceed the minimum required to inhibit the expected microbial contamination. Bean (1972) proposed, however, that a biocidal rather than a biostatic activity would be preferred. The general view at present appears to be that the preservative in a formulation should decrease the microbial count incurred by unavoidable contamination of a product following good manufacturing procedures. Both the preservative and its concentration should be selected carefully for optimal activity on any potential

Preservatives in the Food,
Pharmaceutical and Environmental
Industries

0–632–01727–9

contaminants of a product whilst avoiding toxicity to the user. In addition, the concentration of a preservative within a cosmetic formulation should not perturb the equilibrium of the biozone created by the skin microflora (Tronnier 1968).

Standards of Preservation

A direct relationship exists between the function of a preservative in a preparation and any applicable microbial standards. Specific pharmaceutical products, such as ophthalmic solutions and injections, must be sterile. When these preparations are packaged in multiple-dose forms, a preservative capable of maintaining sterility of the product should be included in the formulation. Microbial spoilage of cosmetic products has been a great concern of manufacturers (Manowitz 1961) as the cosmetic must not infect the user and it must be resistant to microbial spoilage (Smart & Spooner 1972). Guidance issued by the council of the Society of Cosmetic Chemists of Great Britain (Anon. 1970*a*) on the hygienic manufacture and preservation of toiletries and cosmetics recommends that relevant known pathogenic micro-organisms be absent from cosmetics and that those preparations intended for use on newborn infants, for direct installation into the conjuctival sac, or for use on broken skin should be sterile at the time of manufacture. While good manufacturing procedures have undoubtedly diminished the levels of microbial contamination of susceptible cosmetic products, the addition of a preservative to such preparations should at least inhibit the growth of any micro-organisms present in the product at the time of manufacture and provide protection against contamination introduced accidentally during use (Croshaw 1977).

A working party was set up by the Pharmaceutical Society of Great Britain to examine the problems of microbial contamination of pharmaceutical preparations intended for oral and topical use (Sykes 1971). It found that preparations formulated with a preservative were microbiologically satisfactory; in the absence of a suitable preservative, liquid medicines in common use could support significant levels of bacteria. The report thus recommended that a final formulation should be capable of reducing or preventing the increase of such contamination. Cowen & Steiger (1976) also have noted that, unless adequately preserved, most modern topical products provide an environment for bacterial or fungal growth. The level of microbial risk is influenced by the physical state of the product. Aqueous solutions and multiple phase oil-in-water emulsions are highly susceptible to microbial contamination while water-in-oil emulsions are less so because the continuous oil phase bars microbial penetration into the aqueous phase (Tice & Barr 1958).

In order to ensure that a preservative will fulfill its microbiological requirements in a product, the preservative which is selected for use must be evaluated. Antimicrobial properties of a preservative can be ascertained by screening tests. These tests will eliminate unsuitable compounds and broth-dilution or agar-plate diffusion methods will provide information on the antimicrobial activity of a preservative. The results of this preliminary perscrutation will provide an indication of the inhibitory attributes of a preservative (Parker 1982). These initial tests can be followed by a challenge test with the final product. This testing method is essential as each preparation exhibits its own particular problems due to any one of a number of factors which can influence preservative action (Lachman 1968). The challenge test is a means of determining the efficiency of a preservative (Cowen & Steiger 1976). It also provides a useful foundation for assessing the biological availability of preservatives in pharmaceutical and cosmetic products which are susceptible to microbial spoilage (Eriksen 1970).

The first developments of the current challenge test procedures occurred with the artificial inoculation of a preparation with selected microbial contaminants. This type of test was initially proposed by van Abbe (1947) but has been frequently modified. A more comprehensive preservative test system has been described by Rzok *et al.* (1955) and by Kenney *et al.* (1964). For the testing of injectables, ophthalmics, topicals and oral preparations, these tests require artificial contamination of samples of the product with either pure or mixed cultures. Contamination by mixed cultures has been used in an attempt to achieve a more realistic test procedure. The results obtained with such cultures may be less reliable than those obtained with pure cultures (Yablonski 1972). Ideal test strains would seem to be those isolated from a decomposed product of the type to be tested. Due to standardization difficulties, however, strains from culture collections are most frequently employed. These may be supplemented with strains isolated from a decomposed product or with strains, such as an osmophilic yeast, possessing particular properties (Anon. 1980*a*).

The first official pharmacopoeial test for multiple-dose sterile products was described in the United States Pharmacopoeia (USP, Anon. 1970*b*) and a modified test has been retained in subsequent editions (Anon. 1975, 1980*c*). Although the British Pharmacopoeia (BP) has listed suitable bactericides for preservation for about 40 years, only the current edition (Anon. 1980*b*) includes a detailed challenge test. This test exhibits many similarities to the challenge test of the USP.

Many criticisms have been raised against these official pharmacopoeial methods, notably against the selection, number, size and condition of the challenge organisms together with the sampling and recovery methods. The test procedure has been described as inefficient because only five individual cultures are used. The large inoculum size has been criticized as it is probably unrepresentative of the challenge experienced by a product during use (Leitz 1972). Difficulties do occur in relating the test inoculum to practical conditions; storage and growth conditions of the inoculum may modify the sensitivity of the organisms to the test conditions (Al-Hiti & Gilbert 1980).

Under ideal circumstances an adequately preserved preparation should eradicate completely contaminating organisms in the intervals between the withdrawal of doses. This is, however, a difficult standard to achieve. As an alternative, limits which are known to be satisfied by well-preserved products would suffice (Spooner & Croshaw 1981). The BP test is not a mandatory requirement. Nevertheless it does furnish valuable information on the performance of any suitable antimicrobial preservative for inclusion in the BP.

Guidance for the testing of preserved cosmetics have been supplied by the Society of Cosmetic Chemists of Great Britain (Anon. 1970*a*). In addition to a range of bacteria, moulds and yeasts from culture collections, this testing procedure also includes recent isolates from contaminated samples. Variation can occur in the organisms selected for use in this system and this should enhance the test severity. The procedure can be modified for different cosmetic products which are under investigation.

There has been a tendency to design a single capacity test to fulfill all situations. As noted by Parker (1982), however, there is probably a case for simplifying challenge tests as much as possible. These simplified tests would suffice for testing a specific group of products such as cosmetics and non-sterile pharmaceutical preparations which possess similar microbiological requirements. Several publications have noted a need for an alternative test procedure in order to test multiple-dose injections, ophthalmic products and similar preparations containing a rapidly acting preservative. Leitz (1972) has suggested that the sampling time in the test procedure should vary with the type of product.

Chemical and Physical Properties

Several factors exist within a pharmaceutical or cosmetic formulation: pH; oil-water ratio; surfactant concentration and character; colloids and suspended solids; container interaction. Any one of these could influence the activity and the stability of an antimicrobial agent (Lachman 1968). Rosen & Berke (1973) have suggested that the pH of a product should be considered first during the selection of a suitable antimicrobial preservative. Changes in the pH of a formulation may influence the degree of ionization of a preservative, affect partitioning of the preservative and/or influence the ionisation of chemical groups on the microbial cell surface (p. 257).

While the ideal preservative should be effective over a wide pH range, in practice the optimal activity of most preservatives exists over a narrow range (Raiman 1970). Studies carried out to determine optimal pH conditions for preservative activity by Entrekin (1961) and Wickliffe & Entrekin (1964) led to the general observation that the lower the pH the greater the antimicrobial activity of the 17 preservatives included in their investigations. Such a relationship appears to exist for certain weak carboxylic acids such as benzoic or sorbic acid, the degree of dissociation of these compounds being pH dependent (Eklund 1983). The antibacterial and antifungal properties of these preservatives is due predominantly to the undissociated acid. Both acids exhibit much greater activity in acidic than in neutral or alkaline solutions and hence they are normally used at pH 4 or below (Entrekin 1961). A similar relationship exists for esters of *p*-hydroxybenzoic acid (Bandelin 1958; Entrekin 1961). Hydrolysis of these esters occurs at pH 7 (McDavid 1973) and at pH 10 it occurs so rapidly that the esters are completely unsuitable for use as preservatives (Wickliffe & Entrekin 1964). This effect of increasing alkalinity on the hydrolysis rate of esters of *p*-hydroxybenzoic acid is enhanced by heat (Raval & Parrott 1967). The dissociation of preservatives as a function of pH has been reviewed by Wedderburn (1964).

The pH of the treatment medium has also been shown to influence the sensitivity of *Pseudomonas cepaciae* and *Proteus vulgaris* to a combination of chlorhexidine gluconate (0.05% w/v) and cetrimide (0.5% w/v). These organisms grow on this preservative combination under acid conditions up to pH 6, but die under alkaline conditions (Bassett 1971).

Microbial growth occurs in the aqueous phase of a monophase or multiple phase preparation. Thus, for suitable effectiveness, an adequate preservative concentration is required within this phase of a formulation (Croshaw 1977). Aqueous solutions are relatively easily preserved as all the preservative is present throughout the system. In a biphasic oil and water formulation, however, the pH of the aqueous phase can affect the quantity of preservative lost by partitioning into the oil phase. This loss may be highly significant as the preservative concentration remaining in the aqueous phase could be reduced to an ineffective level (Chamburkar & Joslin 1975). Preservation failure for a cream or emulsion is further complicated by the presence of at least one emulsifying agent in these multiple phase systems. Loss of a preservative could then occur due to its partitioning into the oil phase or its interaction with an emulsifying agent (Bean *et al.* 1969, 1970). In order to aid preservation of multiple phase systems, the ideal preservative should have a low oil-water partition coefficient (Bean & Heman-Ackah 1964). Parker (1982) has detailed some of the characteristics of preservatives which are used in creams.

Many emulsifying agents used in both pharmaceutical and cosmetic products can markedly influence the activity of a preservative. Great care is therefore essential in the selection of a suitable preservative in order to avoid interaction between the preservative, the emulsifying agent, and other components in the formulation. Non-ionic surfactants are frequently used to produce stable formulations. Their use has created many problems, however, due to their interaction with and inactivation of preservatives (Schmolka 1973). The extent of this interaction varies with the hydrophile-lipophile balance of the macromolecular substance and with the chemical structure of the preservative. Those non-ionic compounds possessing a greater lipophilic character appear to produce a greater reduction in the effectiveness of a preservative (Wedderburn 1970). This association between these two potential components of a formulation is believed to involve hydrogen bonding (Lachman

1968). The non-ionic surface active agent, polysorbate 80, has been shown to interact with, and thereby decrease, the effectiveness of a variety of preservatives (Bahal & Kostenbauder 1964; Kostenbauder 1968; Yousef *et al.* 1973).

Emulsions which are prepared with anionic are considered to be preserved more easily than those prepared with non-ionic emulsifying agents (Jacobs *et al.* 1975). A useful mathematical formula has been used by Kazmi & Mitchell (1971, 1976) in order to predict the distribution of a preservative between the oil, aqueous and micellar components of an emulsified system.

Gums, suspending and thickening agents are used in a wide variety of pharmaceutical and cosmetic preparations (Young 1970). Interactions between preservatives and these ingredients have been well documented. It would appear to be generally accepted that preservative molecules are rendered inactive when they bind to other molecules within a formulation. The mechanism of these interactions involves either ionic incompatibility, hydrogen bonding, van der Waals forces, physical adsorption or inclusion formation (Coates 1973).

Preservative Activity and Application

The frequency of preservative use in cosmetic formulations has been discussed by Richardson (1977, 1981). Details of the preservatives used in both pharmaceutical and cosmetic formulations (Table 1) have been presented by Wallheuser (1974).

Benzoic and sorbic acids

Benzoic acid is one of the oldest preservatives used in pharmaceutical and cosmetic formulations. It is suitable for the preservation of oral preparations and has the advantages of low toxicity and lack of colour. Due to its greater solubility, the sodium salt is often used (Chichester & Tanner 1972; Kimble 1977) although the non-ionized acid is the form which exhibits good antibacterial and antifungal properties in acid solution (Rahn & Conn 1944). Although the use of benzoic acid in many pharmaceutical formulations has been displaced by other preservatives (Elowe 1955), 0.1% w/v is contained in childrens mixtures of the BP (Anon. 1980*b*) and the British Pharmaceutical Codex (BPC, Anon. 1979). Sodium benzoate (0.1–0.5% w/v) continues to be used extensively in cosmetic formulations (Richardson 1981).

Sorbic acid (2,4-hexadecanoic acid) is a more effective antifungal agent than benzoic acid (Puls *et al.* 1955); the undissociated molecule exhibits greater activity in conditions below pH 4 (Wickliffe & Entrekin 1964). This compound is used in both pharmaceutical and cosmetic preparations at a concentration of 0.1–0.2% (w/v). Indeed its use may be preferred for products containing non-ionic surfactants. Sorbic acid 0.01% (w/v) has been suggested as a suitable preservative for syrups

TABLE 1. *Summary of preservatives and their cosmetic and pharmaceutical applications**

Preservative	Concentration range (% w/v)	Product
Benzoic acid	0.1	Childrens oral medicines
Sorbic acid	0.2	Pharmaceutical ointments
Phenol	0.5	Solutions for injection
Cresol	0.3	Solutions for injection
Chlorocresol	0.1	Solutions for injection
Methyl parabens	0.1	Oral liquid medicines
	0.1–0.2	Eye make-up
Chlorhexidine acetate	0.01	Eye drops
Bronopol†	0.2–0.5	Shampoos
Benzalkonium chloride	0.01	Eye drops
Cetrimide (cetyltrimethyl)-ammonium bromide)	0.1	Creams
Phenylmercuric nitrate	0.001	Solutions for injection
	0.002	Eye drops
Formaldehyde	0.05	Shampoos
Germall 115†	0.1–0.5	Eye make up
	0.1	Hair conditioner
Kathon CG†	0.02–0.1	Shampoos, lotions and creams

* See pp. 207 *et seq.* for additional details.

† Manufacturers' trade name: Bronopol: 2-nitro-2-bromo-1,3-propanediol (Boots Pure Drug Co. Ltd. Nottingham, England); Germall 115: substituted imidazolidinyl urea components (Sutton Laboratories Inc. Rosella, U.S.A.); Kathon CG: mixture of isothiazolin-3-one derivatives (Rohm and Haas Co., Philadelphia, USA).

(Richards 1972). It appears to be a suitable preservative against most common microbial contaminants (Lenart 1977) although some moulds show a marked resistance and some can metabolize it (Melnick *et al.* 1954; Troller 1965).

Phenolic compounds

These weakly acidic compounds have been used extensively in pharmaceutical preparations although their inclusion in many products has declined due to the disadvantages of odour, irritation, volatility, toxicity and other incompatibilities (Elowe 1955). These disadvantages have also restricted the usefulness of these compounds in cosmetic formulations (Wells & Lubowe 1964).

Phenol

An aqueous solution of phenol (0.5% (w/v) is included in the BP (Anon. 1980*b*) as a preservative standard. The inclusion of phenol in cosmetics and toiletries is restricted in the UK under the Cosmetic Products Regulations 1978 (Anon. 1978). As with other phenolic compounds, this compound is more active against Gram positive than Gram negative bacteria while bacterial spores are considered to be resistant (Briggs & Yazdany 1974). Indeed large numbers of *Bacillus subtilis* spores can survive exposure to phenol at a concentration of up to 5% (w/v) for at least 100 h at 25° or 37 °C (Loosemore & Russell 1963; Russell & Loosemore 1964).

o-*Cresol*

This is a mixture of three isomers together with a small quantity of phenol (Anon. 1976). Although this compound possesses similar activity to phenol, a concentration of 0.3% (w/v) is used as a preservative for injections. It is also used as a preservative in lotions and ointments. It is too irritating and odour-producing for use as a cosmetic preservative. At a concentration of up to 0.1% (w/v) it is bacteriostatic against Gram positive and Gram negative bacteria with an approximately identical bacterial sporostatic concentration. This compound exhibits little sporocidal activity (Russell 1982). *o*-Cresol does have the advantage in that its antimicrobial activity is not greatly reduced in the presence of organic matter (Russell *et al.* 1967).

Chlorocresol

Chlorocresol (*p*-chloro-*m*-cresol) has considerably greater anti-microbial activity than phenol or *o*-cresol. It exhibits low toxicity and was the first official ophthalmic preservative used in the UK (Brown 1971). The preservative is widely used in pharmaceutical preparations where a concentration of 0.1% (w/v) is used in aqueous multiple-dose injections BPC (Anon. 1979). Its inclusion in most of the eye drops listed in the BPC (Anon. 1963) has been withdrawn. It is used also as a preservative in creams, topical pharmaceutical preparations and in cosmetic formulations at levels of up to 1% (w/v) at which concentration the odour level is minimal (Gucklhorn 1968). Less than 0.1% (w/v) chlorocresol was required to sterilize broth suspensions (10^8/ml) of three strains of *Ps. aeruginosa*. An identical preservative concentration in ophthalmic preparations killed *Ps. aeruginosa* cell counts of approximately 10^8/ml 'almost instantaneously' (Klein *et al.* 1954). The effectiveness of chlorocresol is greatly suppressed, however, in the presence of oils, fats, or a non-ionic surfactant (Yousef *et al.* 1973).

Chloroxylenol

Chloroxylenol (*p*-chloro-*m*-xylenol) is used in a number of cosmetic formulations (Richardson 1981) and, at normal-use concentrations, it is non-irritating and retains activity at alkaline pH (Judis 1962). It exhibits a relatively broad spectrum of

ethylenediamine tetra acetic acid increases its activity against Gram-negative organisms (MacGregor & Elliker 1958).

Cetrimide

Cetrimide is the generic name given by the BP (Anon. 1980b) and European Pharmacopoeia (EP, Anon. 1971) for a mixture of three alkyltrimethylammonium bromide compounds differing only in their alkyl chain length. This mixture consists mainly of tetradecyltrimethylammonium bromide (C_{14}) together with smaller amounts of the dodecyl (C_{12}) and hexadecyl (C_{16}) derivatives. The USP (Anon. 1980*c*) considers the name cetrimide to represent hexadecyltrimethylammonium bromide (C_{16}) alone. An investigation carried out by Al-Taae *et al.* (1983) indicated that a single alkyltrimethylammonium bromide, preferably the tetradecyl (C_{14}) derivative, would yield greater activity against a range of micro-organisms, rather than a mixture of these compounds. The solubility of these compounds is related to their alkyl chain length with the dodecyl (C_{12}) derivative being highly water soluble. The solubility of the other compounds decreases as the alkyl chain length increases beyond this length (Gucklhorn 1968).

Bronopol

This, a relatively new preservative, is included in various cosmetic and toiletry preparations (Bryce *et al.* 1978). It appears to be too toxic for use in liquid oral medicinal preparations (Parker 1979). Bronopol exhibits a wide spectrum of antibacterial activity, being highly active against Gram-negative organisms including *Ps. aeruginosa* (Bryce *et al.* 1978). It is less active against moulds and yeasts (Gucklhorn 1970). It is active over a pH range from pH 5 to 8; however exposure of aqueous solutions of bronopol to light under alkaline conditions results in slow break down and discolouration (Bryce & Smart 1965). Aqueous suspensions of powders and thickening agents as used in pharmaceutical and cosmetic preparations under alkaline conditions have been shown to affect adversely the antibacterial activity of bronopol (Myburgh & McCarthy 1978). Formaldehyde and nitrites are among the decomposition products of bronopol which has been implicated in nitrosamine formation. Holland (1981) claims, however, that this preservative can only result in the formation of nitrosamines in the presence of readily nitrosated amines.

Organic mercurial compounds

Phenylmercuric salts and thiomersal are powerful biostatic compounds which, although relatively slow acting, are useful preservatives in ophthalmic, parenteral and topical formulations (Eriksen 1970). Phenylmercuric salts are used at a concentration of 0.001% (w/v) for injection solutions and for the preservation of some ophthalmic drops at a concentration of 0.002% (w/v). Thiomersal (0.01%–0.02% w/v) has been used as a preservative in biological products having a low protein content (Wallheuser 1974). The application of these organic mercurial compounds in the UK for cosmetics and toiletries is restricted under the Cosmetic Products Regulations 1978 (Anon. 1978).

The mercurial preservatives exhibit a broad spectrum of antimicrobial activity being both fungistatic and bacteriostatic. A low concentration of phenylmercuric nitrate (0.00002% w/v) is required to inhibit the growth of *Staphylococcus aureus* and spores of *Bacillus subtilis* whereas they exhibit little activity against many strains of *Ps. aeruginosa* (Wallheuser 1974). A disadvantage of the use of these compounds is that continued use of eye-drops containing a mercurial preservative for a prolonged period may result in the intra-ocular deposition of mercury. This condition is, however, slow to develop (Abrams 1963). The use of these mercurial preservatives is under careful investigation and Wallheuser (1974) has suggested that some agencies may prohibit their clinical use.

Phenylmercuric salts are readily lost from ophthalmic solutions packaged in an eye-drop bottle possessing a silicone rubber or Nylon dropper assembly. The loss of the mercurial compound occurs due to its adsorption into the plastic or rubber dropper (p. 217; Eriksson 1967). The antimicrobial activity of organic mercurial compounds can be lost also due to inactivation by organic matter (Wilson 1965), anionic agents and, to a much lesser extent, by non-ionic surfactants (Eckhardt 1970). Unlike thiomersal, which is effective over a broad pH range up to pH 9, phenylmercuric nitrate is only active above pH 6 in an aqueous environment (Wallheuser 1974). Thus, as discussed by Allwood (1978), the latter may have little antimicrobial activity in several injections of the BP as many of these parenteral preparations possess an acidic pH. When used as a preservative in a topical cream, phenylmercuric nitrate is active over the range pH 5 to 8 (Parker 1982).

Formaldehyde

Formaldehyde is frequently used for the preservation of cosmetic products (Richardson 1981) such as shampoos and in topical products applied to the skin for short time periods (Anon. 1970*a*). This compound has good antimicrobial properties with high concentrations exhibiting a lethal action against moulds, yeasts, bacteria, and bacterial spores (Rubbo *et al.* 1967). A low formaldehyde concentration is bacteriostatic (Hamilton 1971).

The indication by the U.K. Health and Safety Executive that the inhalation of formaldehyde vapour may be a carcinogenic risk to humans may have a marked influence on the use of this preservative in cosmetic preparations. Restraint in the use of formaldehyde in many cosmetic products already exists due to the pungent odour, the volatility and the incompatabilities of formaldehyde with several components of cosmetic formulations. In order to overcome the problems of odour and volatility, several products have been developed the activity of which is based upon the gradual release of formaldehyde (Sharpell & Manowitz 1983). Germall 115 is an example of one of the more prominent formaldehyde-releasing compounds recommended for use in cosmetic formulations.

Germall 115 is a member of the imidazolidinyl ureas which were described for use as preservatives by Berke & Rosen (1970). It is an odourless, tasteless and water soluble white powder which, in addition to its stability over a wide pH range, is also compatible with almost all ingredients of cosmetic formulations (Berke & Rosen 1970). This compound exhibits varying activity against moulds and yeasts; it is however effective against both Gram-positive and Gram-negative vegetative organisms including *Pseudomonas* spp. A concentration of 0.3% (w/v) adequately protects most cosmetic formulations (Berke & Rosen 1978).

Kathon CG

This is a very recent addition to the range of preservatives available for use in cosmetic formulations. It is available as an aqueous solution containing 1.5% (w/v) of the active ingredients which are a mixture of 5-chloro-2-methyl-4-isothiazolin-3-one and 2-methyl-4-isothiazolin-3-one together with magnesium salts (Hugo & Russell 1982).

Kathon CG was used initially as a preservative in cosmetic preparations in 1978 and is now used in a wide range of cosmetic products including shampoos and creams, at a concentration of up to 25 μg active ingredient per ml. It is highly water-soluble and, although effective over the entire pH range of cosmetic preparations, it appears more stable in acid than in a mildly alkaline condition. It is active against a wide range of moulds, yeasts and bacteria including *Ps. aeruginosa* (Croshaw 1977). The antimicrobial activity does not appear to be due to the release of formaldehyde. Kathon CG also has the advantages of being compatible with most ingredients of cosmetic formulations including anionic, cationic and non-ionic agents and it is non-irritating at normal in-use concentrations (Hugo & Russell 1982).

Preservative combinations

Any deficiencies in the action of a preservative may be rectified by combining it with another preservative. This could either result in a synergistic effect, an increase in the spectrum of anti-bacterial activity and/or restraining the development of microbial resistance (Garrett 1966). For example, the addition of parabens to Germall 115 increases the antimicrobial spectrum of the latter (Jacobs *et al.* 1975). Hugbo (1976) has shown that combination of preservatives possessing different modes of antimicrobial action can produce a synergistic effect. As an alternative to preservative combination, greater anti-microbial activity can be achieved by the addition of a non-antimicrobial potentiating agent to a preservative (Parker *et al.* 1968). Several examples of preservative potentiating agents, including propylene glycol, have been listed by Parker (1982). With the continuing increased complexity of product formulations, combinations of preservatives may play an important role in routine preservation in the future.

Acknowledgement

I wish to thank Mr S. M. MacManus and Mr R. F. Steel for a useful discussion.

References

AALTO, T. R., FIRMAN, M. C. & TIGLER, N. E. 1953 *p*-Hydroxybenzoic acid esters as preservatives I. *Journal of the American Pharmaceutical Association Scientific Edition (Washington)* **42**, 449–457.

ABRAMS, J. D. 1963 Iatrogenic mercurialentis. *Transactions of the Ophthalmological Society of the United Kingdom* **83**, 263.

AL-HITI, M. M. A. & GILBERT, P. 1980 Changes in preservative sensitivity for the USP antimicrobial agents effectiveness test micro-organisms. *Journal of Applied Bacteriology* **49**, 119–126.

ALLWOOD, M. C. 1978 Antimicrobial agents in single and multi-dose injections. *Journal of Applied Bacteriology* **44**, Svii-Sxvii.

AL-TAAE, A. A. N., DICKINSON, N. A. & GILBERT, P. 1983 Antimicrobial activity of some alkytrimethylammonium bromides. *Journal of Pharmacy and Pharmacology (Supplement)* **35**, 60P.

ANDERSON, K., LILLIE, S. & CROMPTON, D. 1964 Efficacy of bacteriostats in ophthalmic solutions. *Pharmaceutical Journal* **192**, 593–594.

ANON. 1963 *British Pharmaceutical Codex* London: The Pharmaceutical Press.

ANON. 1970*a* Hygenic manufacture and preservation of toiletries and cosmetics. *Journal of the Society of Cosmetic Chemists* **21**, 719–800.

ANON. 1970*b* *United States Pharmacopoeia* Eighteenth Revision. United States Pharmacopoeial Convention, Bethesda, U.S.A.

ANON. 1971 *European Pharmacopoeia* Saint-Ruffine: Maisonneuve S.A.

ANON. 1975 *United States Pharmacopoeia* Nineteenth Revision. United States Pharmacopoeial Convention, Rockville, U.S.A.

ANON. 1976 *Merck Index* Ninth edition. Rahway, U.S.A.: Merck & Co. Inc.

ANON. 1978 *The Cosmetic Products Regulations 1978*. Statutory Instruments 1978 No. 1354. London: HMSO.

ANON. 1979 *British Pharmaceutical Codex* London: The Pharmaceutical Press.

ANON. 1980*a* Report: The test for the effectiveness of antimicrobial preservation of pharmaceuticals. *Pharmaceutica Acta Helvetia* **55**, 40–49.

ANON. 1980*b* *British Pharmacopoeia* London: HMSO.

ANON. 1980*c* *United States Pharmacopoeia* Twentieth Revision. United States Pharmacopoeial Convention, Rockville, U.S.A.

BAHAL, C. K. & KOSTENBAUDER, H. B. 1964 Interaction of preservatives with macromolecules. *Journal of Pharmaceutical Sciences* **53**, 1027–1029.

BANDELIN, F. J. 1958 The effect of pH on the efficiency of various mould inhibiting compounds. *Journal of the American Pharmaceutical Association Scientific Edition (Washington)* **47**, 691–694.

BASSETT, D. C. J. 1971 The effect of pH on the multiplication of a pseudomonad in chlorhexidine and cetrimide. *Journal of Clinical Pathology* **24**, 708–711.

BEAN, H. S. 1972 Preservatives for pharmaceuticals. *Journal of the Society of Cosmetic Chemists* **23**, 703–720.

BEAN, H. S. & HEMAN-ACKAH, S. M. 1964 Influence of oil:water ratio on the activity of some bactericides against *Escherichia coli* in liquid paraffin and water dispersions. *Journal of Pharmacy and Pharmacology* **16**, 58T–67T.

BEAN, H. S., HEMAN-ACKAH, S. M. & THOMAS, J. 1965 The activity of antibacterials in two-phase systems. *Journal of the Society of Cosmetic Chemists* **16**, 15–27.

BEAN, H. S., KONNING, G. H. & MALCOLM, S. A. 1969 A model for the influence of emulsion formulations on the activity of phenolic preservatives. *Journal of Pharmacy and Pharmacology* **21** (Supplement), 173S–181S.

BEAN, H. S., KONNING, G. H. & THOMAS, J. 1970 Significance of the partition coefficient of a preservative in cosmetic emulsions. *American Perfumer and Cosmetics* **85**, 61–65.

BERKE, P. A. & ROSEN, W. E. 1970 Germall, a new family of antimicrobial preservatives for cosmetics. *American Perfumer and Cosmetics* **85**, 55–60.

BERKE, P. A. & ROSEN, W. E. 1978 Imidazolidinyl urea activity against *Pseudomonas*. *Journal of the Society of Cosmetic Chemists* **29**, 757–766.

BOTWRIGHT, W.E. 1952 Quaternary ammonium compounds in cosmetics and related products. *Journal of the Society of Cosmetic Chemists* **3**, 118–123.

BRIGGS, A. & YAZDANY, S. 1974 Resistance of *Bacillus* spores to combined sporicidal treatments. *Journal of Applied Bacteriology* **37**, 623–631.

BROWN, M. R. W. 1971 Inhibition and Destruction of *Pseudomonas aeruginosa*. In *Inhibition and Destruction of the Microbial Cell* ed. Hugo, W. B. pp. 307–367. London: Academic Press.

BROWN, M. R. W., FOSTER, J. H. S., NORTON, D. A. & RICHARDS, R. M. E. 1964 Preservation of ophthalmic solutions. *Pharmaceutical Journal, London* **192**, 8.

BRYCE, D. M. & SMART, R. 1965 The preservation of shampoos. *Journal of the Society of Cosmetic Chemists.* **16**, 187–201.

BRYCE, D. M., CROSHAW, B., HALL, J. E., HOLLAND, V. R. & LESSEL, B. 1978 The activity and safety of the antimicrobial agent Bronopol (2-bromo-2-nitropropan-1,3-diol). *Journal of the Society of Cosmetic Chemists* **29**, 3–24.

BURDON, D. W. & WHITBY, J. L. 1967 Contamination of hospital disinfectants with *Pseudomonas* species. *British Medical Journal* **2**, 153–155.

CHAMBURKAR, P. B. & JOSLIN, R. S. 1975 Effect of flavouring oils on preservative concentrations in oral liquid dosage forms. *Journal of Pharmaceutical Sciences* **64**, 414–417.

CHICHESTER, D. F. & TANNER, F. W. 1972 Antimicrobial food additives. In *Handbook of Food Additives* 2nd edn. ed. Furia, T. E., pp. 115–184. Cleveland, Ohio: CRC Press.

COATES, D. 1973 Preservatives—microbiological and physical factors. *Manufacturing Chemist and Aerosol News* **44**, 35–37.

COWEN, R. A. & STEIGER, B. 1976 Antimicrobial activity—a critical review of test methods of preservative efficiency. *Journal of the Society of Cosmetic Chemists* **27**, 467–481.

CROMPTON, D. O. 1962 Ophthalmic prescribing. *Australian Journal of Pharmacy* **43**, 1020–1028.

CROMPTON, D. O. 1964 Sterility of eye medicaments. *Lancet* **i**, 41.

CROSHAW, B. 1977 Preservatives for cosmetics and toiletries. *Journal of the Society of Cosmetic Chemists* **28**, 3–16.

D'ARCY, P. F. 1971 Inhibition and destruction of moulds and yeasts. In *Inhibition and Destruction of the Microbial Cell*, ed. Hugo, W. B. pp. 614–636. London: Academic Press.

DAVIES, G. E., FRANCIS, J., MARTIN, A. R., ROSE, F. L. & SWAIN, G. 1954 1:6-4′-chlorophenyldiguanidohexane (Hibitane). Laboratory investigation of a new antibacterial agent of high potency. *British Journal of Pharmacology and Chemotherapy* **9**, 192–196.

DELUCA, P. P. & KOSTENBAUDER, H. B. 1960 Interaction of preservatives with macromolecules. IV Binding of quaternary ammonium compounds by nonionic agents. *Journal of the American Pharmaceutical Association, Scientific Edition (Washington)* **49**, 430–437.

DYMICKY, M. & HUHTANEN, C. N. 1979 Inhibition of *Clostridium botulinum* by *p*-hydroxybenzoic acid esters. *Antimicrobial Agents and Chemotherapy* **15**, 798–801.

ECKHARDT, W. A. 1970 Phenylmercuric compounds as preservatives for cosmetics. *American Perfumer and Cosmetics* **85**, 83–85.

ELOWE, L. N. 1955 Preservation of pharmaceuticals Part II. Preservatives used in pharmacy. *Ontario College of Pharmacy Bulletin* **4**, 45–47.

ECKLUND, T. 1983 The antimicrobial effect of dissociated and undissociated sorbic acid at different pH levels. *Journal of Applied Bacteriology* **54**, 383–389.

ENTREKIN, D. N. 1961 Relation of pH to preservative effectiveness I. Acid media. *Journal of Pharmaceutical Sciences* **50**, 743–746.

ERIKSEN, S. P. 1970 Preservation of ophthalmic nasal and otic products. *Drug and Cosmetic Industry* **107**, 36–40.

ERIKSSON, K. 1967 Loss of organomercurial preservatives from medicaments in different kinds of containers. *Acta Pharmaceutica Suecica* **4**, 261.

GARRETT, E. R. 1966 A basic model for the evaluation and prediction of preservative action. *Journal of Pharmacy and Pharmacology* **18**, 589–601.

GERSHENFELD, L. & MILANICK, V. E. 1941 Bactericidal and bacteriostatic properties of surface tension depressants. *American Journal of Pharmacy* **113**, 306–326.

GRAY, G. W. & WILKINSON, S. G. 1965 The action of ethylenediamine tetraacetic acid on *Pseudomonas aeruginosa*. *Journal of Applied Bacteriology* **28**, 153–164.

GUCKLHORN, I. R. 1968 Cosmetic microbiology. *Manufacturing Chemist and Aerosol News* **39**, 32–33.

GUCKLHORN, I. R. 1970 Antimicrobials in cosmetics Part 8. *Manufacturing Chemist and Aerosol News* **41**, (1) 45, 47.

HALL R. 1967 Pharmaceutical medical applications. *Process Biochemistry* **2**, 24–25.

HAMILTON, W. A. 1971 Membrane active antibacterial compounds. In *Inhibition and Destruction of the Microbial Cell.* ed. Hugo, W. B. pp. 77–93. London: Academic Press.

HEARD, D. D. & ASHWORTH, R. W. 1968 The colloidal properties of chlorhexidine and its interaction with some macromolecules. *Journal of Pharmacy and Pharmacology* **20**, 505–512.

HOLLAND, V. R. 1981 BNPD and nitrosamine formation. *Cosmetic Technology* **3**, 31–36.

HUGBO, P. G. 1976 Additivity and synergism in vitro as displayed by mixtures of some commonly employed antibacterial preservatives. *Canadian Journal of Pharmaceutical Sciences* **11**, 117–120.

HUGO, W. B. 1978 Introduction: The scope of the problem. *Journal of Applied Bacteriology* **44**, Siii–Siv.

HUGO, W. B. & RUSSELL, A. D. 1982 Types of antimicrobial agents. In *Principles and Practice of Disinfection, Preservation and Sterilization.* ed. Russell, A. D., Hugo, W. B. & Ayliffe, G. A. J. pp. 8–106. Oxford: Blackwell Scientific Publications.

JACOBS, G., HENRY, S. M. & COTTY, V. F. 1975 The influence of pH, emulsifier, and accelerated ageing upon preservative requirements of o/w emulsions. *Journal of the Society of Cosmetic Chemists* **26**, 105–117.

JUDIS, J. 1962 Studies on the mechanism of action of phenolic disinfectants I. *Journal of Pharmaceutical Sciences* **51**, 261–265.

KAZMI, S. J. A. & MITCHELL, A. F. 1971 Dialysis method for determining preservative distribution in emulsions. *Journal of Pharmaceutical Sciences* **60**, 1422–1424.

KAZMI, S. J. A. & MITCHELL, A. F. 1976 The interaction of preservative and non-ionic surfactant mixtures. *Canadian Journal of Pharmaceutical Sciences* **11**, 10–17.

KENNEY, D. S., GRUNDY, W. E. & OTTO, R. H. 1964 Spoilage and preservative tests as applied to pharmaceuticals. *Bulletin of the Parenteral Drug Association* **18**, 10–19.

KIMBLE, C. E. 1977 Chemical food preservatives. In *Disinfection, Sterilization and Preservation* 2nd edn. ed. Block, S. S. pp. 834–858. Philadelphia, Pennsylvania: Lea & Febiger.

KLEIN, M., MILLWOOD, E. G. & WALTHER, W. W. 1954 On the maintenance of sterility in eye-drops. *Journal of Pharmacy and Pharmacology* **6**, 725–732.

KOHN, S. R., GERSHENFELD, L. & BARR, M. 1963 Effectiveness of antibacterial agents presently employed in ophthalmic preparations as preservatives against *Pseudomonas aeruginosa. Journal of Pharmaceutical Sciences* **52**, 967–974.

KOSTENBAUDER, H. B. 1968 In *Disinfection, Sterilization and Preservation.* ed. Lawrence, C. A. & Block, S. S. pp. 50 Philadelphia: Lea & Febiger.

KURUP, T. R. R. 1981 Microbiological quality in pharmaceuticals. *The Eastern Pharmacist* **24**, 53–69.

LACHMAN, L. 1968 The instability of antimicrobial preservatives. *Bulletin of the Parenteral Drug Association* **22**, 127–144.

LAWRENCE, C. A. 1955 An evaluation of chemical preservatives for ophthalmic solutions. *Journal of the American Pharmaceutical Association, Scientific Edition (Washington)* **44**, 457–464.

LAWRENCE, C. A. 1960 Antimicrobial activity invitro of chlorhexidine. *Journal of the American Pharmaceutical Association, Scientific Edition (Washington)* **49**, 731–734.

LEITZ, M. 1972 Panel discussion: Critique of preservative challenge systems I. Critique of USP microbiological test. *Bulletin of the Parenteral Drug Association* **26**, 212–216.

LENART, G. 1977 Microbiological testing of some nonsterile liquid pharmaceutical preparations. *Gyogyszereszet* **21**, 408–412.

LOOSEMORE, M. & RUSSELL, A. D. 1963 The sporicidal activity of phenol. *Journal of Pharmacy and Pharmacology,* **15**, 558.

LOWBURY, E. J. L. 1951 Contamination of cetrimide and other fluids with *Pseudomonas pyocanea. British Journal of Industrial Medicine* **8**, 22–25.

LUECK, E. 1980 *Antimicrobial Food Preservatives.* Berlin: Springer-Verlag.

MANOWITZ, M. 1961 Preservation of cosmetic emulsions. In *Developments in Industrial Microbiology* ed. Koda, C. **2** 65–71 New York: Plenum Press.

MELNICK, D., LUCKMANN, F. H. & GOODING, C. M. 1954 Sorbic acid as a fungistatic agent for foods. VI Metabolic degradation of sorbic acid in cheese by molds and the mechanism of mold inhibition. *Food Research* **19**, 44–58.

MYBURGH, J. A. & MCCARTHY, T. J. 1978 *Cosmetics and Toiletries* **93**, 47.

MACGREGOR, D. R. & ELLIKER, P. R. 1958 A comparison of some properties of strains of *Pseudomonas aeruginosa* sensitive and resistant to quaternary ammonium compounds. *Canadian Journal of Microbiology* **4**, 499–503.

MCDAVID, J. E. 1973 Characteristics of the parabens as preservatives. International symposium on Preservatives in Biological products, San Francisco, 1973. *Developments in Biological Standardization* **24**, 49–55.

PARKER, M. S., MCCAFFERTY, M. & MCBRIDE, S. 1968 Phenonip a broad spectrum preservative. *Soap Perfumery and Cosmetics* **41**, 647–650.

PARKER, M. S. 1979 Preservation of fluid oral dosage forms. *The Pharmaceutical Journal* **222**, 92.

PARKER, M. S. 1982 The preservation of pharmaceutical and cosmetic products. In *Principles and Practice of Disinfection, Preservation and Sterilization.* ed. Russell, A. D., Hugo, W. B. & Ayliffe, G. A. J. pp. 287–305. Oxford: Blackwell Scientific Publications.

PLOTKIN, S. A. & AUSTRIAN, R. 1958 Bacteremia caused by *Pseudomonas* spp. following the use of materials stored in solutions of cationic surface-active agents. *American Journal of Medical Science* **235**, 621–627.

PULS, D. D., LINDGREN, L. F. & COSGROVE, F. P. 1955 Sorbic acid as a fungistatic agent for certain pharmaceutical preparations. *Journal of the American Pharmaceutical Association. Scientific Edition (Washington)* **44**, 85–87.

RAHN, O. & CONN, J. E. 1944 Effect of increase in acidity on antiseptic efficiency. *Industrial and Engineering Chemistry* **36**, 185–187.

RAIMAN, H. L. 1970 Problems relating to pharmaceutical preservatives including criteria for stability, efficacy and testing methods. In *The Dating of Pharmaceuticals* eds. Windheuser, J. L. & Blockstein, W. L. pp. 55–76. Wisconsin: University Press.

RAVAL, N. N. & PARROTT, E. L. 1967 Hydrolysis of methylparaben. *Journal of Pharmaceutical Sciences* **56**, 274–275.

RICHARDS, R. M. E. 1972 *Stability of Syrup*. Letter: Pharmaceutical Journal **209**, 91.

RICHARDS, R. M. E. & MCBRIDE, R. J. 1973 Enhancement of benzalkonium chloride and chlorhexidine acetate activity against *Pseudomonas aeruginosa* by aromatic alcohols. *Journal of Pharmaceutical Sciences* **62**, 2035–2037.

RICHARDS, R. M. E. & MIZRAHI, L. M. 1978 Differences in antibacterial activity of benzalkonium chloride. *Journal of Pharmaceutical Sciences* **67**, 380–383.

RICHARDSON, E. L. 1977 Preservatives: frequency of use in cosmetic formulas as disclosed to FDA. *Cosmetics and Toiletries* **92**, 85–86.

RICHARDSON, E. L. 1981 Update frequency of preservative use in cosmetic formulas as disclosed to FDA. *Cosmetics and Toiletries* **96**, 91–92.

ROSEN, W. E. & BERKE, P. A. 1973 Modern concepts of cosmetic preservation. *Journal of the Society of Cosmetic Chemists* **24**, 663–675.

RUBBO, S. D., GARDNER, J. F., WEBB, R. L. 1967 Biocidal activities of glutaraldehyde and related compounds. *Journal of Applied Bacteriology* **30**, 78–87.

RUSSELL, A. D., JENKINS, J. & HARRISON, I. H. 1967 The inclusion of antimicrobial agents in pharmaceutical products. *Advances in Applied Microbiology* **9**, 1–38.

RUSSELL, A. D. 1982 *The Destruction of Bacterial Spores*. pp. 168–231. London: Academic Press.

RUSSELL, A. D. & FURR, J. R. 1977 The antibacterial activity of a new chloroxylenol preparation containing ethylenediamine tetraacetic acid. *Journal of Applied Bacteriology* **43**, 253–260.

RUSSELL, A. D. & LOOSEMORE, M. 1964 Effect of phenol on *Bacillus subtilis* spores at elevated temperatures. *Applied Microbiology* **12**, 403–406.

RZOK, E. J., GRUNDY, W. E., KIRCHMEYER, F. J. & SYLVESTER, J. C. 1955 Determining the efficacy of preservatives in pharmaceutical products. *Journal of the American Pharmaceutical Association. Scientific Edition (Washington)* **44**, 613–616.

SCHMOLKA, I. R. 1973 The synergistic effects of nonionic surfactants upon cationic germicidal agents. *Journal of the Society of Cosmetic Chemists* **24**, 577–592.

SHARPELL, F. & MANOWITZ, M. 1983 Preservation of Cosmetics. In *Disinfection, Sterilization and Preservation* 3rd ed. ed. Block, S. S. pp. 589–607. Philadelphia: Lea & Febiger.

SMART, R. & SPOONER, D. F. 1972 Microbiological spoilage in pharmaceuticals and cosmetics. *Journal of the Society of Cosmetic Chemists* **23**, 721–737.

SPOONER, D. F. & CROSHAW, B. 1981 Challenge testing—the laboratory evaluation of the preservation of pharmaceutical preparations. *Antonie van Leeuwenhoek* **47**, 168–169.

SYKES, G. 1971 Microbial contamination in pharmaceuticals for oral and topical use: Society's working party report. *Pharmaceutical Journal* **207**, 400–402.

TENENBAUM, S. 1967 Preservatives for pharmaceuticals. *Journal of the Society of Cosmetic Chemists* **18**, 797–807.

TICE, L. F. & BARR, M. 1958 Factors to be considered in the preservation of cosmetic emulsions. *Journal of the Society of Cosmetic Chemists* **9**, 171.

TROLLER, J. A. 1965 Catalase inhibition as a possible mechanism of the fungistatic action of sorbic acid. *Canadian Journal of Microbiology* **11**, 611–617.

TRONNIER, H. 1968 The preservation of cosmetics. *Soap, Perfumery and Cosmetics* **41**, 871.

VAN ABBE, N. J. 1947 Bacteriostasis in solutions for injections. *Pharmaceutical Journal* **159**, 146.

WALLHEUSER, K. H. 1974 Antimicrobial preservatives in biologics. *Pharmazeutische Industrie* **36**, 716–722.

WEDDERBURN, D. L. 1958 Preservation of toilet preparations containing nonionics. *Journal of the Society of Cosmetic Chemists* **9**, 210–228.

WEDDERBURN, D. L. 1964 Preservation of emulsions against microbial attack. *Advances in Pharmaceutical Sciences* **1**, 195–268.

WEDDERBURN, D. L. 1970 Interactions in cosmetic preservation. *American Perfumer and Cosmetics* **85**, 49–53.

WELLS, F. V. & LUBOWE, I. I. 1964 *Cosmetics and the Skin*, pp. 586–598. New York: Reinhold Publishing Corporation.

WICKLIFFE, B. & ENTREKIN, D. N. 1964 Relation of pH to preservative effectiveness II. *Journal of Pharmaceutical Sciences* **53**, 769–773.

WILSON, G. 1965 Use of disinfectants in hospitals. *British Medical Journal* **1**, 408–413.

YABLONSKI, J. I. 1972 Panel discussion: critique of preservative challenge systems. III. Fundamental concepts of preservation. *Bulletin of the Parenteral Drug Association* **26**, 212–216.

YOUNG, A. E. 1970 Gums and thickeners used in cosmetic formulations. *Manufacturing Chemist and Aerosol News* **41**, 27–29.

YOUSEF, R. T., EL-NAKEEB, M. A. & SALAMA, S. 1973 Effect of some pharmaceutical materials on the bactericidal activities of preservatives. *Canadian Journal of Pharmaceutical Sciences* **18**, 54–56.

Preservation of Mixtures, Suspensions and Syrups

M. C. Allwood

Regional Research & Development Laboratory, East Anglia RHA, Central Pharmacy, Addenbrooke's Hospital, Cambridge UK

and R. J. S. Shaw

Regional Quality Control Laboratory, Norfolk & Norwich Hospital, Norwich UK

Liquid pharmaceutical preparations are generally intended for oral or external use. Oral liquid medication is chosen for its pharmacological advantages or because of ease of administration compared with tablets or capsules. Liquids are especially valuable for administration to children, the elderly and some severely handicapped patients. Viscous liquids that coat the buccal cavity or throat during swallowing provide an important treatment vehicle. Rapid dispersion in the stomach after swallowing is another valuable attribute. Finally the provision of relatively large amounts of bulky powders or oils can be accomplished only by using a liquid vehicle. As a consequence of all these factors, most oral liquid medicines are relatively complex, aqueous preparations containing a wide variety of excipients. They are vulnerable to microbial contamination and will support growth if appropriate precautions are not taken. Indeed the likelihood of contamination is substantial especially as most preparations are presented in a multi-dose form. External liquids are chosen because they offer a vehicle that can be applied over relatively large areas of the body surface. These may be aqueous-based lotions or organic solvent-based spirits. Many lotions are antiseptic preparations and as such they are unlikely to support growth because they contain active antimicrobial compounds. Spirits are also unlikely to support significant microbial growth. It may be concluded, therefore, that the oral liquid medicines are at greater risk of microbial contamination. It is essential that such preparations are protected against microbial abuse by adequate preservation. The sources and likely effects of microbial abuse have a major effect on the source of preservative system to be used to protect the product.

Preservatives in the Food, Pharmaceutical and Environmental Industries

0–632–01727–9

Sources of Contamination

Available evidence confirms that oral liquid preparations are the most likely of all the non-sterile pharmaceutical products to be contaminated by micro-organisms. Many preparations will actively support growth if a preservative is not included in the formulation. The range of products contaminated and the types or organisms isolated have been widely documented (Kallings *et al.* 1966; Anon. 1971; Baird *et al.* 1976, 1979; Baird & Petrie 1981). Aqueous suspensions have been found to be most commonly contaminated. In two studies, more than 20% of samples contained micro-organisms (Baird *et al.* 1979; Baird & Petrie 1981). Syrups, emulsions and solutions were also found on occasions to contain viable micro-organisms. Numbers isolated varied from less than 100 to more than 10^6/ml in some suspensions and liquids. The types of micro-organism isolated have also been found to vary widely. Gram-positive spore-formers are the most commonly occurring contaminants. Gram-negative rods are also recovered from many mixtures. In particular, many mixtures are contaminated with pseudomonads. The ability of preservatives to inactivate pseudomonads is therefore of particular importance in such preparations. The prevalence of pseudomonads and other Gramnegative rods reflects the ability of these organisms to survive and multiply in nutritionally-poor environments.

A major reason why oral liquid medicines are commonly contaminated arises from the many potential sources of micro-organisms associated with their manufacture. Apart from the obvious risks associated with the use of Purified water, many raw materials of natural origin are used. These may contain large numbers of viable spores. The inclusion of low concentrations of sugars, as flavouring agents, or of polysaccharides, as suspending agents, enrich the preparation with growth-supporting substrates.

The manufacturing process contributes also to the microbiological contamination of oral liquid medicines. Many of these products have been prepared traditionally as low cost items, with insufficient care taken to ensure maximum control of the microbiological aspects of manufacture. These preparations are generally packed in relatively large bottles from which many individual doses are taken. This practice enhances the microbiological risks to which a product is exposed.

The summation of all these factors ensures that oral liquid preparations pose some of the greatest difficulties in providing adequate protection against microbial attack. The inclusion of preservatives is essential as part of the measures required to minimize the risk to the patient. Unfortunately many of these products, because of their complex nature, are extremely difficult to preserve adequately.

conclusion

Factors governing the choice of preservative

It is clear that non-sterile aqueous pharmaceuticals must be preserved but the criteria that any acceptable preservative must fulfil severely limits the available choice. The major factors—perhaps unique to these products—are those of safety and lack of toxicity after oral intake. It should be emphasized that these medicines are commonly administered, in some cases for long periods of time, to children and elderly people. The complete freedom of serious toxic side-effects is therefore essential. Many preservatives that can be applied topically, and therefore used in external liquids or creams, are too toxic or irritant when given orally. There is a lack of information about the long term effects of ingestion of many preservatives. It is common to find therefore that only traditional and familiar preservatives are used in oral liquids. Moreover as data on the safety of many recently introduced preservatives are not readily attainable, they are not used in these products. The pharmaceutical aspects of a preservative also govern its value in preserving different oral liquid medications. For example, the compound(s) must be soluble and compatible in the formulation; biological activity should be unaffected by either the active ingredients or other excipients of the preparation. Microbiological properties must also be taken into account. In particular, activity against Gram-negative rods, especially pseudomonads, is a highly desirable attribute.

The British Pharmacopoeia (Anon. 1980) has traditionally included chloroform as the recommended preservative in most official formulae for oral liquid medication. The Pharmacopoeia allows manufacturers to include, however, a suitable alternative preservative, using as a microbiological criterion the Efficacy of Anti-microbial Preservative Test (Anon. 1980).

Choice of preservative

Considerations related to the formula and nature of the preparation

The formulae of many liquid preparations are often relatively complex. Both the nature of the active ingredients and the excipients must be taken into account when choosing a suitable preservative.

Active ingredients

Drugs may be soluble or insoluble in the liquid formulation. It is essential that the preservative is chemically compatible with any soluble drugs. The pH of the preparation may be governed by the active ingredients. The drug may have a direct effect on pH or the formulation may be buffered to maintain solubility or stability. In either case it is important to ensure that the preservative is also soluble, stable and retains its antimicrobial activity within the pH profile of the formulation. Insoluble but active ingredients may also affect pH. For instance, magnesium trisilicate will raise the pH. Many insoluble materials can adsorb preservatives. A number of examples have been reported. Benzalkonium chloride has been shown to be removed from solution by light kaolin (Batuyios & Brecht 1957; Bean & Dempsey 1971) and by magnesium trisilicate (Yousef *et al.* 1971). Beveridge & Hope (1967) showed that the preservative action of benzoic acid in Sulphadimidine Mixture Paediatric BPC was impaired due to adsorption of the preservative from solution by the undissolved sulphonamide. Horn *et al.* (1971) found that the concentration of benzoic acid decreased significantly for twelve out of fifteen suspensions containing commonly used pharmaceutical materials. Clarke & Armstrong (1972), who studied the influence of pH on the adsorption of benzoic acid by kaolin, noted that maximal adsorption occurred between pH 4.0 and 5.0. Yousef *et al.* (1973) examined the effect of eleven different materials on the bactericidal activities of benzyl alcohol, chlorbutol, chlorhexidine diacetate, chlorocresol, methyl *p*-hydroxybenzoate (methyl paraben) and then phenylmercuric nitrate. Magnesium trisilicate eliminated the activity of chlorbutol and methyl paraben and considerably reduced that of benzyl alcohol and chlorhexidine diacetate; polysorbate 80 antagonized the bactericidal action of all the agents. Horn *et al.* (1971) have also reported varying degrees of inactivation of methyl paraben and phenylmercuric nitrate solutions in the presence of fifteen insoluble solids. Allwood (1982), who investigated the effect of pH on the adsorption of parabens esters by magnesium trisilicate, showed that there was a correlation between the degree of ionization and adsorption. The ionized ester was adsorbed preferentially by the solid. In contrast, the paraben esters were not adsorbed by kaolin.

The effect of thirteen different particulate solids on the anti-microbial activity of methylbenzathonium chloride (Hyamine 10X), 2-bromo-2-nitro propane-1, 3-diol (Bronopol), 2.4.4′-trichloro-2′-hydroxydiphenyl ether (Irgasan DP 300) and 5-chloro-2-methyl-4-isothiazolin-3-one (Kathon CG) was studied by Myburgh & McCarthy (1980*a,b*). Hyamine 10X was adsorbed strongly by nearly all the materials; only light kaolin adsorbed significantly all four preservatives. Other antimicrobial agents studied for susceptibility to adsorption include cetylkonium chloride (Horn *et al.* 1971), cetylpyridinium chloride (Batuyios & Brecht 1957; Horn *et al.* 1971) and *o*-chlorbenzoic acid (Horn *et al.* 1971).

Chloroform does not appear to adsorb insoluble solids. Lynch *et al.* (1977) and Purkiss (1978) concluded, however, that sorption was responsible for the unexpectedly low supernatant concentrations of this preservative in a magnesium trisilicate mixture. Their evidence must be set against the normal methods of preparing such materials. Firstly, the method of manufacture of magnesium trisilicate and kaolin mixtures can allow losses of chloroform by volatilization greater than those that would occur during the preparation of a simple solution. Thus the low initial concentration of chloroform in the sedimented mixtures could be due, at least in part, to losses during manufacture rather than adsorption. Also, its loss from a magnesium trisilicate mixture following removal of portions might be expected to be reduced by the presence of a 'reservoir' of adsorbed chloroform when compared with a simple solution. Lynch *et al.* (1977) noted some loss from Ammonia and Ipecacuanha mixture and Sodium Chloride Mixture Compound BPC stored in unopened containers but none from a double-strength chloroform-water mixture. As both preparations are solutions, it is possible that some process other than sorption was involved.

Influence of excipients on preservatives

Many different types of excipients may be included in liquid medicines. A number of excipient-preservative interactions have been described.

Suspending and viscosity-adjusting agents

Suspending and thickening agents can reduce the antimicrobial activity of preservatives. The activity of the parabens esters, for instance, is reduced by Veegum (aluminium magnesium silicate/bentonite—Yousef *et al.* 1973) and sodium alginate (Myburgh & McCarthy 1980*b*). It has been shown also that gum tragacanth can suppress the activity of chlorbutol, methyl or propyl *p*-hydroxybenzoate (Eisman *et al.* 1957). This effect is very dependent on concentration (Taub *et al.* 1958). Interaction is difficult to quantify (Miyawaki *et al.* 1959). Alginates are anionic and may be incompatible with cationic preservatives. As the substances noted above are extracted from plants, they may be contaminated with micro-organisms some of which may have the potential to hydrolyze plant polymers.

Oils and emulsifying agents

Some oral liquids are formulated as oil-in-water emulsions, using suitable emulsifying agents. Many surface-active materials used for this purpose can form micelles. The preservative may migrate preferentially both into the oil phase and into micelles thereby becoming ineffective. This is a well recognized problem of phenolic preservatives and parabens esters. Migration into the oil phase can be calculated using oil-water distribution coefficients (Mitchell & Kazmi 1975). Some examples are given in Table 1. In general, migration into vegetable oil is far greater than into mineral oils. Unfortunately, the former are used in oral liquid emulsions. It should be noted that migration into the oil phase will depend on pH if the compound is a weak acid—phenolics for example. Only the non-ionized compound will be lipophilic. It is also true that this is generally the active form of the compound. Therefore migration into the oil phase is paralleled by loss of activity.

TABLE 1. *Typical vegetable oil-and-water partition coefficients for some preservatives commonly used in liquid medicines*

Preservative	Partition Coefficient
Chlorocresol	190.0
Methylparabens	5.8
Propylparabens	66.0
Butylparabens	280.0
Benzyl alcohol	1.3
Benzalkonium chloride	<1.0
Chlorhexidine acetate	0.04
Bronopol	0.11

Flavouring Agents

Sucrose in the form of a 67% syrup is the most common flavouring additive used to improve the palatability of many liquid medicines. Syrup BP is a suitable vehicle for many drugs used to treat throat and upper-respiratory tract conditions. Such a high concentration of sugar inhibits the growth of many micro-organisms because of the depression in the water activity coefficient (a_w). The a_w of Syrup BP is <0.8 which is less than the minimum (0.88–0.90) tolerated by facultatively halophilic micro-organisms such as micrococci or staphylococci. Such an environment is not protected totally from microbial attack, especially by osmophilic yeasts and xerophilic moulds. In addition, poor storage conditions with hot/cold cycling can cause condensation around the neck of a container and dilution of the syrup at the top of the contents. Although some additional safeguard is necessary even with concentrated syrups, the concentration of any additional preservative can usually be less than that used in an aqueous base with an a_w around 1.0. Thus it is common to add approximately half the normal strength of methyl parahydroxybenzoate to Syrup BP. If soluble sodium salts of parabens are used for this purpose, then a substantial increase in the pH of the weakly buffered Syrup may occur. Such an alkaline drift can cause loss of compatibility of drugs such as weak bases, especially the alkaloids. The precipitation of alkaloids in some commercial sources of Syrup BP, which has been observed in the authors' laboratory, was attributed to the raised pH of parabens syrup in which the sodium salt had been used.

As the inhibition of growth of micro-organisms is a function of water activity, the substitution of sucrose by other sweet tasting humectants such as sorbitol and glycerol does not compromise microbial safety. Indeed these alternatives are often preferred because of the association of dental caries with sucrose intake. Saccharin or aspartame as well as other flavouring agents are commonly included in oral liquid formulations to improve palatability. In general such additives have no appreciable effect on preservative efficacy.

Stabilizers

Small concentrations of specific stabilizers may be included in formulations, for example the anti-oxidant, sodium metabisulphite, is used to protect oxidizable drugs. It may have a small potentiating effect on the antimicrobial activity of some preservatives but generally provides no substantial protection against microbial attack. Likewise the small concentrations of ethanol used to dissolve a drug may enhance preservative efficacy. The pH of the preparation may also be important especially if a product has either a marked acid or alkaline reaction. As noted above, however, pH may have other effects relating to the chemistry of the preservative.

Containers

Dean (1978) has reviewed all aspects of the interactions between packaging and products. Glass containers do not interfere with the activity or availability of preservatives and glass is the preferred material for packing oral and topical liquid preparations. Plastic ones may interact with particular preservatives. Rubber stoppers may allow certain preservatives to escape from containers (p. 217). McCarthy (1971) reported that polyethylene containers caused substantial losses of some phenolics but not esters of parahydroxybenzoate. Polyvinylchloride and polypropylene appeared to be far more suitable for packaging pharmaceutical products.

Preservatives used in oral liquid medicines

Esters of parahydroxybenzoate (parabens)

The esters of parahydroxybenzoic acid are perhaps the most commonly used preservatives in liquid pharmaceuticals. As a result of their low systemic toxicity (Mathews *et al.* 1956) and common use for more than 50 years without any substantial adverse effects—there are rare reports of skin sensitivity—they are now used in oral liquid preparations also. The methyl ester alone or a combination of methyl and propyl compounds are in common use, the concentration [0.15–0.20% (w/v) total ester] depending on the formulation. The concentration (Table 2) may be reduced in a preparation with a pH below *ca.* 5.0 or in one with low a_w (<0.9). Activity is reduced substantially at pH values above *ca.* 8.0 and in addition the esters (pKa 8.4) will hydrolyze in alkaline conditions following ionization (Blaug & Grant 1974). Loss of antimicrobial activity is associated with sorption onto solids (p. 217). Because of the low solubility of the parabens in water, it is common practice to use the sodium salt. As this salt can cause a marked alkaline drift in poorly buffered products, it must be neutralized. It needs to be noted that parabens esters are incompatible with iron salts.

TABLE 2. *Parabens and preservative efficacy**

Preparation	Ester(s) concentration (% w/v)	Result
Kaolin mixture	methyl, 0.08† propyl, 0.02 (+chloroform)	Pass
	methyl, 0.16† propyl, 0.04	Pass
Magnesium trisilicate	methyl, 0.16† propyl, 0.04	Pass
Peppermint water	methyl, 0.16† propyl, 0.04	Fail
2% Methylcellulose gel	methyl, 0.2	Pass

*BP (Anon. 1980) Test.

The antimicrobial spectrum of the parabens esters is relatively wide. In general they are more active against Gram-positive bacteria than Gram-negative ones or moulds. There are some reports of resistance among *Pseudomonas* spp. (Hugo & Foster 1964; Beveridge & Hart 1970; Close & Nielsen 1976; Rosen *et al.* 1977).

Chloroform

The use of chloroform as an antimicrobial agent was first reported by Barrett (1874). In the same year, chloroform was introduced into the British Pharmacopeia as a solution in water [equivalent to 0.5% (v/v)]. Subsequently, the standard BP strength chloroform water became 0.25% (v/v) and, in 1911, the British Pharmaceutical Codex, BPC (Anon. 1973) cited a formula for Aqua Chloroformi Fortior [0.5% (v/v)] which was to become known as 'Duplex' or Double Strength Chloroform Water. The stated solubility of chloroform in water is 0.815% w/w (0.55% v/v) or 0.00123 mole fraction at 20 °C (Horvarth 1982). The BP (Anon. 1980) quotes a figure of 1 in 200 parts by volume. Double Strength Chloroform Water BP contains 0.5% (v/v) chloroform and is therefore at the limit of solubility. Chloroform spirit BP contains 5% (v/v) chloroform in 90% (v/v) ethanol, with which it is miscible. Considerable agitation is required to achieve complete dissolution of chloroform in an aqueous medium at a concentration intended for preservation, normally 0.25% (v/v).

Since its introduction as a preservative, chloroform has been included in many official formulae in the UK, particularly mixtures, linctuses and emulsions. The effectiveness of chloroform as a preservative against specific organisms was studied by Westwood & Pin-Lim (1972) and Richards *et al.* (1972). They concluded that chloroform was an effective antimicrobial agent at a concentration of 0.25% (v/v) providing one of its major drawbacks, volatility, was recognized. Sykes (1965) pointed out that evaporation from products at ambient temperatures could lead to the chloroform concentration falling below *ca.* 0.1% and to bacterial growth.

Lynch *et al.* (1977), who simulated 'in-use' conditions, estimated the rate of loss of chloroform from BPC mixtures, and measured the antimicrobial effect of chloroform against large and small inocula of four different genera of bacteria. They found that the rate of chloroform loss varied with the size of the container, the frequency of opening, and the amount of preparation present. Losses of between 30 and 60% of the original chloroform content were found to occur in bulk and dispensing containers which were opened regularly and portions of the contents removed during a 30-day period. The effect of removing portions of Magnesium Trisilicate Mixture BPC over a period of time was investigated by Purkiss (1978) and he too found that loss of chloroform varied with the volume of mixture removed from and the frequency of opening of a container. Loss was influenced also by the 'form' of chloroform used, namely Chloroform Water Double Strength BPC, chloroform emulsion or chloroform spirit. A reduction of up to 47% of initial chloroform content was found following removal of 20 × 100 ml samples of Magnesium Trisilicate Mixture from a 2 l container. A smaller loss (33%) was recorded when 6 × 300 ml portions were removed over a 2-month period. Losses of a similar order of magnitude occurred with Chloroform Water BPC and Chloroform Water Double Strength BPC. In mixtures containing chloroform emulsion or chloroform spirit, higher initial concentrations were attained and losses over a period of time were less than in those containing Double Strength Chloroform water.

The toxic nature of chloroform was recognized in 1945, when the nephrotoxicity to mice was first observed (Taylor 1977). He noted that the metabolites formed during chloroform breakdown are probably responsible for this toxicity. Lowry & Banzer (1978) analysed samples of commercial chloroform for the presence of chloromethyl ether (CME), a recognized carcinogen (Laskin *et al.* 1975). Lowry & Banzer (1978) surmised that CME formed during the production of methyl chloride would contaminate chloroform during the final stage of manufacture. If CME were formed, its contamination of chloroform was less than 1 ppm. Reports of tumour formation in the kidney and liver of rats and mice respectively led to the prohibition of chloroform as an ingredient of drugs and cosmetics by the Food and Drugs Administration (Anon. 1976) in the United States of America. The Committee on Safety of Medicines advocated another policy. It concluded in a report of the UK Medicines' Commission (Anon. 1979) that there was no evidence of chloroform being harmful to man when used in low dosage levels. Nevertheless the concern about carcinogenicity and toxicity, led to the recommendation that sale or supply would be restricted to medicinal products containing 0.5% (w/w or v/v) or less of chloroform.

The authors found that chloroform was not adsorbed by kaolin, magnesium trisilicate, light magnesium carbonate or Fuller's earth in aqueous suspension. In contrast substantial adsorption occurred with charcoal. Chloroform may also be salted-out by electrolytes such as sodium bicarbonate. Using a head-space analysis technique, it was proved that the major reason for poor preservative efficiency was due to losses by volatilization. This is illustrated in Fig. 1 which shows likely losses from a medicine bottle during repeated opening and closing to remove a dose. Losses during the manufacturing of pharmaceutical products could be substantial also if steps are not taken to minimize evaporation (Shaw 1984).

In conclusion, the susceptibility of chloroform to adsorption is low, when compared with many other preservatives. In this regard it is an appropriate choice for pharmaceutical products containing the substances noted above but not for those containing activated charcoal. When the BP (Anon. 1980) introduced discretionary use of antimicrobial agents, some manufacturers included esters of *p*-hydroxybenzoic acid into their formulations of official mixtures. As parabens have relatively low antimicrobial activity under alkaline conditions (Allwood 1982; p. 185), they have limited application. Although chloroform is largely unaffected by the concentration of hydrogen-ions and does not interact with solids, its effectiveness can be minimized by volatilization. Other manufacturers have used parabens in combination with chloroform. A preparation containing these two manifests a high initial antimicrobial activity and, even if chloroform is lost by evaporation, bacteriastasis is maintained by the other agent. This approach would appear to have

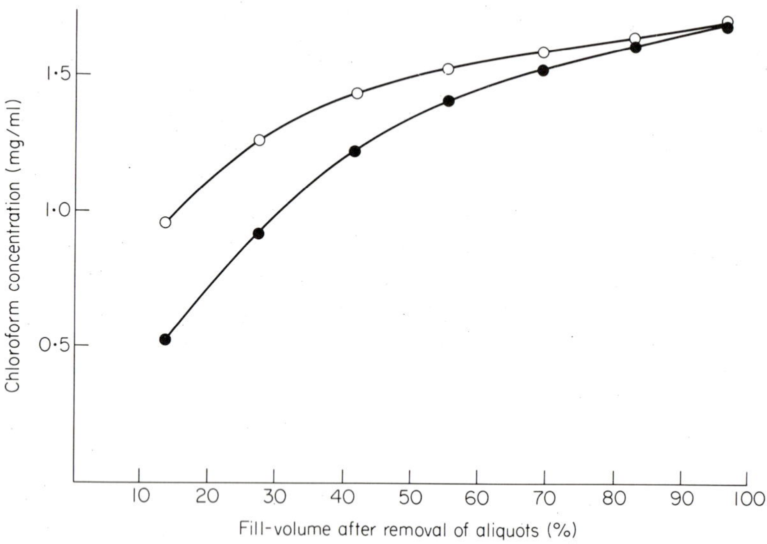

FIG. 1. Predicted liquid mixture chloroform concentrations with or without head-space extraction, after consecutive removal of medicine doses. (○) Medicine doses only removed, (●) medicine dose removed and head-space purged.

some merit. Chloroform has useful carminative and flavouring properties also. Some formulations would need to be reviewed, therefore, in terms of these aspects if chloroform was to be substituted by alternative antimicrobial agents.

Benzyl alcohol

Benzyl alcohol, a weak antimicrobial agent (MIC 0.3–0.65%), has been included in paediatric formulations at a concentration of 1–2% (v/v). It is slightly less active against Gram-negative compared with typical Gram-positive bacteria. It is most active at pH below 5.0 and least above pH 8.0. Benzyl alcohol is oxidized slowly in the presence of air or other oxidizing agents. Antimicrobial activity is reduced in the presence of non-ionic surfactants but the effect is far less than in the case of the parabens. Benzyl alcohol is not bound to methylcellulose. It is sorbed slowly by rubber and by polyethylene. Losses to PVC or polypropylene are negligible. While relatively free from toxicity after oral ingestion, parenteral use is not recommended, especially in infants, because of the likely occurrence of the well established toxic syndrome of benzyl alcohol and related compounds.

Benzoic acid and sodium benzoate

Benzoic acid is active only in the undissociated form. As the pKa of the acid is *ca.* 4.2, this particular acid is only active below pH 5.0. It is used, however, in some mixtures for children at a concentration of 0.1% (w/v). Its weak antimicrobial properties are exhibited against bacteria and moulds. At or below pH 5.0, typical MICs are in the range of 100 μg for Gram-positive bacteria to 1000–1600 μg for Gram-negative bacteria and between 400 and 2000 μg/ml for moulds. Benzoic acid is relatively stable but is incompatible with ferric, calcium and heavy metal salts. The activity is reduced substantially by non-ionic surfactants. It is absorbed by kaolin, depending on the pH of the suspension, with maximum absorption at *ca.* pH 5.0. Oral administration of large amounts of benzoic acid may cause severe gastric irritation. It may also cause allergic reaction and is an irritant to the skin, eyes and mucous membranes. It is, however, generally recognized as safe to use as a preservative in relatively low concentrations.

Conclusions

The preservation of liquid preparations, especially those for oral administration, is one of the more difficult microbiological problems facing the formulator. The relative complexity of these products, based on an aqueous vehicle intended primarily for administration to children and the elderly, introduces many difficulties in the path to acceptable levels of protection against microbial abuse. The mechanisms by which preservative efficacy is reduced or inhibited are many and varied. In addition such preparations can be heavily contaminated, either from raw materials or during the manufacturing process. The major problems that must be considered in the choice of a suitable preservative system may be summarized as follows:

1 Adsorption by insoluble solids in the preparation.
2 Inactivation due to reaction with drugs, excipient or adverse pH.
3 Insolubility or partitioning into the micellar or oil phase.
4 Inappropriate pH affecting antimicrobial activity.
5 Losses by evaporation from a multi-dose form.
6 Nutritional value of the formulation to micro-organisms, since many excipients are subject to microbial attack.
7 Lack of toxicity for oral administration, to children and the elderly in particular.

This formidable list explains why the choice is both narrow and imperfect. Certainly an ideal preservative system for these preparations does not exist and the task of the microbiologist is to balance antimicrobial efficacy against safety.

Acknowledgements

Mrs S. Negus and Mrs G. Griggs for technical assistance.

References

ALLWOOD, M. C. 1982 The adsorption of esters of *p*-hydroxybenzoic acid by magnesium trisilicate. *International Journal of Pharmaceutics* **11**, 101–107.

ANON. 1971 Report of Public Health Laboratory Service Working Party. Microbial contamination of medicines administered to hospital patients. *Pharmaceutical Journal* **207**, 96–103.

ANON. 1973 *British Pharmaceutical Codex*. London: Pharmaceutical Press.

ANON. 1976 *Food and Drugs Administration*. Federal Register 41, (126). pp. 268426.

ANON. 1979 Medicines Act Information Letter No. 24, London: Department of Health and Social Security.

ANON. 1980 *British Pharmacopoeia*. London: Pharmaceutical Press.

BAIRD, R. M., BROWN, W. R. L. & SHOOTER, R. A. 1976 *Pseudomonas aeruginosa* in hospital pharmacies. *British Medical Journal* **i**, 511.

BAIRD, R. M., CROWDEN, C. A., O'FARRELL, S. M. & SHOOTER, R. A. 1979 Microbial contamination of pharmaceutical products in the home. *Journal of Hygiene* **83**, 277–283.

BAIRD, R. M. & PETRIE, P. S. 1981 A study of microbiological contamination of oral medicaments. *Pharmaceutical Journal* **226**, 10–11.

BARRETT, F. J. 1874 untitled. *The Pharmaceutical Journal and Transactions*. December 5th.

BATUYIOS, N. H. & BRECHT, E. A. 1957 An investigation of the incompatibilities of quaternary ammonium germicides in compressed troches I. The adsorption of cetylpyridinium chloride and benzalkonium chloride by talc and kaolin. *Journal of the American Pharmaceutical Association* **46**, 524–531.

BEAN, H. S. & DEMPSEY, G. 1971 The effect of suspensions on the bactericidal activity of *m*-cresol and benzalkonium chloride. *Journal of Pharmacy and Pharmacology* **23**, 699–704.

BEVERIDGE, E. G. & HART, I. A. 1967 Inactivation of benzoic acid in sulphadimidine mixture for infants B.P.C. *Pharmaceutical Journal* **198**, 457–458.

BEVERIDGE, E. G. & HOPE, I. A. 1970 The utilization for growth and the degradation of *p*-hydroxybenzoate esters by bacteria. *International Biodeterioration Bulletin* **6**, 9–12.

BLANG, S. M. & GRANT, D. E. 1974 Kinetics of degradation of the parabens. *Journal of the Society of Cosmetic Chemists* **25**, 495–506.

CLARKE, C. D. & ARMSTRONG, N. A. 1972 Influence of pH on the adsorption of benzoic acid by kaolin. *Pharmaceutical Journal* **209**, 44–45.

CLOSE, J. & NEILSON, P. A. 1976 Resistance of a strain of *Pseudomonas cepacia* to esters of *p*-hydroxybenzoic acid. *Applied and Environmental Microbiology* **31**, 718–722.

DEAN, D. A. 1978 Some recent advances in the packaging of pharmaceuticals. *Drug Development and Industrial Pharmacy* **4**, v–vi.

EISMAN, P. C., COOPER, J. & JACOMIA, D. 1957 Influence of gum tragacanth on the bactericidal activity of preservatives. *Journal of the American Pharmaceutical Association* **46**, 144.

HORN, N. R., MCCARTHY, T. J. & PRICE, C. H. 1971 Interaction between preservatives and suspension systems. *American Perfumer and Cosmetics* **86**, 37–40.

HORVARTH, A. L. 1982 *Halogenated hydrocarbons*, p. 410. New York: Marcel Decker.

HUGO, W. B. & FOSTER, J. H. S. 1964 Growth of *Pseudomonas aeruginosa* in solutions of esters of *p*-hydroxybenzoic acid. *Journal of Pharmacy and Pharmacology* **16**, 209.

KALLINGS, L. O., RINGERTZ, O., SILVERSTOLPE, L. & ERNERFELDT, F. 1966 Microbial contamination of medical preparations. *Acta Pharmaca Suecica* **3**, 219–228.

LOWRY, S. R. & BANZER, J. D. 1978 Analysis of commercial chloroform for chloromethyl methyl ether by Fourier Transform Infrared Spectometry. *Analytical Chemistry* **50**, 1187–1189.

LASKIN, S., DREW, R. T., CAPPIELLO, V., KUSCHNER, M. & NEILSON, N. 1975 Inhalation carcinogenicity of *alpha* halo ethers. *Archives of Environmental Health* **30**, 70–72.

LYNCH, M., LUND, W. & WILSON, D. A. 1977 Chloroform as a preservative in aqueous systems. *Pharmaceutical Journal* **219**, 507–510.

MATHEWS, C., DAVIDSON, J., BAUER, E., MORRISON, J. L. & RICHARDSON, A. P. 1956 *p*-Hydroxybenzoic acid esters as preservatives II Acute and chronic toxicity in dogs, rats and mice. *Journal of the American Pharmaceutical Association* **45**, 260–267.

MCCARTHY, T. J. 1971 Aspects of preservative inactivation. *South African Pharmaceutical Journal* **6**, 21–28.

MIYAWAKI, G. M., PATEL, N. K. & KOSTENBAUDER, H. B. 1959 Interactions of preservatives with macromels. III Parahydroxybenzoate esters in the presence of hydrophilic polymers. *Journal of American Pharmaceutical Association* **48**, 315–318.

MITCHELL, A. G. & KAZMI, S. J. A. 1975 Preservative availability in emulsified systems. *Canadian Journal of Pharmaceutical Sciences* **10**, 67–68.

MYBURGH, J. A. & MCCARTHY, T. J. 1980*a* Inactivation of preservatives in the presence of particulate solids. *Pharmaceutisch Weekblad* **2**, 137–142.

MYBURGH, J. A. & MCCARTHY, T. J. 1980*b* The influence of suspending agents on preservative activity in aqueous solid/liquid dispersions. *Pharmaceutisch Weekblad* **2**, 143–148.

PURKISS, R. 1978 Loss of chloroform from magnesium trisilicate mixture under simulated in-use conditions. *Journal of Clinical Pharmacy* **2**, 163–170.

RICHARDS, R. M. E., MCBRIDE, R. J. & GUNN, M. A. 1972 Preliminary investigation of the preservative properties of 3-phenylpropanol. *Journal of Pharmacy and Pharmacology* **24**, 158–159.

ROSEN, W. E., BERKE, P. A., MATZIN, T. & PETERSON, A. F. 1977 Preservation of cosmetic lotions with imidazolidinyl urea plus parabens. *Journal of the Society of Cosmetic Chemists* **28**, 83–87.

Shaw, R. J. S. 1984 *Interaction of chloroform with insoluble solids and its consequences for preservation.* M. Phil Thesis. University of Bradford.

Sykes, G. 1965 *Disinfection and sterilization.* 2nd edn. p. 456 London: E. & F. N. Spon.

Taub, A. W., Meer, A. & Clausen, L. W. 1958 Conditions for the preservation of gum tragacanth. *Journal of American Pharmaceutical Association* **47**, 235–237.

Taylor, D. C. 1977 Chloroform as a preservative. *Pharmaceutical Journal* **219**, 541.

Yousef, R. T., El-Nakeeb, M. A. & Labib, S. 1971 Effect of magnesium trisilicate on the antibacterial activity of benzalkonium chloride. *Acta Pharmaca Suecica* **8**, 303–310.

Yousef, R. T., El-Nakeeb, M. A. & Salama, S. 1973 Effect of some pharmaceutical materials on the bactericidal activities of preservatives. *Canadian Journal of Pharmaceutical Sciences* **18**, 54–56.

Westwood, N. & Pin-Lim, B. 1972 Survival of *Escherichia coli, Staphylococcus aureus, Pseudomonas aeruginosa* and spores of *Bacillus subtilis* in B.P.C. mixtures. *Pharmaceutical Journal* **208**, 153–154.

Preservation of Sterile Pharmaceutical Products

J. R. Furr and D. T. Rogers

Welsh School of Pharmacy, University of Wales Institute of Science and Technology, PO Box 13, Cathays Park, Cardiff, UK

Theoretical Considerations

Microbial spoilage of pharmaceutical products has been known for many years. Spoilage may result in deterioration of the product due to loss of potency or induction of an infection in the user (Beveridge 1975, 1983; Ringertz & Ringertz 1982; Stephenson *et al.* 1984).

Certain pharmaceutical products require to be sterile on manufacture. These can be classified into two groups:

1 Single dose preparations where any unused material is discarded immediately after use.

2 Multiple dose preparations which permit successive doses to be withdrawn from the same container.

The various types of products that require to be sterile are given in Table 1.

Re-usable products, particularly those in an aqueous vehicle, are at risk from accidental contamination during use. Products of this type are multidose injections, eye drops, domicillary eye lotions, contact lens solutions, ear drops, topical ointments, creams and lotions. It is therefore necessary to include, as part of the product formulation, chemical compounds which are designed to prevent the multiplication of any contaminating micro-organisms (p. 215). In this context, these agents are often referred to simply as preservatives. Before considering these agents it is necessary to outline briefly the methods available for the sterilization of pharmaceuticals.

The British Pharmacopoeia (BP; Anon. 1980*a*) recognizes the following processes and the method used will depend on the stability and nature of the ingredients in the product.

Preservatives in the Food,
Pharmaceutical and Environmental
Industries

0–632–01727–9

TABLE 1. *Classification of sterile products*

Parenteral products	Single dose injections Multidose injections Large volume intravenous fluids products Total parenteral nutrition Immunological products
Eye preparations	Drops (single dose/multidose) Ointments (single dose/multidose) Lotions (first aid/domicillary) Contact lens solutions
Miscellaneous products	Surgical powders/equipment Irrigation solutions Syringes, needles etc. Topical preparations (creams, lotions, ointments etc.)

Heating in an autoclave

Sterilization by heating in saturated steam under pressure should be used wherever possible for aqueous preparations (Alder & Simpson 1982). The recommended combinations of temperatures and times to be used are given in Table 2.

TABLE 2. *Recommended times and temperatures to be used in the autoclaving process*

Holding temperature (°C)	Minimum holding time (min)
115–118	30
121–124	15
126–129	10
134–138	3

Heating with a bactericide

This process may be used for aqueous preparations the ingredients of which are unstable at the temperatures achieved in the previous process (Pharmaceutical Codex, PC; Anon. 1979). It involves heating the whole of the container to a temperature of 98–100 °C for a holding time of 30 min. The preparation contains a suitable concentration of a chemical agent (bactericide) to enable the product to be rendered free from any viable micro-organisms. The nature of these chemical agents depends on the type of product to be sterilized (p. 215).

Filtration

This process is used for aqueous products the ingredients of which are unstable at the temperatures achieved in the other two processes. It involves the passage of the liquid through a filter or combination of filters designed to remove micro-organisms (Denyer *et al.* 1982).

Dry heat

This is used for heat-stable non-aqueous preparations, powders and certain impregnated dressings. The product should be maintained at the recommended temperatures for the appropriate time (Alder & Simpson 1982) (Table 3).

TABLE 3. *Recommended minimum holding times for dry heat sterilisation at different temperatures**

Holding temperature (°C)	Holding time (min)†
160	60
170	40
180	20

* Anon. (1962). † Holding times are for loads that have been brought to the required temperature before the sterilizing time commences.

Sterilization by ionizing radiation or ethylene oxide

These are two alternative processes which are recognized for specific pharmaceutical products where heat sterilization or filtration are not possible. The principles and practice of these methods have been discussed by Christensen *et al.* 1982; Russell 1982; Christensen & Kristensen 1982; Davies & Hobbs 1983; Richards *et al.* 1984. The requirements for the preservation of the various types of sterile aqueous products is given in Table 4.

It is proposed to consider each of these types of product individually and to examine the factors that govern the choice of the anti-microbial agents to be used. It is first necessary, however, to clarify and standardize the various terms used to describe these agents when consulting different pharmacopoeias (Table 5).

The term antimicrobial preservative (AP) will be used in this chapter unless the agent is part of the sterilization process in which case the term bactericide will be used. It should be noted that the bactericide will act as an AP subsequent to the sterilization process.

TABLE 4. *Preservative requirements for sterile products*

Type	Form		Preservation
Injections	Single dose		No*
	Multidose		Yes
	Large volume intravenous		No
Eye preparations	Drops	(single dose/	No*
		multidose)	Yes
	Lotions	(first aid	No
		domicilary)	Yes
	Ointments	(single dose/	No
		multidose)	Yes
	Contact lens solutions		
	(single dose/		No*
	multidose)		Yes
Solutions	Irrigation (single dose)		No

* Included if part of the sterilization process or if the product is prepared aseptically.

TABLE 5. *Various terms used in Pharmacopoeias to describe antimicrobial agents*

Source	Terminology	Specified uses
British Pharmacopoeia (Anon. 1980*a*)	Antimicrobial preservative	Inhibition of microbial growth
Pharmaceutical Codex (Anon. 1979)	Bactericide	Inhibition of microbial growth Part of sterilization process
United States Pharmacopoeia (Anon. 1980*b*)	Antimicrobial preservative	Inhibition of microbial growth
European Pharmacopoeia (Anon. 1975)	Antimicrobial agent	Inhibition of microbial growth

Factors influencing the choice of antimicrobial preservatives

The substance chosen should have: a broad spectrum of activity against micro-organisms; be non-toxic at the dose to be administered; be active over a wide pH range; be uninfluenced by the containers or closures used, and be stable in aqueous solution at high temperatures and during storage of a product. No one substance possesses all these requirements in relation to every injection and hence a choice must be made depending on the nature of the injection. Most Pharmacopoeias recommend certain APs together with suitable concentration (Table 6).

TABLE 6. *APs recommended for inclusion in multidose injections and aseptically prepared aqueous injections*

	Concentration (% w/v) recommended in			
Antimicrobial preservative	BP* Anon. (1980*a*)	PC† Anon. (1979)	EP‡ Anon. (1975)	USP§ Anon. (1980*b*)
Phenol	–	0.5¶	–	0.5
o-Cresol	0.3	0.3	0.3	0.5
Chlorocresol	0.1	0.1	0.1	0.5
Chlorbutol	–	0.5	–	0.5
Phenylmercuric nitrate (PMN)	0.002	0.001	0.002	0.01
Phenylmercuric acetate (PMA)	0.002	0.001	–	0.01
Phenylmercuric borate (PMB)	0.002	–	0.002	0.01
Benzyl alcohol	1.0	–	1.0	–

* British Pharmacopoeia; † Pharmaceutical Codex; ‡ European Pharmacopoeia, and § United States Pharmacopoeia. ¶ Maximum Concentrations.

Injectable Preparations

The BP (Anon. 1980*a*) defines injectable preparations as sterile products containing one or more medicaments intended for administration into body tissues. It further classifies these preparations into:

1 Injections: Sterile solutions or suspensions in water for injections, a non-aqueous liquid or a mixture of the two if they are miscible. They may also be emulsions.
2 Intravenous infusions: Solutions or emulsions with water as the continuous phase.
3 Powders for injection: Sterile solid substances which are reconstituted with the prescribed volume of the appropriate sterile liquid prior to use.

The BP (Anon. 1980*a*) recognizes that certain products require the addition of an AP and, in addition to multidose injections, states that aqueous preparations that cannot be sterilized in their final containers and are prepared by using aseptic precautions may also contain a suitable AP. For example, single dose injections may under these conditions contain an AP even though it is not part of the sterilization process.

Antimicrobial Activity and Toxicity

The PC (Anon. 1979) states that any other substance used must have a bactericidal activity at least equivalent to that of an aqueous solution containing 0.5% (w/v) of phenol. There are certain limitations on the use of APs in injections because of potentially toxic side-effects. The BP (Anon. 1980*a*) and PC (Anon. 1979) recommend that the maximum volume of an injection containing an AP administered at any one time should be 15 ml. The United States Pharmaeopeia (USP; Anon. 1980*b*) recommends 5 ml. Because the meninges are readily irritated and prone to inflammation, the addition of any AP to injections which give access to the cerebrospinal fluid is precluded. Intra-cardiac and intra-arterial injections also must not contain an AP.

It is clear that there is considerable variation in both agents and concentrations recommended by various standard texts. The final concentration of any substance is governed by its potential toxicity linked to its antimicrobial activity. The Addendum (Anon. 1982) to the BP (Anon. 1980*a*) Appendix XVIC, specifies a test for the Efficacy of Antimicrobial Preservatives in Pharmaceutical Products. This should interrelate toxicity and antimicrobial activity (see later).

Effects of pH

As most drugs in common use are bases and as the pH of the injections ranges between 3.5 and 6.0, an AP must be active in acid conditions. All of the phenolics are active in the unionized state, pK_a of *ca.* 10. Their activity should not be affected therefore by pH values below 9.0.

Knowledge of the pK_a of a preservative allows the calculation of the fraction undissociated at any given pH and hence the protection afforded at that pH:

$$\text{Fraction of undissociated preservative} = \frac{1}{1+\text{antilog}\,(\text{pH}-\text{p}K_a)}$$

Chlorbutol is less active above pH 5.0 and unstable above pH 6.0. Phenylmercuric nitrate, however, is only active above pH 6.0 (Wallhausser 1974). In contrast Thiomersal is more active in acid conditions.

Effect of concentration

Kinetic studies involving the effect of concentration on the activity of antimicrobial substances have used a symbol which is termed the concentration exponent (p. 281). It is a measure of the effect of changes in concentration on cell death rate. It can be determined by measuring the time necessary to produce a comparable degree of death of a bacterial suspension at two different concentrations of the antimicrobial agent. For example, if C_1 and C_2 represent the two concentrations and t_1 and t_2 the respective time to reduce the population to a similar degree, then:

$$C_1 t_1 = C_2 t_2 \quad \text{or} \quad \eta = \frac{\log t_2 - \log t_1}{\log C_1 - \log C_2}$$

Substances with high η values require a marked increase in time to achieve a comparable kill if their concentration is decreased. Substances with low η values are much less influenced (Table 7). A knowledge of η is essential in ensuring adequate preservative levels in pharmaceutical products and to ensure that the agent has been inactivated in the sterility testing of such products (p. 227).

TABLE 7. *Concentration exponents (η values) of antimicrobial preservatives used in injections**

Substance	η value	Increased time factor when concentration is reduced to 50%
Phenols	6	2^6 i.e. 64 ×
Mercury compounds	1	2^1 i.e. 2 ×
Alcohols	10	2^{10} i.e. 1024 ×

* Based, in part, on Bean (1967)

Effect of containers and closures

All the recommended agents react in some way or another with the container systems used for multi-dose injections. Because of leaching of hydroxyl ions from the glass, the main effect is on the pH of the product. This can be overcome by using containers that comply with a 'limit test' for alkalinity. If the pH of the product is very low initially, however, and either little or no AP has been included because of the preserving action of the low pH, then leaching could have a significant effect on the product.

Multidose containers are sealed by a rubber closure to exclude micro-organisms. Both natural rubber and synthetic polymers absorb many compounds, including APs (Weiner 1955; Wing 1955, 1956*a*, *b*). Substances may also diffuse through the rubber and be volatilized from the surface of the closure exposed to the atmosphere (p. 215). After allowing for equilibration between natural rubber and AP solutions, which takes up to one month at room temperature, Royce & Sykes (1957) found that in all cases a proportion of the AP was absorbed into the rubber. Using the BP (Anon. 1980*a*) recommended concentrations, they found that the amount of agent taken up by 1 g of natural rubber from 3 ml solution was (%): phenol 25, cresol 33, chlorocresol 85, chlorbutol 80–90 and phenylmercuric nitrate 95. Similar results were obtained with other types of rubber. The time taken for equilibrium to be reached was reduced by a rise in temperature. It should occur within 30 min at 115 °C. This may not be a sufficient treatment to saturate the rubber closure. In order to minimize absorption from an injection, the BP (Anon. 1980*a*) recommends that rubber closures should be autoclaved at 115 °C for 30 min in double the concentration of the AP to be included in the injection and that at least 2 ml of solution be supplied per g of rubber. The closure should be stored in the AP solution until used. This treatment still does not prevent entirely loss of the AP from injections. For example 40% of phenol and 55% of chlorocresol were lost after one year from a solution despite prior saturation of the rubber closure. As both these substances have high values, such losses pose a serious problem from the viewpoint of adequate maintenance of preservative activity in products. Even though both agents were lost through vapourisation from the outer surface of the saturated closure, attempts to prevent loss by covering this surface with wax, coated foils or metal overseals were unsuccessful. The use of closures made from materials with low affinity for an AP (Lachman *et al.* 1964), increasing the thickness of and/or decreasing the surface area of closures are alternative approaches to this problem. The logical compromise is to relate AP loss with time and to print a suitable expiry date on the product. It is also wise to minimize the number of doses in a container. Indeed the BP (Anon. 1980*a*) recognizes this and suggests that a container should not contain an excessive number of doses. This advice inter-relates with the danger of accidentally contaminating an injection when it is withdrawn from the container.

Other factors influencing the choice of antimicrobial preservatives

In addition to the factors already mentioned, several others may influence the choice of the AP.

The problem of compatibility with ingredients

The BP (Anon. 1980*a*) states that 'the chosen agent should not interfere with the therapuetic efficiency of the drug'. Incompatibility data have been published by Carter (1974), Martindale (1982) and Remington (1985).

Suspensions

The AP may be adsorbed onto the insoluble material of a suspension thereby reducing its concentration in the aqueous phase (Armstrong & Clarke 1972). Bacterial suspensions, as used in vaccines, may exhibit this phenomena. A 'reaction' between an AP and the components on the cell surface may also diminish the concentration of the 'free' AP in a product.

Surface-active agents

Non-ionic surface-active agents are known to reduce markedly the antimicrobial action of many compounds (Evans & Dunbar 1965). Interactions may result in micellar solubilization or complex formation between molecules. Low concentrations of polysorbates, tritons and tergitols have been shown to enhance the antimicrobial activity of certain agents (Allwood 1973).

Partitioning between oil and water

As micro-organisms reside/multiply in the aqueous phase of a multiphasic system, an adequate concentration of AP must be maintained in this phase. Bean (1972) derived the following equation whereby the concentration in the aqueous phase can be calculated:

$$C_w = C(\Psi + 1)/(K^{\circ}{}_w\Psi + 1)$$

C_w, concentration AP in aqueous phase; C, total preservative concentration, Ψ oil/water ratio, and $K^{\circ}{}_w$, partition coefficient. No allowance is made for the nature and effect of the emulgent in the system in this formula.

Metal ions

The activity of APs may be reduced, enhanced or remain unchanged by the presence of cations (Galbraith & Miller 1973). The antibacterial activity of many APs is potentiated by ethylenediame tetra acetic acid (Wilkinson 1975; Russell & Furr 1977).

Oily products in multiple-dose containers may include (% w/v): phenol, 0.5%; cresol, 0.3%, or chlorocresol 0.1%. The PC (Anon. 1979) recognizes that 'these bactericides may afford some protection from contamination with vegetative micro-organisms but are ineffective against sporing organisms'.

Practical Aspects

Although their use in British hospitals is now discouraged, multiple-dose injections have certain practical advantages for the user when compared with single-dose ampoules. Many pharmacopoeias, however, condone the use of the former and the PC (Anon. 1979), BP. (Anon. 1980*a*) and USP (Anon. 1980*b*) now set specific standards regarding the effectiveness of the antimicrobial preservative to be included in the product. For example the PC (Anon. 1979) makes the general statement that the substance should be capable of sterilizing an injection within 3 h after inoculation with 10^6 vegetative cells of unspecified organisms/ml.

The Addendum (1982) to the BP (Anon. 1980*a*), Appendix XVIC specifies a test for the 'Efficacy of Antimicrobial Preservatives in Pharmaceutical Products'.

This monograph does emphasise, however, that the inclusion of an AP should not be used as a substitute for good manufacturing procedures:

> 'The testing procedure described in this Appendix is intended as a means whereby, during the development of a formulated product, the efficacy of any antimicrobial preservative included in that product may be assessed. It is intended for application to those products that may support the growth of microbial contaminants in the absence of a preservative system. The organisms specified for use in the testing procedure are intended to be representative of those that might be expected to be found in the environment in which the preparation is manufactured, stored and used. The designated organisms should be supplemented by other strains or species, especially those likely to be found in the conditions under which a particular product is made or used, or which might offer a particular challenge to the type of product being tested'.

Single strain challenges (rather than mixed cultures) should be used throughout.

Media

The cultivation media used for the revival of the specified organisms should be those described in the text but additional media may also be used or alternatives used where organisms other than the specified strains are employed. The recommended media are:

Soya tryptone agar. This contains (g/l): Pancreatic digest of casein, 15; papain digest of soya beans, 5; NaCl, 5; agar, 15; pH 7.2–7.4.

Sabouraud agar. This contains (g/l): D-glucose, 40; peptone, mycological or commercial, 10; agar, 15; pH 5.4–5.6.

Peptone water (0.1%). This contains (g/l): peptone, 1; NaCl, 8.9; pH 7.2–7.4.

The medium used in the preparation of pour plates should be that stated in the text, except that additional media may be used or alternatives used where organisms other than the specified strains are employed. The organisms should be maintained in the laboratory as recommended by the Curator of the appropriate culture collection.

Organisms

The following organisms are used in the test: *Aspergillus niger* IMI 149 007 (ATCC 16404); *Candida albicans* NCPF 3179 (ATCC 10231); *Pseudomonas aeruginosa* NCTC 6750 (NCIB 6750; ATCC 19429); *Staphylococcus aureus* NCTC 10788 (NCIB 9518; ATCC 6538).

Inoculum

For each bacterial strain, inoculate the surface of soya tryptone agar from a suitable, recently-prepared stock culture; for yeasts and moulds, use Sabourand agar.

Several subcultures may be needed after revival before an organism is in its optimal state. Incubate bacterial cultures at 30–35 °C for 18–24h, the culture of *Candida albicans* at 20–25 °C for 48 h, and the culture of *Aspergillus niger* at 20–25 °C for 7 d. Harvest the bacteria and *C. albicans* using 0.1% peptone water to wash the surface growth into a suitable receptacle and then dilute with the same liquid to reduce the microbial count to *ca.* 1×10^8 organisms per ml. To harvest *A. niger*, use 0.1% peptone water containing 0.05% (w/v) of polysorbate 80, filter through glass wool if necessary, and adjust to a microbial count of *ca.* 1×10^8 organisms per ml using 0.1% peptone water. Suspensions should be used immediately.

Test procedure

Introduce into portions of the product being examined sufficient of a suspension of the micro-organism to achieve a final concentration of *ca.* 1×10^6 organisms per g or per ml; use separate portions for each micro-organism. At the same time, introduce the same volume of inoculum into separate equivalent quantities of 0.1% peptone water for each organism (control). Ensure that the micro-organisms are evenly dispersed throughout the portion of the product being examined. The volume of the inoculum should not exceed 1% of the sample weight or volume. Store the inoculated product at 20–25 °C, and protect from light.

Remove 1 g or 1 ml quantities of the inoculated products at 0, 6, 24, and 48 h; 7, 14 and 28 d, as appropriate. The control preparations should be examined at the same time as the zero-hour sample of the inoculated product and, using the same procedure, in order to determine the viable counts of the cultures used and the suitability of the media used for their growth.

Introduce each portion into 9 ml of 0.1% peptone water. This may contain up to 10% (w/v) of polysorbate 80 but, when concentrations of the surface agent in excess of 1% are used, the ability of the medium to support the growth of viable organisms should be confirmed. Specific antagonists or neutralizing agents for preservatives present in the product may also be incorporated but, where such materials are used, the ability of the system to support the growth of the test organism should be confirmed. Disperse the material, using sterile glass beads if necessary. Dilute to an appropriate extent using 0.1% peptone water (which may contain preservative antagonists or neutralizing agents), and use 1 ml portions of these dilutions to prepare (in duplicate) pour-plates using soya tryptone agar (which may also contain preservative antagonists or neutralizing agents). If the preservative system present in the product cannot be suitably antagonized or neutralized, suitable dilutions should be made to prevent carry-over of inhibitory concentrations of preservative into the nutrient agar. Due allowance should be made for the reduced sensitivity in the recovery of small numbers of viable organisms when dilution procedures are employed. Incubate the pour plates at 30–35 °C for 3 d for bacteria and at 20–25 °C for 5 d for moulds and yeasts.

Determine the average number of colony-forming units of each test organism recovered after incubation for 3 d and, for moulds and yeasts, at 5 d also, and estimate the total number of viable test organisms per ml or per g of the original product taking into account the subsequent dilutions. A separate note should be made of results obtained with organisms other than those specified under 'Organisms'. A separate note should also be made of any colony-forming units not having the usual morphological characteristics of the test organism. Such organisms should be identified. The repeated introduction of inocula of the same organism into the same sample of the product should be considered in appropriate cases for example creams.

In all cases, the antimicrobial activity of the total product (including container) should be investigated. Such investigations may be carried out on samples removed from the container immediately prior to testing. In choosing a container it should be borne in mind that preservatives may be adsorbed by the container or closure.

Antimicrobial precautions

Challenge tests should be conducted under conditions that prevent accidental contamination of the product during the test. The precautions taken to prevent contamination should not affect the survival of organisms in the product being examined.

Interpretation

A product may be considered to be effectively preserved if laboratory and production batches satisfy the relevant criteria (p. 223) over the claimed shelf-life and the likely period of use. While effective product preservation, as defined by these criteria, is highly desirable, it is only one aspect of pharmaceutical formulation. It is recognized that for a number of products, e.g. certain preparations of biological origin, the criteria are unlikely to be achieved except at the expense of some other property of equal or greater importance. The following criteria of acceptance should therefore be applied after due account has been taken of any special considerations relevant to specific products.

Injectable preparations intended for use on more than one occasion

Bacteria. The number of organisms recovered per ml is reduced by a factor of not less than 10^3 within 6 h of challenge, and no organism is recovered from 1 ml at 24 h and thereafter.

Moulds and yeasts. The number of organisms recovered per ml is reduced by a factor of not less than 10^2 within 7 d of challenge, and there is no increase thereafter.

The test period should extend normally over not less than 28 d. In practice, it is

often found that repeated testing for much longer periods is necessary to assess adequately the preservative efficacy of a particular formulation. A period of 3 months for such testing is not excessive. In any case, it is advisable to examine the product for maintenance of preservative efficacy at the end of its claimed shelf-life to ensure that such activity has not been impaired by storage.
The USP XX (Anon. 1980*b*) states:

> 'Antimicrobial preservatives are substances added to dosage forms to protect them from microbial contamination. They are used primarily in multiple-dose containers to inhibit the growth of micro-organisms that may be introduced inadvertently during or subsequent to the manufacturing process. Antimicrobial agents should not be used solely to reduce the viable microbial count as a substitute for good manufacturing practice. Situations may arise, however, where their use may be required to minimize proliferation of micro-organisms. It should be recognized that the presence of dead micro-organisms or the metabolic by-products of living micro-organisms may cause adverse reactions in sensitized persons.
>
> Any antimicrobial agent may exhibit the protective properties of a preservative. However, all useful antimicrobial agents are toxic substances. For maximum protection of the consumer, the concentration of the preservative shown to be effective in the final packaged product should be considerably below the concentrations of the preservative that may be toxic to human beings.'

Until the publication of USP XVIII (Anon. 1970) there were no official methods for determining the efficacy of APs in pharmaceutical formulations. The USP XX procedure requires the product to be challenged with each of five organisms: *Staphylococcus aureus* ATCC 6358; *Pseudomonas aeruginosa* ATCC 9027; *Escherichia coli* ATCC 8739; *Aspergillus niger* ATCC 1 64 4, and *Candida albicans* ATCC 10231. The number of organisms set as the challenge is between 10^5 and 10^6 ml of the product and the inoculated formulations are incubated at 20–25 °C or at the storage temperature specified on the label. The containers are examined at 7, 14, 21 and 28 d after inoculation and the number of viable organisms remaining determined by a plate count procedure in conjunction with an observation regarding the appearance of the product.

The effectiveness of the preservative is determined by the following three criteria:

1. The concentrations of viable bacteria are reduced to not more than 0.1% of the initial concentrations by the fourteenth day.
2. The concentrations of viable yeasts and moulds remain at or below the initial concentrations during the first 14 d.
3. The concentration of each test organism remains at or below these designated levels during the remainder of the 28 d test period.

The European Pharmacopoeia (Anon. 1975) states that:
'multidose injections must contain a suitable antimicrobial agent'. There is no statement regarding expected levels of activity of any chosen preservative.

Ophthalmic preparations

Eye drops

The BP (Anon. 1980*a*) defines eye drops as sterile solutions or suspensions of one or more medicaments intended for instillation into the conjunctival sac for diagnostic or therapeutics purposes. It also specifies that:

> 'aqueous preparations supplied in containers intended for use on more than one occasion contain suitable APs at appropriate concentrations except when the preparation itself has adequate antimicrobial properties'.

Eye drops intended for use in surgical procedures should not contain any added AP and must be supplied in containers intended for use on ONE occasion only.

The BPC (Anon. 1979) states that the APs included in multiple containers should have bactericidal and fungicidal activity and that suitable agents are PMN and PMA (Table 6) at a concentration of 0.002% (w/v), benzalkonium chloride and chlorhexidine acetate at 0.01% (w/v). The choice of the AP used depends on its compatibility with the medicament and the duration of the period during which the eye drops are to be used. In local anaesthetic eye-drops, which reduce the blink reflex of the eye, or in eye-drops for patients with exposure keratitis, benzalkonium chloride is an unsuitable preservative because of its surface activity on the precorneal film.

The significant difference between eye drops and injections with respect to APs lies in the strength of the AP. Only one strength is recommended irrespective of the method of sterilization as this reflects the increased chance of contamination 'in use' of eye drops compared with injections. The BP (Anon. 1980*a*) also allows the use of thiomersal 0.01% (w/v). Like the rubber closures used in multidose injections (p. 217), the rubber teats used in eye drops should be impregnated and stored in solutions of the selected AP. Care must also be taken that the chosen AP is compatible with both the medicament, the container, the plastic closures and the teats used. As benzalkonium chloride is incompatible with natural rubber, silicone rubber should be used. As the latter is permeable to water vapour, and as a steady loss of water can occur from the product, the storage of the filled container is limited to 3 months. If eye-drops utilizing these teats are required to be stored for longer than 3 months they should be sealed initially with a plain phenolic cap fitted with a suitable liner. The complete dropper assembly should be supplied separately as a wrapped, sterilized unit. This unit should be exchanged for the cap immediately before use. Only a limited number of APs can be used in eye drops as it is essential that, in addition to having adequate antimicrobial properties at the concentration used, they must be non-irritant to the eye.

Eye lotions

These are defined in the PC (Anon. 1979) as '. . . sterile aqueous solutions used, usually undiluted, for bathing the eyes'. Two types of eye lotions are recognized:

1 Sterile aqueous solutions which contain no AP. These are to be used once only for first aid or for a period of treatment not longer than 24 h.
2 Aqueous solutions (sterile when issued) containing an AP. These are designed for intermittent domicillary use for up to 7 d.

Eye ointments

The BP (Anon. 1980*a*) described these as '. . . sterile semi-solid preparations of homogenous appearance intended for application to the conjunctiva'. They may be supplied as single dose units supplied in containers that facilitate their administration without contamination or in collapsible tubes of metal or plastic which can be reused. The BP (Anon. 1980*a*) states that an AP may be included in the latter type of product. Both with eye lotions and eye ointments, all the pharmacopoeias are vague and do not specify any preservatives. When a preservative is selected, irritancy to the eye must be a major consideration.

Contact lens solutions

The corneal contact lens are made of plastic materials and are designated 'hard' (hydrophilic) or 'soft' (hydrophobic) lenses according to their physical characteristics. Whatever type is used, a high standard of hygiene is required to minimize the possibility of infection to the eye due to lens insertion. Specially formulated solutions have been developed for users of contact lenses and each solution has been formulated for a specific purpose. A list of these solutions and their purposes is given in Table 8.

A wide range of APs are used in these solutions, frequently in combination and in conjunction with 0.1% (w/v) ethylenediamine tetra acetic acid (EDTA). Benzalkonium chloride (0.004–0.01% w/v) is the antimicrobial agent used in many formulations, often in conjunction with chlorhexidine (0.0025–0.006% w/v) or with EDTA. Thiomersal is also used widely but has the disadvantage of being absorbed rapidly from solution by certain types of plastic containers. It is also incompatible with EDTA when the pH is below 7.2. Some solutions use chlorbutol (0.24–0.5% w/v). Hydrophilic soft lenses absorb the AP and care must be taken to add the minimum concentration of the AP that is required to be effective. Such a concentration minimizes any irritancy produced by the release of the agent when the lenses are worn.

TABLE 8. *Function of different types of solutions used in association with contact lenses*

Type	Function	Antimicrobial preservative
Soaking & cleaning solutions*	Remove ocular debris, protein deposits, dust particles etc.	Yes
Storage & disinfecting solutions*	Storage for lenses when not in use	Yes
Wetting & cushioning solutions*	To enable lachrymal fluid to wet rapidly the contact lens surface permitting comfortable wear.	Yes
Hydrating and preserving solution†	Hydration of lens	Yes
Cleansing solution	Cleansing and hydration after removal of lens from the eye	Yes

* Hard lenses. † Soft lenses.

Interpretation of efficacy test

The PC (Anon. 1979) states that in the absence of published standard the onus rests on the manufacturer to demonstrate by adequate testing procedures that the preparations can efficiently combat a level of contamination and within a period of time which is consistent with contact lens practice. The BP Addendum (Anon. 1982), however, does specify that opthalmic preparations must comply with a specific challenge with appropriate micro-organisms.

Opthalmic preparations and preparations for use in the ear intended for use on more than one occasion

Bacteria. The number of organisms recovered per ml or per g is reduced by a factor of not less than 10^3 within 6 h of challenge, and no organism is recovered from 1 ml or 1 g at 24 h and thereafter.

Moulds and yeasts. The number of organisms recovered per ml or g is reduced by a factor of not less than 10^2 within 7 d of challenge, and there is no increase thereafter.

Otic Preparations—Ear Drops

The BP (Anon. 1980*a*) defines these as:

> 'suspensions, emulsions or solutions of one or more medicaments in vehicles suitable for instillation into the ear. Ear drops may also contain suitable APs'.

These preparations do not require to be sterile on manufacture but the product must comply with the same standards with respect to the 'Efficacy Test' for APs as opthalmic preparations when they are supplied as preparations intended for use on more than one occasion.

Topical preparations

There are many problems associated with the preservation of sterile topical preparations such as creams, emulsions, aqueous and oily suspensions and ointments. Despite theoretical approaches to the selection of preservatives, the challenge test is the most practicable approach. Apart from preservation, there is also the problem of the recovery of the micro-organisms from the oily phase of certain of these products. The efficacy of the AP used in topical preparations is as stated below.

Bacteria. The number of organisms recovered per ml or per g is reduced by a factor of not less than 10^3 within 48 h of challenge, and no organisms are recovered from 1 ml or 1 g at 7 d and thereafter.

Moulds and yeasts. The number of organisms recovered per ml or per g is reduced by a factor of not less than 10^2 within 14 d of challenge and there is no increase thereafter.

Sterility Testing

After the manufacture of sterile pharmaceutical products it is necessary to confirm that the process was carried out correctly. One confirmatory method is to test samples from the batch and establish whether or not there are micro-organisms which can proliferate in suitable nutrient media. As the sample may contain an AP, a suitable inactivating agent needs to be added (Russell *et al.* 1979).

It is essential that the inactivating agent does not have any antimicrobial activity. Although this can be tested with small numbers of physiologically fit cells, those that have survived a sterilization process may be damaged and hence more susceptible to an inactivating agent. A list of suitable methods for inactivating APs commonly used in pharmaceutical preparations is shown in Table 9.

TABLE 9. *Recommended methods of inactivating various non-antibiotic APs used in pharmaceutical preparations*

Preservatives	Inactivation
Phenolics	Dilution
Mercury compounds	Sodium thioglycollate
Alcohols	Dilution
Quaternary ammonium compounds (QACs)	Letheen Broth Lecithin & Tween 80
Biguanides	Letheen Broth Lecithin & Tween 80
EDTA	Dilution: Mg^{++}

Comments and Conclusions

This paper has described various antimicrobial preservatives that may be incorporated into sterile pharmaceutical preparations, especially parenteral and opthalmic products. The properties of such microbial inhibitors and a consideration of the factors influencing their activity was considered also.

The preservative effectiveness tests of the USP (Anon. 1980*b*) and BP Addendum (Anon. 1982) are described. These tests provide invaluable information about the activity of antimicrobial preservatives in various types of sterile (or non-sterile) pharmaceutical products when challenged by specific organisms. However, Al-Hiti & Gilbert (1980) have criticised the use of complex nutrient media for growing the test organisms because of inter-laboratory variations in the preparation of such media. They stated also that there was unlikely to be any significant inter-laboratory variation when chemically-defined media were employed. Al-Hiti & Gilbert (1980) also demonstrated that Gram-positive and Gram-negative bacteria, *Candida albicans* and *Aspergillus niger* grown in synthetic or semi-synthetic media under conditions of various nutrient depletions showed varying responses to thiomersal, chlorhexidine and benzalkonium chloride. Due attention should be paid to these findings when pharmacopoeial test methods are next being reviewed.

References

Alder, V. G. & Simpson, R. A. 1982 Sterilisation and disinfection by heat methods. In *Principles and Practice of Disinfection, Preservation and Sterilisation*, ed. Russell, A. D., Hugo, W. B. and Ayliffe, G. A. J. pp. 433–453. Oxford: Blackwell Scientific Publications.

Al-Hiti, M. M. A. & Gilbert, P. 1980 Changes in preservative sensitivity for the USP antimicrobial agents effectiveness test micro-organisms. *Journal of Applied Bacteriology* **49**, 119–126.

Allwood, M. C. 1973 Inhibition of *Staphylococcus aureus* by combinations of non-ionic surface-active agents and antibacterial substances. *Microbios* **7**, 209–214.

Anon. 1962 Medical Research Council *The Sterilization, Use and Care of Syringes.* MRC Memorandum No. 41. London: HMSO.

Anon. 1970 *United States Pharmacopoeia* XVIII Rockville, Md: United States Pharmacopoeial Convention Inc.

Anon. 1975 *European Pharmacopoeia* Volume 3, Sainte-Rufine: Maisonneuve S.A.

Anon. 1979 *Pharmaceutical Codex* London: The Pharmaceutical Press.

Anon. 1980*a* *British Pharmacopoeia* London: HMSO.

Anon. 1980*b* *United States Pharmacopoeia* United States Pharmacopoeial Convention Inc., Rockville.

Anon. 1982 *British Pharmacopoeia Addendum* London: HMSO.

Armstrong, N. A. & Clarke, C. D. 1972 The influence of pH on the adsorption of benzoic acid by kaolin. *Pharmaceutical Journal*, **209**, 44–46.

Bean, H. S. 1967 Types and characteristics of disinfectants. *Journal of Applied Bacteriology*, **30**, 6–16.

Bean, H. S. 1972 Preservatives for pharmaceuticals. *Journal of the Society of Cosmetic Chemists* **23**, 703–720.

BEVERIDGE, E. G. 1975 The microbial spoilage of pharmaceutical products. In *Microbial Aspects of Deterioration of Materials* ed. Lovelock, D. W. & Gilbert, R. J. pp. 213–235 London: Academic Press.

BEVERIDGE, E. G. 1983 Microbial spoilage and preservation of pharmaceutical products. In *Pharmaceutical Microbiology* 3rd Edn, ed. Hugo, W. B. & Russell, A. D. pp. 334–353, Oxford: Blackwell Scientific Publications.

CARTER, S. J. 1974 Incompatibilities of bactericides In *Dispensing for Pharmacy Students* 12th edn. ed. Cooper, J. W. & Gunn, G. Appendix 3, p. 557. London: Pitman Medical.

CHRISTENSEN, E. A. & KRISTENSEN, H. 1982 Gaseous sterilisation. In *Principles and Practice of Disinfection, Preservation and Sterilisation* ed. Russell, A. D., Hugo, W. B. & Ayliffe, G. A. J. pp. 548–568, Oxford: Blackwell Scientific Publications.

CHRISTENSEN, E. A., KRISTENSEN, H. & SEHESTED, K. 1982 Ionising radiation. In *Principles and Practice of Disinfection, Preservation and Sterilisation* ed. Russell, A. D., Hugo, W. B. & Ayliffe, G. A. J. pp. 513–533, Oxford: Blackwell Scientific Publications.

DAVIES, D. J. G. & HOBBS, R. J. 1983 Principles of sterilisation. In *Pharmaceutical Microbiology* ed. Hugo, W. B. & Russell, A. D. pp. 366–385, Oxford: Blackwell Scientific Publications.

DENYER, S. P., RUSSELL, A. D. & HUGO, W. B. 1982 Filtration sterilisation. In *Principles and Practice of Disinfection, Preservation and Sterilisation* ed. Russell, A. D., Hugo, W. B. & Ayliffe, G. A. J. pp. 569–609, Oxford: Blackwell Scientific Publications.

EVANS, W. F. & DUNBAR, S. F. 1965 The effect of surfactants on germicides and preservatives. In *Society of Chemical Industry: Surface Activity and the Microbial Cell.* London: Society for Chemical Industry.

GALBRAITH, H. & MILLER, T. B. 1973 Effect of metal cations and pH on the antibacterial activity and uptake of long chain fatty acids. *Journal of Applied Bacteriology* **36**, 635–646.

LACHMAN, L., SHETH, P. B. & URBANYI, T. 1964 Lined and unlined rubber stoppers for multiple-dose vial solutions. 1. Sorbtion of preservatives and leaching of extractives. *Journal of Pharmaceutical Science* **53**, 211–219.

MARTINDALE, W. 1982 Disinfectants and antiseptics. In *The Extra Pharmacopoeia* 28th edn, ed. Reynolds, J. E. F. pp. 547–578. London: Pharmaceutical Press.

REMINGTON, J. P. 1985 Antimicrobial drugs. In *Pharmaceutical Sciences* 17th edn, ed. Osol, A., pp. 1158–1169. Easton, Pennsylvania: Mack Publishing Company.

RICHARDS, C., FURR, J. R. & RUSSELL, A. D. 1984 Inactivation of micro-organisms by lethal gases. In *Cosmetic and Drug Preservation. Principles and Practice* ed. Kabara, Jon J. pp. 209–222, New York and Basel: Marcel Dekker Inc.

RINGERTZ, O. & RINGERTZ, S. 1982 The clinical significance of microbial contamination in pharmaceutical and allied products. In *Advances in Pharmaceutical Sciences* Vol. 5 ed. Bean, H. S., Beckett, A. H. & Carless, J. E. pp. 201–226, London: Academic Press.

ROYCE, A. & SYKES, G. 1957 Losses of bacteriostats from injections in rubber closure containers. *Journal of Pharmacy and Pharmacology* **9**, 814–822.

RUSSELL, A. D. 1982 Ultraviolet radiation. In *Principles and Practice of Disinfection, Preservation and Sterilization* ed. Russell, A. D., Hugo, W. B. & Ayliffe, G. A. J. pp. 534–547, Oxford: Blackwell Scientific Publications.

RUSSELL, A. D., AHONKAI, I. & ROGERS, D. T. 1979 Microbiology applications of the inactivation of antibiotics and other antimicrobial agents. *Journal of Applied Bacteriology* **46**, 207–245.

RUSSELL, A. D. & FURR, J. R. 1977 The antibacterial activity of a new chloroxylenol preparation containing ethylenediamine tetracetic acid. *Journal of Applied Bacteriology* **45**, 253–260.

STEPHENSON, J. R., HEARD, S. R., RICHARDS, M. A. & TABAQCHALI, S. 1984 Outbreaks of septicaemia due to contaminated mouthwash. *British Medical Journal* **289**, 1584.

WALLHAUSSER, K. H. 1974 Antimicrobial preservatives in biologics. *Die Pharmazeutische Industrie* **36**, 716–722.

WIENER, S. 1955 The interference of rubber with the bacteriostatic action of thiomersalate. *Journal of Pharmacy and Pharmacology* **7**, 118–125.

Wilkinson, S. G. 1975 Sensitivity to ethylene diaminetetraacetic acid. In *Resistance of* Pseudomonas aeruginosa, ed. Brown, M. R. W., pp. 145–188 London: John Wiley & Sons.

Wing, W. T. 1955 An examination of rubber used as a closure for container of injectable solutions. Part I. Factors affecting the absorption of phenol. *Journal of Pharmacy and Pharmacology* **7**, 648–658.

Wing, W. T. 1956*a* An examination of rubber used as a closure for containers of injectable solutions. Part II. The absorption of chlorocresol. *Journal of Pharmacy and Pharmacology* **8**, 734–737.

Wing, W. T. 1956*b* An examination of rubber used as a closure for containers of injectable solutions. Part III. The effect of the chemical composition of the rubber mix on phenol and chlorocresol absorption. *Journal of Pharmacy and Pharmacology* **7**, 738–743.

Preservation of Cosmetic and Toiletry Products

CAROLE DURANT AND P. HIGDON

Johnson & Johnson Ltd., Southampton Road, Cosham, Portsmouth, Hants, UK

Preservatives are used primarily in cosmetic/toiletry products to prevent product spoilage and to avoid infection of the consumer. A cosmetic is defined in Anon. (1976) as:

> '... any substance or preparation to be placed in contact with external parts of the human body or with the teeth and oral cavity with a view to exclusively or principally perfume them, protect them, or keep them in good condition, to change their appearance or to correct body odours.'

Contamination and Spoilage

Spoilage can occur through oxidation, reduction, hydrolysis, change in colour, pH, odour or emulsion breakdown in the product. Many of these effects can be caused by micro-organisms. Improperly preserved products could cause infection.

Article 2 in the 1976 EEC Cosmetics Directive (Anon. 1976) states that:

> 'Cosmetic products put on the market within the community must not be liable to cause damage to human health when they are applied under normal conditions of use.'

The distinctions between a non-pathogen and a pathogen and potential infection are very tenuous. It will vary with the age and health of the individual and the conditions encountered. Many of the constituents that are used in the manufacture of cosmetics and toiletries can support the growth of a wide variety of micro-organisms. This is particularly the case with the contemporary use of proteins, vitamins, gums, beer and herbs. Indeed these can be regarded as additives to the nutritive detergents, water and soft paraffins of cosmetics.

Preservatives must be chosen to prevent microbial growth, to render the product safe for use and to extend its shelf-life. Cosmetic and toiletry products are likely to

Preservatives in the Food,
Pharmaceutical and Environmental
Industries

0-632-01727-9

be severely challenged in the home, typically with micro-organisms in saliva, on dirty hands and in tap water used to dilute a product. Typical contaminants that can be encountered under these conditions are: *Enterobacter* spp., *Pseudomonas* spp., *Klebsiella* spp., *Alcaligenes* spp. and *Serratia* spp. The effect of heat and humidity from storage in bathrooms will influence also the chances of survival and growth of the contaminants. Consequently the preservatives chosen for these products must counteract such insults whilst being compatible both with the formulation and the packaging (p. 217). Aspects such as pH, light, temperature, water activity, presence of proteins and surfactants must be taken into account also. Thus, for example, with emulsions, an oil-in-water emulsion is very susceptible to microbial contamination but a water-in-oil one is less so. Should the latter be unstable, the aqueous phase will be exposed to micro-organisms.

Incidents of cosmetic/toiletry products causing microbiological problems to consumers are very rare. One of the earliest and most dramatic incidents occurred in New Zealand in 1946. The death of four babies caused by *Clostridium tetani* infections was attributed to the use of talcum powder contaminated with this organism (Tremewan 1946).

There are also a few reports of wounds being infected by the use of products manufactured in hospitals and some hospital infections following the use of commercially-produced non-cosmetic products. Likewise there are few, if any, reports of authenticated cases of infection of the consumer from use of cosmetic/toiletry products outside the hospital environment. Two papers, however, one by Cooke *et al.* (1976) and the other by Morse *et al.* (1967), are often quoted as evidence of risk to the consumer. Over the years, there have been a number of surveys on cosmetic/toiletry products. These are summarized below. These surveys show that contaminated products have occurred because of inadequate or no preservation:

1 A survey (1966–68) of 23 Food & Drug Administration recalls showed contamination (Tenenbaum 1969).

2 Wolven & Levenstein (1969) found 24% of retail cosmetics to be contaminated.

3 Dunnigan & Evans (1970) found contamination in 20% of topical drugs and cosmetics in hospitals in the UK.

4 Wilson *et al.* (1971) reported contamination in ocular cosmetics in the US.

5 Jarvis *et al.* (1974) found that 10% of cosmetics and toiletries contained 1000 micro-organisms/g in the UK.

6 In a second survey, Wolven & Levenstein (1972) found contamination in only 3.5% of the products tested in the US.

7 In Canada, Myers & Pasutto (1973) tested used and unused cosmetics; they found a higher incidence of contamination in the former.

8 Bruch (1976) published the results of a microbial study of eye products in the US.

9 Baird (1977) found *Pseudomonas* spp. in 4% of unused cosmetics and toiletries in the UK.

10 A study by the Cosmetic, Toiletry and Fragrance Association (CFTA) (Anon. 1977) carried out in 1972–75 reported that 100% of the baby and eye products and 99.5% of other toiletries/cosmetics tested in the US were within the CTFA limits.

The need for Microbiological Limits and Good Manufacturing Practice

Preservatives should not be used as biocides to compensate for poor manufacturing practices. They are intended to maintain the microbiolgical quality of a product and to protect it against contamination during use. As discussed by Chapman (p. 177) there are few preservatives that can be used in cosmetic and toiletry products. Indeed the whole philosophy of ensuring that products are safe relies on Good Manufacturing Practices (GMP) and effective preservation. The aim of GMP is not to produce products totally free of micro-organisms but merely to keep them to an acceptably low level. The purpose of the preservative system is to maintain this level. As a consequence of these objectives, GMP covers: control of raw materials; treatment of water; design and ease of cleaning of equipment; the training of personnel; effective cleaning regimes; and a satisfactory working environment.

The level of microbial contamination of raw materials must be minimized so that the total load in the finished product is small. Some raw materials of natural origin will be very contaminated. Others will support the growth of contamination *viz.* pigments, surfactants, natural gums, waxes, detergents, oils, proteins and water. Indeed it may be that a raw material awaiting use has to be preserved. This would need to be taken into account in the formulation of the finished product so that excessive concentrations of preservatives were avoided.

Cleanliness during manufacture is important in order to reduce the number of micro-oragnisms which may attack a product. Adaptation of micro-organisms to a preservative system could occur if the product comes into contact with the environment or is diluted. Packaging materials too must be monitored during the development phase. They may affect the stability of a product and thereby have a deleterious effect on the preservative system. As this effect may become evident during storage, it should be taken into account when interpreting data obtained from challenge tests (p. 205).

Microbial limits have been chosen by the industry as the best quantifiable estimate of the point at which a definite risk of spoilage to the product, or hazard to the consumer, might exist (Table 1). Major pathogenic micro-organisms which would cause concern are: *Staphylococcus aureus*, *Pseudomonas aeruginosa*, *Candida albicans* and *Clostridium* spp. It must be recognized also that pathogenicity depends on the site of infection, the resistance of an individual and the numbers of micro-organisms involved. The micro-organisms causing spoilage are less well defined but it is well known that many Gram-negative bacteria and fungi are involved. Estimating if a potential dynamic situation leading to microbial growth can occur

TABLE 1. *Trade association guidelines on microbial limits for cosmetic toiletries products*

Trade Association	Limits for eye and baby products	Limits for 'others'	Comments
Cosmetic, Toiletry and Perfumery Association Ltd., UK	<100 micro-organisms/g	<1000 micro-organisms/g	Where there are substantial numbers, the presence of harmful micro-organisms should be excluded.
Cosmetics, Toiletry & Fragrance Association Inc., USA	$\leqslant 500$ micro-organisms/g	$\leqslant 1000$ micro-organisms/g	In addition to microbial limits no product shall have a microbial content recognized as harmful to the user.

depends on a knowledge of the contaminants and experience of the product including the packaging materials and the challenge test data.

Preservatives used in Cosmetics and Toiletries

The EEC Cosmetics Directive, Annex VI (Anon. 1984), lists 12 preservatives which cosmetic products may contain and another 60 which are provisionally allowed. A preservative in the latter category is one that can be used until it is no longer listed. The substances on the provisional list will only be allowed permanently after safety data have been submitted to clear them and to ensure that they can be used safely in cosmetic products. Maximum authorized concentrations are detailed, along with the limitations, requirements and the conditions of use and warnings which must be printed on the label of the product.

A preservative is listed in the 'Directive' (Anon. 1984) as a substance which may be added to cosmetic products for the primary purpose of inhibiting the development of micro-organisms in such products. Some of the listed preservatives may be added to cosmetic products at other concentrations for other specific purposes, e.g. anti-dandruff agents in shampoos. It is recognized also that other substances within a formulation may exhibit some anti-microbial properties, e.g. some alcohols, and thereby contribute to the preservation of the product. Such substances are not included in the lists. The complete list of allowed and provisionally allowed preservatives at the time of writing is given in the Appendix (pp. 244–251). The data are taken direct from the EEC Cosmetics Directive, 1984 updated (Anon. 1984).

The preservatives available for use in cosmetic and toiletry products are limited by two major factors: the compatability of the preservative with the formulation, and the toxicological properties of the preservative. A summary of the properties, advantages and disadvantages of the preservatives most commonly used for cosmetic and toiletry products is given in Table 2.

The most important factors influencing the activity of preservatives in formulations are: pH; the presence of oil, and the presence of surfactants. The pH of the formulation affects the activity of most preservatives to a greater or lesser extent. Examples include the virtual loss of activity of benzoic and sorbic acids in alkaline solutions and the loss of activity of 4-hydroxybenzoic acid salts and esters and 2-bromo-2-nitropropan-1,3-diol at $pH > 8.0$. The partition coefficient of a preservative in an oil/water system is another important determinant of preservative efficacy. Preservatives with a high oil/water partition co-efficient concentrate in the oil and thus become unavailable to any micro-organisms present elsewhere in the product. Examples of preservatives affected in this way are the salts and esters of 4-hydroxybenzoic acid. The third most important factor influencing preservative efficiency is the presence of surfactants, particularly non-ionic ones, in the formulation. These compounds are known to decrease the efficiency of many preservatives including the salts and esters of 4-hydroxybenzoic acid and phenols. Other factors, e.g. absorption of the preservative by the primary packaging material, must also be considered. In general these are of lesser importance than the factors described above.

TABLE 2. *Major preservatives used in cosmetics and toiletries products**

Preservative	Antimicrobial spectrum	pH	Possible inactivators
Benzoic acid detergent	Broad	<4.5	Non ionic
Benzyl alcohol	Antibacterial	<7.0	–
Boric acid	Broad	<7.0	Alkali
Bronidox	Broad	<9	–
Bronopol	Antibacterial	5–8	–
Chlorhexidine proteins	Broad	<8	Anionics, some non ionics
Dehydroacetic acid	Broad	<6.0	Non ionics
Dowicil 2000	Broad	<9.0	Urea, ammonia, proteins
Formaldehyde	Broad	<9.0	Urea, ammonia, proteins
Germall 115	Antibacterial	<10	–
Giv Gard DXN	Antibacterial	<9	–
Irgasan DP300 proteins	Broad (ineffective against *Pseudomonas aeruginosa*)	<9	Non ionics, oil phase
Kathon CG	Broad	<9	Amines, mercaptan
Parabens	Broad	<9	Non ionics
PCMX	Broad	<9	Non ionics, oil, proteins
Quaternary ammonium compound	Broad (some Gram-negatives are resistant)	<9	Anionics, Non ionics
Sorbic acid	Antifungal	<6	Non ionics

* See also Table 1 (p. 183) & Table 1 (p. 202) for additional details.

Challenge Testing in Cosmetic and Toiletry Products

A resumé of challenge tests for cosmetics and toiletries used in the UK and US is given in Table 3. The data taken from the trade associations of Britain and the United States are compared with those of the pharmacopoeias of these two countries. As might be expected, the UK Trade Association (The Cosmetic, Toiletry and Perfumery Association Ltd; CTPA) Guidelines (Anon. 1983) and the BP (Anon. 1980*a*) Method are comparable. The US Trade Association (The Cosmetic, Toiletry & Fragrance Association Inc.; CFTA) Guidelines (Anon. 1973) are comparable with those in the USP (Anon. 1980*a*). The major difference between the two countries is that the UK define acceptance criteria at a 48 h period, whilst the US methods do not require any acceptance criteria until after a 7 d period.

Important points to consider when setting up challenge tests are:

1 The range of micro-organisms used in challenge tests should include those that have been associated with spoilage problems in the past or those which might be envisaged to cause problems in the future. Use of such a range of micro-organisms gives good background data for in-house control.

2 Adapted micro-organisms can be used. If the micro-organisms have been isolated from a product, they may be adapted already; otherwise adaptation can be achieved by growing micro-organisms in a medium containing diluted or unpreserved product.

3 In practice, repeated challenge testing for periods longer than 28 d may be thought necessary to assess the preservative efficacy. It is also valuable to assess preservative activity at different temperatures and under different storage conditions.

4 It is valid to challenge-test the product in its container as the packaging material may modify preserving activity.

5 Repeated challenges may be desirable. When in use, products will be challenged continually. A method which includes re-challenging is obviously more demanding than a single challenge. It is particularly suitable for those products which will be subjected to much abuse and challenge due to the way in which they are used or packaged. Even if the product does not fall into this category, it may be useful to re-challenge in order to provide indications of potential problems.

6 An effective recovery system should be used. The diluent must be formulated so that it inactivates the preservative and protects injured cells from further damage. The recovery media which generally contain agar may also contain substances that inactivate a preservative; of course they must facilitate the optimal recovery of the test micro-organisms.

TABLE 3. *Challenge testing in cosmetics and toiletries products*

	Sample size	Final inoculum/ml of product	Micro-organisms	Count at	Re-challenge	Criteria for acceptance
BP* (Anon. 1980*a*)	Not specified	10^6	*Aspergillus niger* (ATCC 16404) *Candida albicans* (ATCC 10231) *Pseudomonas aeruginosa* (ATCC 19429) *Staphylococcus aureus* (ATCC 6538) + others if thought necessary for product	0, 6, 24 & 48 h 7, 14, & 28 d	–	*Topical* Bacteria reduced by a factor of no less than 10^3 in 48 h and no recovery at 7 d. Fungi/yeasts reduced by no less than 10^2 in 7 d and no recovery at 28 d. *Ophthalmic* Bacteria reduced by no less than 10^3 in 6 h and nil recovery in 24 h and moulds and yeasts reduced by no less than 10^2 in 7 d and 7 d +.
USP† (Anon. 1980*b*)	20 ml	$10^5/10^6$	*Aspergillus niger* (ATCC 16404) *Candida Albicans* (ATCC 10231) *Pseudomonas aeruginosa* (ATCC 9027) *Staphylococcus aureus* (ATCC 6538) *Escherichia coli* (ATCC 8739) + others if required	7,14, 21 & 28 d	–	Bacteria reduced to no more than 0.1% of initial concentration in 14 d. Moulds/yeasts to remain at or below initial concentration for first 14 d. Concentration to remain at or below those for rest of 28 d.
CTPA‡ (Anon. 1983) General toiletries	In pack. Amount not specified	$10^5/10^6$	Variety of spoilage micro-organisms. Isolates from naturally contaminated products are preferred.	Variable periods	At intervals over several months	Not defined

TABLE 3.—*cont*

	Sample size	Final inoculum/ml of product	Micro-organisms	Count at	Re-challenge	Criteria for acceptance
Eye cosmetics	Not specified	10^6	As BP + micro-organisms representative of microflora of raw materials/water/environment and eye cosmetic isolates.	0, 6, 24, & 48 h 7, 14 & 28 d	Should be considered	Bacteria reduced by no less than 10^3 in 48 h. Nil recovery in 1 ml at 7 d+. Moulds and yeasts reduced by no less than 10^2 in 7 d. Nil recovery at 28 d+
CTFA§ Cosmetics and Toiletries (Anon. 1973)	≮20 ml or g	≮10^6	*Aspergillus niger* (ATCC 9642) *Penicillium luteum* (ATCC 9644) *Candida albicans* *Pseudomonas aeruginosa* (ATCC 15442/13388) *Staphylococcus aureus* (ATCC 6538) *Escherichia coli*/*Klebsiella*/*Enterobacter*/*Proteus* sp.	0 h 1, 2, 7, 14 & 28 d	If required	If there is relatively high recovery after 7 d it may be considered inadequately preserved. But final judgement made at 28 d +.
Aqueous Liquid and Semi-liquid Eye Cosmetics (Anon. 1975)	≮20 ml or g	10^6 bacteria, 14 yeasts and moulds	*Aspergillus fumigatus* *Candida parapsilosis* *Fusarium solari* *Pseudomonas aeruginosa* *Staphylococcus epidermidis* *Proteus* sp. *Bacillus subtilis* var *globigii*	0 h 1, 2, 7, 14 & 28 d	Suggest re-challenging after two consecutive negative recoveries.	>99.9% reduction of negative bacteria in 7 d after each challenege and continual reduction to 28 d. >90% reduction of yeasts/fungi in 7 d after each challenege and continual reduction to 28 d. Spores—no growth.

* British Pharmacopoeia; † United States Pharmacopoeia XX; ‡ The Cosmetic, Toiletry & Perfumery Association Ltd; § The Cosmetic, Toiletry & Fragrance Association Inc.

The Occurrence of Sub-lethally Damaged Micro-organisms in Cosmetic Preparations

A differential plating procedure, involving subculture onto a selective and non-selective medium, is generally used to both detect and enumerate sub-lethally damaged micro-organisms. Methods devised for the isolation of damaged micro-organisms have been reviewed by Ray (1979), Hurst (1977), Mossell & Corry (1977) and van Schothorst (1976). In the case of cosmetic and toiletry formulations, isolation procedures are complicated by the special requirements of the diluents needed for such products. The diluents commonly need to contain not only inactivators for particular preservatives present in a product but also surfactants to ensure adequate suspension of the product. Although there is a range of inactivators of various preservatives (MacKinnon 1974; Russell 1982), it must be stressed that many of these, e.g. thiosulphate, are known to inhibit certain micro-organisms, Gross & Huff (1973).

We have done two series of experiments to improve our techniques for isolating sub-lethally damaged bacteria, and hence to aid our assessment of the practical significance of such micro-organisms in products. *Pseudomonas aeruginosa* was chosen as the test bacterium in both series. In the first series the effect of diluent composition on the recovery of sub-lethally damaged bacteria was measured. In the second, a simple model system was studied in order to facilitate interpretation of the results.

Effect of diluent composition

Preliminary experiments were made to establish conditions that caused reproducible injury of *Ps. aeruginosa* in a particular product. The pseudomonad was grown overnight in Tryptone Soya broth (TSB) at 30 °C. The culture was centrifuged, washed once and then resuspended in distilled water. A sample (1 ml) of the final suspension was used to inoculate 100 ml of product. This resulted in an inoculum of between 10^6–10^7 micro-organisms/ml of product. At intervals, samples were removed and diluted serially in a suitable diluent containing an inactivating agent. Spread plates of each dilution were then prepared on Tryptone Soya Agar (TSA) and incubated at 32 °C for 3 d. From these experiments an exposure time which resulted in approximately 1% survival of the original inoculum was established.

The following protocol was used to test the effect of various additives to the diluent. An overnight culture of *Ps. aeruginosa* was added to the product as described above. After a known time, portions of the product were weighed into the diluent containing a particular additive. After shaking to obtain a good suspension, 1 ml samples were pipetted into sterile Petri dishes and mixed with 18 ml molten TSA. The medium was incubated at 32 °C for 3 d before colonies were counted. A factorial designed experiment was found to be the most appropriate experimental design. With 3 samples of each diluent formula and 5 pour plates of each diluent, it was possible to test 3 different factors in each experiment.

Model study

In the model study, the effect of bronopol on *Ps. aeruginosa* was tested under defined conditions. *Pseudomonas aeruginosa* was grown to mid-exponential phase in a minimal medium (Boggis *et al.* 1977) of the following composition [g/l of water, pH 7.5]: $(NH4)_2SO_4$ 0.02 g; $(NH_4)_2HPO_4$, 4.54; $(NH_4)H_2PO_4$, 0.64; NaCl 0.058; KCl 0.076, $FeSO_{47}H_2O$ 0.003; glucose 0.69, and $MgSO_{47}H_2O$ 0.041.

The culture was grown at 37 °C using vortex aeration until an OD_{490} of 0.1 was achieved. Sufficient of a 0.1% (w/v) bronopol solution was added to give a final preservative concentration of 100 μg/ml. Incubation and aeration at 37 °C were continued. Samples were taken at intervals and diluted in a neutralizing diluent of the following composition (g/l of water): $(NH_4)SO_4$ 0.026; $(NH_4)HPO_4$ 4.54; $(NH_4)H_2PO_4$ 0.64; NaCl 0.058; KCl 0.76 and $Na_2S_2O_3$ 5.0.

Further serial dilutions were prepared, in diluent from which thiosulphate had been omitted. A Spiral Systems Plater Model C was used to estimate the viable count. The control plating medium for these experiments was the minimal medium (specified above) solidified with 1.5% (w/v) Oxoid agar No. 3. Plates were incubated for 3 d at 32 °C.

Discussion of results

Results from a typical factorial experiment are shown in Table 4. In this experiment *Ps. aeruginosa* was exposed to a detergent-based product preserved with bronopol. The basal diluent had the following formulation: Tryptone Soya Broth (Oxoid) + 5% (w/v) Tween + 5% (w/v) thiosulphate. The effects of 3 alterations, addition of extra thiosulphate, 0.1% lecithin, or 0.1 mm MgCl, were tested singly and in combination. Analysis showed that the combination of magnesium and extra thiosulphate tended to decrease the recovery of viable organisms whereas the presence of magnesium greatly increased recovery. The presence of both lecithin and extra thiosulphate also increased recovery. The overall effect was that the addition of 0.1 mmol of magnesium yielded the largest increase in recovery. Other experiments have shown that significant increase in the recovery of bacteria could be obtained by adding 5 g of product to 95 ml of diluent, rather than 10 g to 90 ml. The most important results of the model study have been the finding that the addition of 200 units catalase to the plating medium increased the recovery of bronopol-treated bacteria (Fig. 1). Other findings from this study confirmed that: the addition of magnesium to the plating medium improved recovery (Fig. 2), and bronopol-damaged bacteria were sensitized to Cetrimide and potassium sulphate (Fig. 3). An interesting adjunct of this study was the discovery that log phase cultures exposed to bronopol recovered viability to a greater extent on minimal agar than they did on TSA.

TABLE 4. *Influence of substances in diluent on recovery of* Pseudomonas aeruginosa *exposed to bronopol*

Additive	Average recovery (%)
None	100
Thiosulphate, 0.5%	100
Lecithin, 0.2%	82.5
Thiosulfate 0.5% and Lecithin, 0.1%	119
$MgCl_2$ 0.1 m	152
$MgCl_2$ 0.1 m and thiosulphate, 0.5%	127
$MgCl_2$ 0.1 m and lecithin 0.1%	139
$MgCl_2$ 0.1 m; lecithin 0.1% and thiosulphate +0.5%	157*

* Not significantly different from response obtained for addition of magnesium chloride alone.

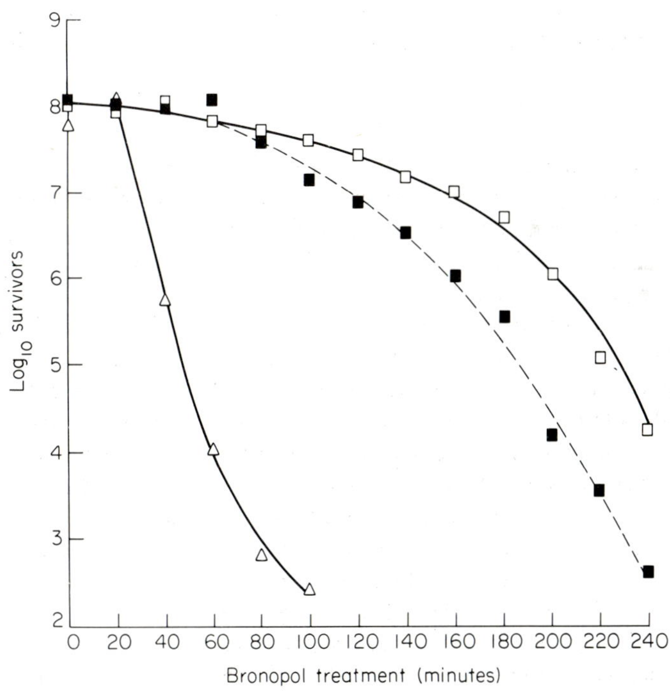

FIG. 1. Effect of catalase on recovery of bronopol-treated *Pseudomonas aeruginosa*. Bronopol-treated *Ps. aeruginosa* were plated out on TSA containing 200 units/ml catalase (■); minimal agar containing 200 units/ml catalase (□), and minimal agar (△).

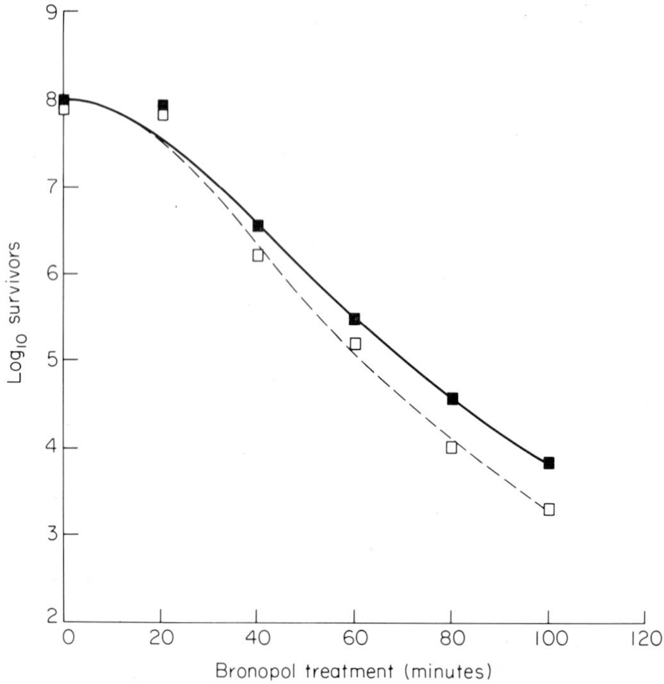

FIG. 2. The effect of magnesium on *Pseudomonas aeruginosa* treated with bronopol. Viable counts of bronopol-treated *Ps. aeruginosa* on TSA (□) and TSA with 0.14% (w/v) magnesium (■). Viable counts based on the average of 2 samples.

These findings parallel closely the results which have been obtained in other studies of micro-organisms sub-lethally damaged by physical agents such as heat (Andrews & Martin, 1979; Ray & Speck 1973; McDonald *et al.* 1983) and preservatives (Nadir & Gilbert 1982). It is thus demonstrated clearly that sub-lethally damaged micro-organisms can occur in cosmetic and toiletry products. The occurrence and significance of the presence of these damaged micro-organisms in such formulations requires further investigation.

Appendix

List of Preservatives which Cosmetic Products may contain

Preamble

1 Preservatives are substances which may be added to cosmetic products for the primary purpose of inhibiting the development of micro-organisms in such products.

2 The substances marked with the symbol (*) may also be added to cosmetic products in concentrations other than those laid down in this Annex for other

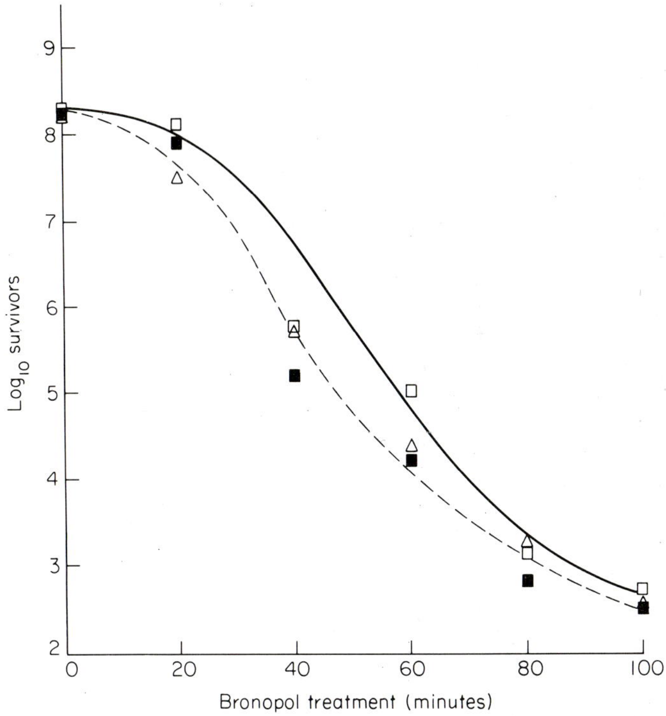

FIG. 3. The effect of potassium sulphate and CTAB on bronopol-treated *Pseudomonas aeruginosa*. Survivor curves of bronopol-treated *Ps. aeruginosa* on TSA (□); TSA containing 1.0% (w/v) potassium sulphate (△), and TSA containing 0.03% (w/v) CTAB (■). Viable counts based on the average of 2 samples.

specific purposes apparent from the presentation of the products, e.g. as deodorants in soaps or as anti-dandruff agents in shampoos.

3 Other substances used in the formulation of cosmetic products may also have anti-microbial properties and thus help in the preservation of the products, as, for instance, many essential oils and some alcohols. These substances are not included in this Annex.

4 For the purposes of this list

—'Salts' is taken to mean: salts of the cations sodium, potassium, calcium, magnesium, ammonium and ethanolamines; salts of the anions chloride, bromide, sulphate, acetate.

—'Esters' is taken to mean: esters of methyl, ethyl, propyl, isopropyl, butyl, isobutyl, phenyl.

EEC Cosmetics Directive
1984 update
Annex VI (Anon. 1984)

ANNEX VI—PART 1†. *List of preservatives allowed*

Colipa no.	Ref. no.	Substance	Maximum authorized concentration	Limitations and requirements	Conditions of use and warnings which must be printed on the label
	a	b	c	d	e
P 2	1	Benzoic acid, its salts and esters (*)	0.5% (acid)		
P 13	2	Proprionic acid and its salts (*)	2% (acid)		
P 14	3	Salicyclic acid and its salts (*)	0.5% (acid)	Not to be used in preparations for children under 3 years of age, except for shampoos	Not to be used for children under 3 years of age (solely for products which might be used for children under 3 years of age and which remain in prolonged contact with the skin)
P 15	4	Sorbic acid (hexa-2,4-dienoic acid) and its salts (*)	0.6% (acid) If mixed with sorbic acid esters, the maximum total concentration of the acids, salts and esters remains fixed at 0.6% (acid)		
P 39	5	Formaldehyde and paraformaldehyde (*)	0.2% (except for products for oral hygiene) 0.1% (products for oral hygiene) expressed as free formaldehyde	Prohibited in aerosol dispensers, except for foams	Contains formaldehyde (only if in a concentration of more than 0.05%)
P 42	6	Hexachlorphen (INN) (*)	0.1%	Prohibited in products for children under 3 years of age, and intimate hygiene products	Not to be used for children under three years of age. Contains hexachlorophene
P 47	7	Biphenyl-2-ol (*o*-phenylphenol) and its salts (*)	0.2% expressed as the phenol		

P 81	8	Pyrithione zinc (INN) (*)	0.5%	Authorized in products rinsed off after use, forbidden in products for oral hygiene	
P 51	9	Inorganic sulphites and hydrogensulphites (*)	0.2% expressed as free SO_2		
P 66	10	Sodium iodate	0.1%	Only for products rinsed off after use	
P 68	11	Chlorobutanol (INN)	0.5%	Prohibited in aerosols, except for foams	Contains chlorobutanol
P 82	12	4-hydroxybenzoic acid and its salts and esters except benzyl ester (*)	0.4% (acid) for 1 ester, 0.8% (acid) for mixtures of esters		

† Directive 76/768 EEC—February, Annex VI/pp. 2–3 (Anon. 1984).

Annex VI—Part 2. *List of preservatives provisionally allowed*

Colipa no.	Ref. no.	Substance	Maximum authorized concentration	Limitations and requirements	Conditions of use and warnings which must be printed on the label
	a	b	c	d	e
P 1	1	2,6-Dimethyl-1, 3-dioxan-4-yl acetate (Dimethoxane)	0.2%		
P 3	2	Boric acid (*)	(a) 0.5%	(a) Products for oral hygiene	
			(b) 3%	(b) Other products	
P 4	3		Chlorphenesin (INN) (*)	0.5%	
P 5	4	3-Acetyl-6-methylpyran-2,4 (3H)-dione (Dehydroacetic acid) and its salts	0.6% (acid)		
P 6	5	Formic acid (*)	0.5% (acid)		
P 7	6	4-Hydroxybenzoic acid benzyl ester	0.1% (acid)		
P 8	7	Hexamidine (INN) and its salts (including isethionate (INN) and 4-hydroxybenzoate (*))	0.1%		
P 9	8	3,3′-Dibromo-4,4′-hexamethylene-dioxydibenzamidine (Dibromohexamidine) and its salts (including isethionate)	0.1%		
P 10	9	Dibromopropamidine (INN) and its salts (including isethionate (INN))	0.1%		

P 12	10	Thiomersal (INN)	0.007% (of Hg) If mixed with other mercurial compounds authorized by this Directive, the maximum concentration of Hg remains fixed at 0.007%	For eye make-up and eye make-up remover only	Contains thiomersal
P 48	11	Phenylmercuric salts (including borate)	Ditto	Ditto	Contains phenylmercuric compounds
P 83	12	Esters of sorbic acid (Hexa-2, 4-dienoic acid) (*)	0.6% (acid) If mixed with sorbic acid and/or its salts, the maximum total concentration of the acids, salts and esters remains fixed at 0.6% (acid)		
P 16 P 17	13	Undec-10-enoic acids: salts, esters, the amide, the mono- and bis (2-hydroxethyl) amides and their sulphosuccinates (*)	0.2% (acid)		
P 18	14	2,6-Diacetyl-1,2,3,9b-tetrahydro-7,9-di-hydroxy-8,9b-dimethyldibenzolfuran-1,3,-dione (usnic acid) and its salts (including the copper salt (*)	0.2%		
P 20	15	Hexetidine (INN) (*)	0.2%		
P 21	16	Benzylformal (a 1:1 mixture of benzyloxymethanol and (benzyloxymethoxy) methanol)	0.2%		
P 22	17	Clorofene (INN)	0.2%		

ANNEX VI—PART 2.—*cont.*

Colipa no.	Ref. no.	Substance	Maximum authorized concentration	Limitations and requirements	Conditions of use and warnings which must be printed on the label
P 23	18	5-Bromo-5-nitro-1,3 dioxane (*)	0.1%	Only for products rinsed off after use	
P 24	19	Bronopol (INN) (*)	0.1%		
P 25	20	6,6-Dibromo-4,4-dichloro-2,2′-methylene-diphenol (Bromochlorophen) (*)	0.1%		
P 26	21	Tetrabromo-*o*-cresol (*)	0.3%		
P 27	22	2-Chloroacetamide	0.3%		Contains chloroacetamide
P 28	23	3,4-Dichlorobenzyl alcohol (*)	0.15%		
P 74	24	2,4-Dichlorobenzyl alcohol (*)	0.15%		
P 29	25	Triclocarban (INN) (*)	0.2%		
P 30	26	4-Chloro-*o*-cresol (*)	0.2%		
P 31	27	Halocarban (INN) (*)	0.3%	Maximum concentration in aerosols: 0.2%	
P 32	28	Triclosan (INN) (*)	0.3%		
P 33	29	Dichlorophen (INN)	0.2%		Contains dichlorophen
P 34	30	N-(Trichloromethylthio) cyclohex-4-ene-1,2-dicarboximide (Captan—(ISO)) (*)	0.5%		
P 35	31	Chlorhexidine (INN) and its digluconate, diacetate and dihydrochloride (*)	0.3%		

P 37	32	4-Chloro-3,5-xylenol (*)	0.5%		
P 38	33	2,4-Dichloro-3,5-xylenol (*)	0.1%		
P 40	34	Quinolin-8-ol and its salts (*)	0.3%	Not to be used in products applied after sunbathing, or in talcum powders for children under 3 years of age	Not to be used for children under three years of age
P 41	35	1,3,5-Tris (2-hydroxyethyl) hexahydra-1,3,5-triasine	0.3%		Contains 1,3,5-Tris (2-hydroxyethyl) hexahydro-1,3,5-triazine
P 43	36	3,3′-Bis(1-hydroxymethyl-2,5 dioxoimidazolidin-4-yl)-1, 1′ methyl-enediurea ('Imidazolidinyl urea') (*)	0.6%		
P 44	37	4-Isopropyl-*m*-cresol	0.1%		
P 45	38	2-Chlora-N-(hydroxymethyl) acetamide	0.3% for the chloracetamide	For products rinsed off after use	
P 46	39	1-Hydroxymethyl-5,5-dimethyl-hydantoin (*)	0.2% expressed as free formaldehyde or theoretically available formaldehyde	For products rinsed off after use	Contains formaldehyde (only if the concentration is greater than 0.05%)
P 49	40	Pyrithione sodium (INN) (*)	0.5%		
P 50	41	2,2′-Dithiobis (pyridine 1-oxide), addition product with magnesium sulphate trihydrate (*)	0.5%		
P 52	42	Poly(1-hexamethylenebiguanide hydrochloride (*))	0.3%		
P 53	43	2-Phenoxyethanol (*)	1%		
P 55	44	Hexamethylenetetramine (*) (methenamine) (INN)	0.2% expressed as free formaldehyde or theoretically available formaldehyde		Contains formaldehyde (only if the concentration is greater than 0.05%)

Colipa no.	Ref. no.	Substance	Maximum authorized concentration	Limitations and requirements	Conditions of use and warnings which must be printed on the label
P 56	45	Mixture of 5-Chloro-2-methyl-isothiasol-3(2H)-one and 2-methylisothiazol-3(2H)-one with magnesium chloride and magnesium nitrate	0.005% (of a mixture in the ratio 3:1 of 5-chloro-2-methyl-isothiazol 3(2H) one and 2-methylisothiazol-3 (2H)-one		
P 58	46	Pyridin-2-ol 1-oxide (*)	0.5%	Only for products rinsed off after use	
P 60	47	Pyrithione aluminium camsilate (INNM)	0.2%		
P 63	48	Methanamine 3-chloroallylo-chloride (INNM)	0.2%		
P 64	49	1-(4-Chlorophenoxy)-1-(imidazol-l-yl)3,3-dimethyl-butan-2-one (*))	0.52%		
P 65	50	1,3-Bis (hydroxymethyl)-5,5-dimethylimidazolidine-2,4-dione (*)	0.2% expressed as free formaldehyde or theoretically available formaldehyde		Contains formaldehyde (only if the concentration is greater than 0.05%)
P 67	51	Benzyl alcohol (*)	1%		
P19	52	1-Dodecylguanidinium acetate (Dodine-ISO) (*)	0.5%	For products rinsed off after use	
			0.1%	For other uses	
P 70	53	Benzethonium chloride (INN) (*)	0.1%		
P 71	54	Benzalkonium chloride (INN), 1-2-Benzisothiazol-3-(2H)-one 1,1 dioxide) (*), its bromide and saccharinate	0.5%		

P 72	55	Alkyl (C12-C22) trimethyl-ammonium bromide and chloride (including Cetri-monium bromide (INN) (*)	0.1%		
P 54	56	1-Phenoxypropan-2-ol	1%		
P 59	57	1-hydroxy-4-methyl-6(2,4,4-trimethylpentyl) 2-pyridon and its monoethanolamine salt	1.0%		For products rinsed off after use
			0.5%		For other uses

† Directive 76/768/EEC—February; Annex VI/pp. 4–9 (Anon. 1984).

References

ANON. 1973 *Microbiological Limit Guidelines for Cosmetics and Toiletries.* Cosmetic, Toiletry & Fragrance Association Guidelines. USA: CTFA.

ANON. 1975 *Microbial Content.* Sub-Committee of the Cosmetics, Toiletry & Fragrance Association Microbiological Committee. USA: CTFA Technical Guidelines Microbiology 3.

ANON. 1976 *EEC Cosmetics Directive 1976.* Luxembourg: Office for Official Publications of the European Communities.

ANON. 1977 The Cosmetic, Toiletry and Fragrance Association National survey of cosmetics and toiletries. *SFTA Cosmetics Journal* **9**, 3.

ANON. 1980*a* *British Pharmacopoeia.* London: HMSO.

ANON. 1980*b* *United States Pharmacopoeia.* Twentieth Revision. USA: United States Pharmacopoeial Convention: Rockville.

ANON. 1983 *Cosmetic, Toiletry & Perfumery Association Recommended Microbiological Limits & Guidelines to Microbiological Quality Control.* London: CTPA.

ANON. 1984 *EEC Cosmetics Directive 1984.* Luxembourg: Office for Official Publications of the European Communities.

Andrews, G. P. & MARTIN, S. E. 1979 Heat inactivation of catalase from *Staphylococcus aureus* MF-31 *Applied and Environmental Microbiology* **37**, 1180–1185.

BAIRD, R. M. 1977 The occurrence of pathogens in cosmetics and toiletries. *Journal of the Society of Cosmetic Chemists* **28**, 17–20.

BOGGIS, W., KENWARD, M. A. & BROWN, M. R. W. 1979 Effects of divalent metal cations in the growth medium upon sensitivity of batch grown *Pseudomonas aeruginosa* to EDTA or polymyxin B. *Journal of Applied Bacteriology* **47**, 477–488.

BRUCH, C. W. 1976 Microbiological quality assurance of eye products. *Drug and Cosmetic Industry* **118**, 49–53; 161–162.

COOKE, E. M., SHOOTER, R. A., O'FARRELL, S. M. & MARTIN, D. R. 1976 Faecal carriage of *Pseudomonas aeruginosa* by new born babies. *Lancet* **11**, 1045.

DUNNIGAN, A. P. & EVANS J. R. 1970 Report of a special survey on microbiological contamination of topical drugs and cosmetics. *Toilet Goods Association CTFA Cosmetic Journal* **2**, 39–41.

GROSS, A. & HUFF, M. B. 1973 Iodine inactivating agent in surgical scrub testing. *Archives of Surgery* **106**, 175–178.

HURST, A. 1977 Bacterial injury: A review. *Canadian Journal of Microbiology* **23**, 935–944.

JARVIS, B. REYNOLDS, A. J., RHODES, A. C. & ARMSTRONG , M. A. 1974 Survey of microbiological contamination in cosmetics and toiletries in the UK. *Journal of the Society of Cosmetic Chemists* **25**, 563–575.

MACKINNON, I. H. 1974 The use of inactivators in the evaluation of disinfectants. *Journal of Hygiene* **74**, 189–195.

MCDONALD, L. C., HACKNEY, C. R. & RAY, B. 1983 Enhanced recovery of injured *Escherichia coli* by compounds that degrade hydrogen peroxide or block its formulation. *Applied and Environmental Microbiology* **45**, 360–365.

MORSE, L. J., WILLIAMS, H. L., GREEN, F. P., ELDRIGE, E. E. & ROTTA, J. R. 1967 Septicaemia due to *Klebsiella pneumoniae* originating from a hand cream dispenser. *New England Journal of Medicine* **277**, 472.

MOSSEL, D. A. A. & CORRY, J. E. L. 1977 Detection and enumeration of sublethally injured pathogenic and index bacteria in foods and water processed for safety. *Alimenta Zurich* (Special issue on microbiology) **16**, 19–34.

MYERS, G. E. & PASUTTO, F. M. 1973 Microbial contamination of cosmetics and toiletries. *Canadian Journal of Pharmaceutical Sciences* **8**, 19.

NADIR, M. T. & GILBERT, P. 1982 Injury and recovery of *Bacillus megaterium* from mild chlorhexidine treatment. *Journal of Applied Bacteriology* **52**, 111–115.

ORTH, D. S. 1981 Principles of preservative efficacy. *Cosmetics and Toiletries* **96**.

RAY, B. & SPECK, M. L. 1973 Freeze injury in bacteria. *CRC Critical Reviews in Clinical Laboratory Sciences* **4**, 161–213.

RAY, B. 1979 Methods to detect stressed organisms. *Journal of Food Protection* **42**, 346–355.

RUSSELL, A. D. 1982 Factors influencing the efficacy of antimicrobial agents. In *Principles and practice of disinfection, preservation and sterilisation.* ed. Russell, A. D. *et al.* pp. 107–133 Oxford: Blackwell Scientific Publications.

SMITH, J. L. 1981 Preserving Cosmetics. *Cosmetics and Toiletries* **96**.

TENENBAUM, S. 1969 The microbial content of cosmetics and toiletries. *Toiletries Goods Association Journal* **1**, 19.

TREMEWAN, H. C. 1946 Tetanus Neonatorium in New Zealand. *The New Zealand Medical Journal* **45**, 312–313.

VAN SCHOTHORST, M. 1976 Resuscitation of injured bacteria in foods. In *Inhibition and inactivation of vegetative microbes.* pp. 317–328. ed. Skinner, F. A. & Hugo W. B. Society of Applied Bacteriology No. 5 New York: Academic Press.

WILSON, L. A., KUEHNE, J. W., HALL, W. & Ahearn, D. G. 1971 Microbial contamination of ocular cosmetics. *American Journal of Opthamology* **71**, 1298.

WOLVEN, A. & LEVENSTEIN, I. 1969 Cosmetics—Contaminated or not. *Toilet Goods Association CTFA Cosmetic Journal* **1**, 34.

WOLVEN, A. & LEVENSTEIN, I. 1972 Microbiological examination of cosmetics. *American Perfumery and Cosmetics* **87**, 63.

Non-plasmidic Resistance towards Preservatives of Pharmaceutical Products

P. GILBERT AND NAOMI WRIGHT

Department of Pharmacy, University of Manchester, Oxford Rd, Manchester, UK

The effectiveness of antimicrobial agents as preservatives is not constant. The physiological state of micro-organisms and the nature of the habitats in which they have grown can lead to considerable variations in their susceptibility towards chemical inactivation. Similarly, exposure to sub-lethal concentrations of these agents can lead to phenotypic adaptation and selection, resulting in a resistant cell population. Resistance is a comparative term and can be defined as the temporary or permanent ability of an organism and its progeny to remain viable and/or multiply under conditions that would destroy or inhibit others.

To be effective, antimicrobial agents must be capable of achieving effective concentrations at their biochemical sites of action. Direct action upon the outermost surface of bacteria is rare, more commonly lesions within the cell envelope or within the cell itself are implicated. Antimicrobial agents must therefore be able to interact with and penetrate the cell envelope in order to exert their antimicrobial action. Resistance may thus be innate and reflect fundamental differences in cellular morphology which render one group of organisms sensitive to certain agents relative to another or it might be acquired and reflect modification of target sites and/or barriers to drug permeation external to those sites, such as the cell wall. In this respect Gram-negative micro-organisms might be regarded as being innately resistant to many chemical agents relative to Gram-positive bacteria (Table 1). Innate resistance such as this is invariably due to the nature of the cell envelopes. For example, sphaeroplasts of *Escherichia coli* and *Staphylococcus aureus* are equally sensitive towards the membrane-active agent tetrachlorosalicylanilide (TCS), yet intact cells differ in terms of sensitivity by a factor of 200.

Whilst the range of chemical substances which find application as preservatives of pharmaceutical products is wide and their antimicrobial action often multiple, the primary target is usually the cytoplasmic membrane and associated enzymes. Most chemical antibacterial agents have differing modes of action dependent upon

Preservatives in the Food,
Pharmaceutical and Environmental
Industries

0–632–01727–9

TABLE 1. *Inate resistance of Gram-positive and Gram-negative bacteria towards some non-antibiotic antimicrobial agents*

	Minimum Growth Inhibitory Concentrations (μg/ml)	
Agent	*Staphylococcus aureus*	*Escherichia coli*
Chlorhexidine	0.3	1.2
Dequalinium	0.6	8.0
Propamidine	8.0	64.0
Dibromopropamidine	1.0	4.0
8-hydroxyquinoline	2.5	6.2
Tetrachlorosalicylanilide	0.15	30.0
Cetyltrimethylammonium bromide	0.39	12.5
Trichlorocarbanilide	0.08	100.0
Triphenylmethane dyes	1.30	13.0
Hexachlorophane	0.05	100.0

concentration. Often bacteristatic and bactericidal effects brought about at their minimal effective concentrations are mediated through precise biochemical lesions in the cytoplasmic membrane rather than through their action as 'general cytoplasmic poisons'. It is towards concentrations and effects such as these that adaptation and selection of resistant phenotypes is possible, rather than towards the much higher concentrations used in infection control regimens.

It is the intention of this article to examine the passage of antimicrobial agents across the cell wall to the cytoplasmic membrane and cytoplasm, and to discuss the extent to which modifications of the cell envelope can result in resistance towards those used as preservatives of pharmaceutical products.

Modulation of Drug Effectiveness by the Cell Envelope

Bacterial cells differ from their mammalian counterparts in that they must interact with an ever-changing and often hostile environment. The effects of such changes upon the cell can often be minimized through compensatory changes in the cell envelope (i.e., cytoplasmic membrane, wall and, if present, slime layers; Fig. 1). The outermost layers of some bacteria, both Gram-negative and Gram-positive, are dispensable extracellular slime-layers or capsule. These are generally carbohydrate in nature and their occurence is subject to the conditions of culture. Often they may occur as a diffuse slime surrounding and containing a number of microbial cells.

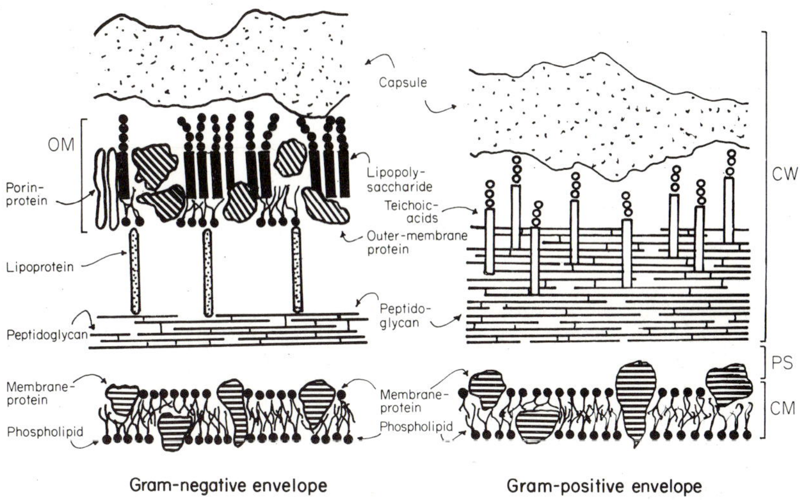

FIG. 1. Diagrammatic representation of the Gram-positive and Gram-negative cell envelopes. CW, cell wall; PS, periplasmic space; CM, cytoplasmic membrane; OM, outer-membrane.

The physiological role of such materials remains unclear, but they do confer advantages upon the cells with respect to their ability to adhere to and colonize surfaces and to resist phagocytosis and desiccation (Costerton *et al.* 1978). Additionally they may represent considerable barriers to drug penetration (Brown & Scott-Foster 1971) and have been found in resistant isolates obtained from stock solutions of antiseptics (Favero *et al.* 1983). Thus mucoid variants of *Pseudomonas aeruginosa* can often be selectively cultured by utilizing their enhanced resistance to penicillins (Govan 1976).

The cell wall, an integral part of the cell, modulates access of nutrients to the underlying cytoplasmic membrane and also to some extent the entry of antimicrobial agents. Two distinct mechanisms are involved in such action:

1 The cytoplasmic membrane is bathed in a hydrophilic medium, continuous with that of the extracellular environment. The peptidoglycan matrix in Gram-positive bacteria (Scherrer & Gerhardt 1971) and in conjunction with the outer-membrane in Gram-negative ones (Nikaido 1976; Nikaido & Nakae 1979) functions as a molecular sieve thereby preventing the passive diffusion of large hydrophilic molecules to the underlying cytoplasmic membrane.

2 When such ready access to the cytoplasmic membrane is denied, then a drug must interact with the cell wall in order to traverse it. With Gram-negative bacteria this may well involve sequential partitioning across the lipophilic layers that constitute the outer membrane.

The Passage of Antimicrobial Agents through the Gram-positive Cell Wall

The Gram-positive cell wall is composed of a complex, crosslinked network of peptidoglycan incorporating teichoic and teichuronic acids (Hess & Lagg 1958; Gerhardt & Judge 1964). Its function as a molecular sieve for large hydrophilic molecules depends upon the thickness of the matrix and the degree of crosslinking (Mitchell 1959; Scherrer & Gerhardt 1971). Both of these properties are variable and depend on the growth-rate and physiological status of the cells (Soloman 1968). Decreases in the degree of crosslinking inevitably increase the porosity of the cell wall and enhances penetration of larger hydrophilic molecules (100 000 Daltons), whilst increases in the degree of crosslinking result in enhanced cell wall stability and apparent resistance towards chemical agents. Thus Gilbert & Brown (1980) showed that growth of *Bacillus megaterium* under conditions of magnesium-limitation, carbon-limitation and phosphate-limitation at different specific growth-rates, led to significant changes in the sensitivity of intact cells, but not sphaeroplasts, towards the bactericidal effects of 2-phenoxyethanol and chlorhexidine (Fig. 2). Both agents act primarily at the cytoplasmic membrane (Hugo & Longworth 1965; Gilbert *et al.* 1977*a, b*). Resistance in this instance was associated with increased

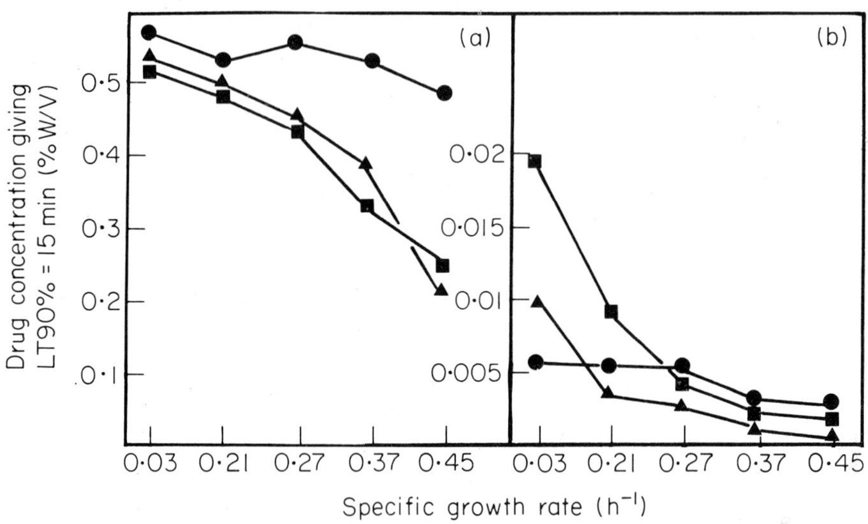

FIG. 2. Influence of nutrient-limitation and cell-growth rate upon the sensitivity of *Bacillus megaterium* suspensions towards bactericidal concentrations of (a) 2-phenoxyethanol and (b) chlorhexidine ●, phosphate-limited cultures, ■, magnesium-limited cultures and ▲ carbon-limited cultures (from Gilbert & Brown 1980).

stability of the cells in the presence of lysozyme and was suggestive of an alteration in the degree of crosslinking influencing drug permeation. The peptidoglycan is able to exclude only large (100 000 Daltons) polymeric molecules (Scherrer & Gerhardt 1971). Molecular sieving can be enhanced, however, by the presence of charged species within the peptidoglycan matrix (Marquis 1968). This, and the large surface area afforded by the cell wall, allows it to function as an ion-exchange column, actively removing strongly charged drug molecules from solution as they diffuse through the matrix. Incoming molecules, including preservatives, would have to saturate all unoccupied binding sites on the peptidoglycan surface before reaching the cytoplasmic membrane (Wagman *et al.* 1975). Subtle changes in the nature of the cell wall can markedly influence such adsorption phenomenon. For example, alteration of the interactive surface of peptidoglycan by cation adsorption has been reported to influence drug permeation by, and sensitivity towards antimicrobial agents in Gram-positive cells. Presence of potassium ions significantly decreased both the growth inhibitory and bactericidal activities of chlorhexidine. This was associated with decreased adsorption of chlorhexidine by intact cells, but not by isolated cell wall preparations (Nadir & Gilbert 1980). This suggested that in the presence of salts, contraction of the peptidoglycan matrix (Marquiś 1968) and competition for binding had significantly reduced permeation by the drug.

Magnesium-limitation has been shown to cause increased affinity of *B. subtilis* cell walls for Mg^{2+} adsorption. These changes correlate with alteration in wall teichoic and teichuronic acid content (Meers & Tempest 1970). Similarly, Tempest *et al.* (1968) showed that phosphate and magnesium-limited *B. subtilis* had walls that were substantially different. Effects such as these, brought about by modification of an organism's growth environment, might therefore be expected to affect sensitivity towards chemical agents.

Little work has been undertaken in this area of induced resistance in Gram-positive bacteria and the practical significance is as yet unknown. There is sufficient evidence, however, to suggest that exclusion of antimicrobial agents from the cytoplasm occurs under certain circumstances (Hamilton 1970; Gilbert & Brown 1980).

The Passage of Antimicrobial Agents through the Gram-negative Cell Wall

Permeation of the cell wall of Gram-negative bacteria by drugs is much more complex. The molecular detail of the outer membrane (OM) of such organisms is now understood and has been the subject of several comprehensive reviews (Costerton 1977; Nikaido & Nakae 1979; Lugtenberg & van Alphen 1983). The cell wall consists typically of an asymetric lipid membrane bilayer attached via Braun's lipoproteins and porin proteins to a thick matrix of peptidoglycan (Nikaido

& Nakae 1979; Osborn & Wu 1980). The exterior surface of the lipid bilayer is composed largely of closely packed lipopolysaccharide (LPS) orientated so that the hydrophilic polysaccharide portion is exposed to the surrounding medium and envelopes the cell surface. The polysaccharide chains attract various cations, such as Mg^{2+}, Ca^{2+} and Zn^{2+}, which can be used to overcome natural electrostatic repulsions between adjacent chains which would otherwise tend to disrupt the membrane. These cations stabilize the chains by neutralizing and bridging between negatively-charged groups (Nikaido & Nakae 1979). LPS therefore acts in a similar fashion to the anionic cell wall polymers of Gram-positive cells, in that it scavenges ionic material from the environment and participates in ion-exchange, thereby controlling the passage of material to the cytoplasm. The lipophilic regions of the LPS are directed inwards and interact with the fatty-acid chains of a phospholipid monolayer which constitutes the inner surface of the OM. Stability of the OM/peptidoglycan complex is maintained by divalent cation bridges (De Pamphilis 1971; Schnaitman 1971*a*, *b*) and lipoprotein (Costerton *et al.* 1974). In *Ps. aeruginosa* there is some evidence that Braun's lipoprotein is much reduced or even absent (Meadow 1975). Presumably divalent cation bridges play a relatively more important stabilizing role in this organism. Hence the potential of ion-chelators such as ethylenediamine tetra acetic acid (EDTA) in preservative combinations (e.g. benzalkonium chloride/EDTA).

The stabilized polysaccharide chains of the LPS form a tightly clustered, highly hydrophilic covering to the cell. This hydrophilic nature, together with a lack of membrane fluidity caused by Lipid A, restricts severely dissolution and diffusion of hydrophobic molecules into and through the outer-membrane (Nikaido & Nakae 1979).

Wild-type strains of *Esch. coli, Salmonella typhimurium* and related Gram-negative organisms possess long antigenic side chains to the LPS. These increase membrane hydrophilicity and make the cells characteristically sensitive to a number of hydrophilic antibiotics such as cephalothin, ampicillin, neomycin and cycloserine. Mutational reduction in LPS polysaccharide chain length has no effect upon sensitivity to these agents, but does increase sensitivity to more hydrophobic ones such as novobiocin, rifamycin and erythromycin (Tamaki *et al.* 1971; Leive 1974; Roantree *et al.* 1977). As the OM is the only part of the cell containing LPS in significant quantities, and the target for these agents is the cytoplasmic membrane or cytoplasm, then the data suggest that at least two diffusion pathways function across the OM, one for hydrophobic drugs affected by LPS structure and at least one other for hydrophilic molecules (Nikaido 1976; Nikaido & Nakae 1979).

Hydrophilic drugs

Together the phospholipid and LPS form a classical lipid bilayer and constitute a formidable barrier to the penetration of hydrophilic molecules. These are able to traverse the cell envelope by two distinct mechanisms, both involving OM proteins. Some material may be transported by highly specific transport proteins, analogous to those of the cytoplasmic membrane (Sabet & Schnaitman 1973; Hantke 1976; Hancock *et al.* 1976; Wookey & Rosenberg 1978), whilst others permeate the OM in a less specific manner through the action of porin proteins (Payne & Gilvarg 1968; Decad & Nikaido 1976; Nikaido & Nakae 1979). An average cell contains 10^5 porin molecules (Rosenbusch 1974) of which there are several types. Evidence suggests that these are aggregated (Palva & Randall 1978) within the OM as trimers and that each trimer forms a hydrophilic water-filled pore allowing the passage of small hydrophilic molecules (Nikaido & Nakae 1979; Yu *et al.* 1979). Their function and properties have been reviewed by Osborn & Wu (1980) and Han & Silhavy (1981). Specificity and rate of transpore diffusion depends upon molecular size (Nikaido & Nakae 1979), charge (Stock *et al.* 1977) and, for molecules approaching the maximum size limit (molecular weight *ca.* 600), lipophilicity (Nakae 1976*a*, *b*; Nikaido 1976; Zimmerman & Rosselet 1977). On the basis of transpore diffusion rates Nikaido (1976) suggested that the diameter of these pores was in the order of 1.2 nm.

Passage of materials through porin–protein pores is non-specific (Nikaido *et al.* 1980); the extent and nature of porin–protein distribution in the cell envelope varies markedly, however, as a function of the physiological state of the cell (Bassford *et al.* 1977; Turnowsky *et al.* 1983). In *Ps. aeruginosa* the pore diameter has been suggested to be significantly larger than in other Gram-negative bacteria (Hancock & Nikaido 1978). This is unexpected since the organism is particularly resistant to most chemical agents (Brown 1975). The available evidence suggests, however, that at any one time a proportion of these pores might be closed and that the organism is thus able to regulate transpore diffusion (Benz & Hancock 1981).

Hydrophobic drugs

Most biological membranes allow transmembrane diffusion of small hydrophobic molecules (Danielli 1952). The OM of many wild-type Gram-negative bacteria is probably exceptional in that such compounds are excluded partly by the surface hydrophilic polysaccharides and partly by the close stacking of the LPS which hinders access to the hydrophobic centre of the OM. This stacking is aided by the relative lack of phospholipid in the outer leaflet of the OM (Nikaido & Nakae 1979). The OM of 'deep rough' Gram-negative micro-organisms resemble typical biological membranes more closely. Hydrophobic compounds are able to traverse the layered structure of the OM through the combined action of oil/water partitioning and passive diffusion processes (Nikaido & Nakae 1979). Hansch & Fujita (1964), Lien *et al.* (1968) and Hansch (1973) argued that the differences in biological activity of chemically related antimicrobial agents, acting on similar sites within the cell, would be a reflection of their relative ease of penetration through the various lipid barriers of the cells and of their ability to react when at that site. The outer components of the cell were considered as a series of aqueous and lipophilic barriers. Substances of low water solubility would be unable to penetrate the aqeous layers and would accumulate within the lipid region, whilst those with low oil-solubility would be unable to cross the lipid regions. Compounds between these two extremes must exist which possess the optimum balance between hydrophilicity and lipophilicity for traversing the cells barriers. Therefore if log P (log oil/water partition coefficient) were plotted against the biological activity of a series of compounds then a parabolic relationship ought to be obtained, log P_0 being the value of log P for the compound with optimal activity. Many such parabolic relationships have been demonstrated (Hansch & Fujita 1964; Daoud *et al.* 1983). In this fashion it is possible to determine the optimal lipophilicity of a drug for cell envelope permeation. The average value for several series of drugs was found to be approximately 4.0 for Gram-negative and 6.0 for Gram-positive cells (Lien *et al.* 1968). This difference was attributed to the increased lipophilicity of the Gram-negative cell wall. Clearly, therefore, changes in OM lipophilicity must be accompanied by changes in the activity of individual agents.

Figure 3 represents a compartmental model for drug disposition in the Gram-negative cell wall. Changes in the polysaccharide chain length of the LPS, changes in the relative amounts of polar and non-polar lipids in the OM or cytoplasmic membrane, alteration of fatty-acid chain length and degree of saturation in LPS/phospholipids or in porin–protein composition and content will influence the rate constants for one or more of these processes. All of these parameters are able to change as the cells adapt to different environmental conditions and physical environments and thus contribute towards preservative and antiseptic sensitivity.

Variation in Preservative Sensitivity with Modification of the Cell Envelope

Under optimal growth conditions, bacterial cells are able to convert exogenous nutrients into new cytoplasmic material with a high level of efficiency. This will enable cultures of, for example, *Esch. coli*, to grow with a generation time as short as 20 min. Such rapid division and growth is unlikely to occur for extended periods of time in natural environments if, indeed, it occurs at all. More likely the rate of growth is controlled by the availability of nutrients. The response of cells to deprivation of particular nutrients is to:

1 economize upon the usage of that particular nutrient by using alternative substrates, by modification of cell composition or by reduction in the amounts of macromolecules containing these substances;

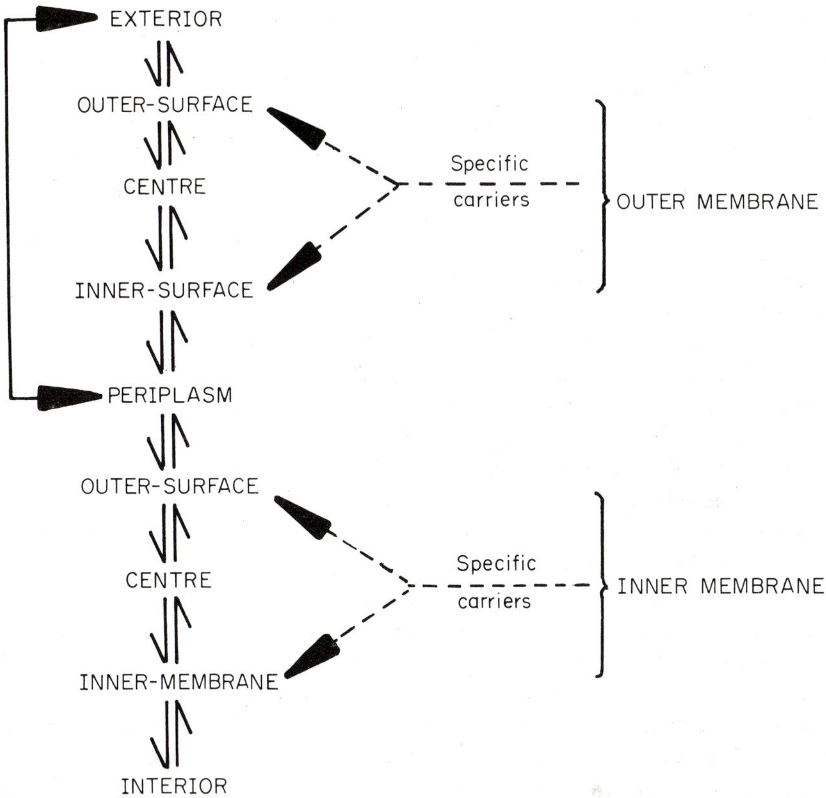

FIG. 3. A compartmental model for drug deposition in the Gram negative cell wall. ↔, diffusion through hydrophilic pores; ⇌, diffusion and partitioning; ← - - - - →, transport by specific carrier proteins.

2 increase the affinity of the cell surface for that material, thereby making its uptake into the cytoplasm more efficient and competitive, or

3 reduce the rate of cell growth and division to the maximum permissible (given 1 and 2).

The effect of different nutrient depletions, in addition to effects on general biochemistry, is to give rise to cells with radically different cell envelopes (Holme 1972; Ellwood & Tempest 1972). This is particularly evident with respect to the OM of Gram-negative bacteria (Brown & Melling 1969*a*, *b*; Brown 1975). The importance of the cell envelope in drug permeation has been discussed above; changes in its composition radically influence the activity of chemical agents, either by mechanisms of exclusion or, for membrane-active agents, by modification of the target site.

Effects of Nutrient-depletion on Preservative Sensitivity

The nutritional status of micro-organisms *in vivo* is difficult to assess (Brown 1977). One can only speculate about the nature of the growth-limiting nutrient in various localized sites, because little, if any, information is available. Broughton *et al.* (1968) reported on falls in serum magnesium levels in patients suffering from burns sufficient to make magnesium levels in the burns growth-limiting. Similarly there is evidence that iron is usually growth-limiting in serum (Weinberg 1978; Bullen *et al.* 1974) and that organisms isolated from infected cystic-fibritic lungs are iron-depleted (Brown *et al.* 1984). A similar *in vivo* growth limiting possibility has been suggested for inorganic phosphate or even zinc (Weinberg 1974). When one considers environments other than animal hosts, then the possibilities are more diverse and less subject to generalization. Various nutrient depletions may be created in chemically-defined minimal salts media in the laboratory by deliberately supplying the cells with a known excess of each elemental requirement, except one. This becomes the growth-limiting nutrient when the culture is approaching or is in the stationary phase of growth (Klemperer *et al.* 1979). The nutrients usually chosen for such studies are phosphorus (P-dep), magnesium (M-dep), a carbon source (C-dep), iron (Fe-dep) or nitrogen (N-dep).

Tables 2 and 3 give a gross chemical analysis of whole cell-preparations of *Esch. coli* grown under conditions of different nutrient-depletion. P-dep resulted in a decrease in cellular phospholipid content compensated for by increases in fatty and neutral lipid. This situation is analogous to the findings with *B. subtilis* where P-dep led to a decreased phospholipid and an increased phosphate-free polar lipid content (Minnikin *et al.* 1972). This suggests a duality of function between the two (Minnikin *et al.* 1971*a*, *b*). The decrease in phospholipid content for *Esch. coli* was found to originate solely from reduction in the amounts of diphosphatidylglycerol and phosphatidylethanolamine. Changes in the relative proportions of neutral and

TABLE 2. *Gross composition of whole cell preparations of* Escherichia coli *K12 W3110 under conditions of different nutrient-depletions.**

Nutrient depletion	Total REL† (% dry wt.)	PHB‡ (% dry wt.)	Phospholipid (% dry wt.)	FAN§ (% dry wt.)	KDO¶ (μg/mg)
Carbon	11.03 ± 2.17	0.223 ± 0.45	4.61 ± 0.39	6.20 ± 0.39	11.16 ± 0.24
Magnesium	10.68 ± 1.82	0.672 ± 0.29	4.69 ± 0.75	5.30 ± 0.75	9.11 ± 0.19
Phosphate	11.78 ± 1.37	1.226 ± 0.21	1.36 ± 0.23	9.20 ± 0.23	8.44 ± 0.42

* Data from Gilbert & Brown (1978*a*); † Readily Extractable Lipid; ‡ β-hydroxybutyrate; § Fatty and neutral lipid; ¶ 2-keto-3-deoxyoctonic acid.

TABLE 3. *Phosphopholipid composition of whole cell preparations of* Escherichia coli *K12 W3110 under conditions of different nutrient depletions*

Nutrient depletion	Composition (% dry weight)			
	DPG†	PE‡	PG§	PC¶
Carbon	1.63 ± 0.06	2.09 ± 0.19	0.58 ± 0.06	0.18 ± 0.03
Magnesium	1.77 ± 0.19	2.30 ± 0.04	0.40 ± 0.12	0.23 ± 0.04
Phosphate	0.60 ± 0.09	0.37 ± 0.04	0.34 ± 0.06	0.16 ± 0.02

* Data from Gilbert & Brown (1978*a*); † Diphosphatidylglycerol; ‡ Phosphatidylethanolamine; § Phosphatidylglycerol; ¶ Phosphatidylcholine.

phosphatide lipids such as these have been implicated in gentamicin resistance (Pechey *et al.* 1974) and also in the resistance of *Esch. coli* towards the polymeric biguanide Vantocil. The latter has been shown to interact specifically with acidic phospholipids (Ikeda *et al.* 1984; Broxton *et al.* 1984). Both of these resistance mechanisms involve modification of the target site. Other workers have been unable to demonstrate direct associations between changes in sensitivity towards antiseptic compounds and individual membrane components. Thus Gilbert & Brown (1978*b*) report that C-dep *Esch. coli* are particularly sensitive towards substituted phenols and 2-phenoxy ethanol (Fig. 4). Activity was assessed in this instance as an increase in the permeability of the cells towards protons. Ismail *et al.* (1977) assessed directly the bactericidal activity of a number of antiseptics and reported that M-dep and P-dep led to increased resistance of *Esch. coli* towards chlorhexidine relative to C-dep; P-dep decreased resistance towards both cetrimide and phenol (Fig. 5). In this instance they were able to attribute changes in resistance to alteration of the LPS moieties of the cell envelope as indicated by changes in cell surface charge. Klemperer *et al.* (1977) obtained similar results with *Proteus mirabilis*; M-dep and P-dep increased resistance towards cetrimide and phenol relative to C-dep cultures. By far the greatest change noted by these authors was for chlorhexidine resistance under P-dep. The latter effect was similar to that observed with the related bibiguanide Vantocil in *Esch. coli* and hence might also be explained in terms of an altered acidic phospholipid content (Broxton *et al.* 1984).

Divalent cations serve to stabilize the outer membrane of Gram-negative bacteria. In *Ps. aeruginosa* they play an especially important role, since this organism lacks Braun's lipoprotein. Kenward *et al.* (1979) and Boggis *et al.* (1979) evaluated the combined effects of M-dep or C-dep and divalent cation availability in the medium. They concluded that under conditions of M-dep, cations such as Mn^{2+} and Ca^{2+} were able to replace Mg^{2+} in the envelope, and that this led to a partial recovery of sensitivity lost under M-dep towards ion-chelating agents such as

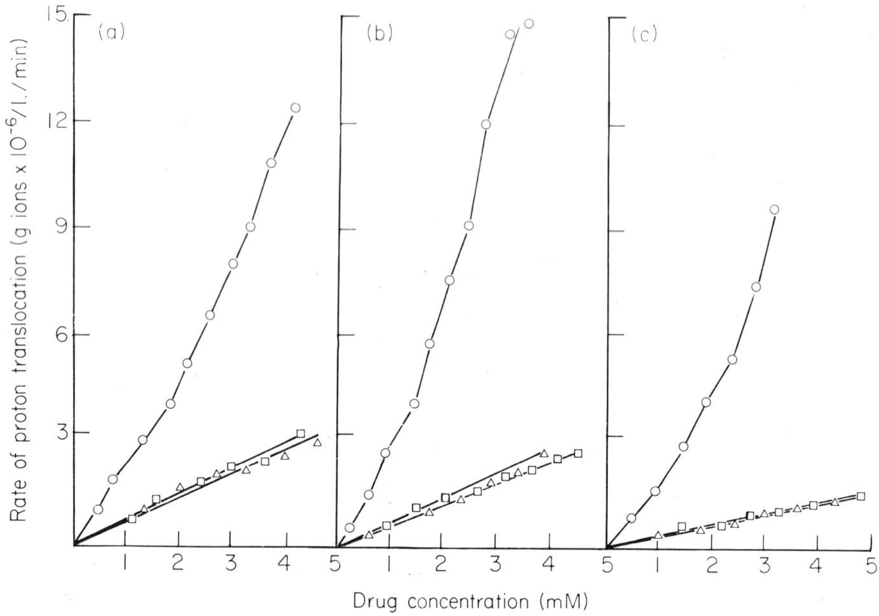

FIG. 4. Effect of nutrient depletions upon the sensitivity of *Escherichia coli* K12 W3110 to (a) 4-chlorophenol, (b) 3-chlorophenol and (c) 2-phenoxyethanol ○ carbon-depleted; □, phosphate-depleted and △, magnesium-depleted (from Gilbert & Brown 1978*b*).

EDTA. Resistance under M-dep was attributed to replacement of Mg^{2+} bridges in the OM by polyamides (Wilkinson 1975). Such effects would have implications to the sensitivity of cells towards other ion-chelating agents such as 8-hydroxyquinoline, tetracycline and EDTA. Failure to attribute antiseptic resistance to particular envelope components in many studies is probably a reflection of the multiplicity of steps involved in drug permeation to its target sites and also of the scarcity of data concerning polysaccharide composition of LPS and OM porin–protein composition under conditions of nutrient-depletion. Effects of nutrient-depletion upon the sensitivity of a wide range of organisms to chemical inactivation have been demonstrated. Al-Hiti & Gilbert (1980) evaluated such effects with the United States Pharmacopoeial Preservative-Efficacy test micro-organisms and several preservative/antiseptic compounds (*United States Pharmacopoeia* 1980). Results are summarized in Table 4. With the exception of *Aspergillus niger*, the preservative sensitivity of the micro-organisms varied significantly with nutrient depletion. Least variation in sensitivity was observed towards thiomersal, and the greatest towards benzalkonium chloride. Notably *Ps. aeruginosa* was the most resistant to all the agents and also showed the greatest variation with nutrient-depletion, justifying its notoriety as an organism particularly resistant to chemical inactivation (Brown

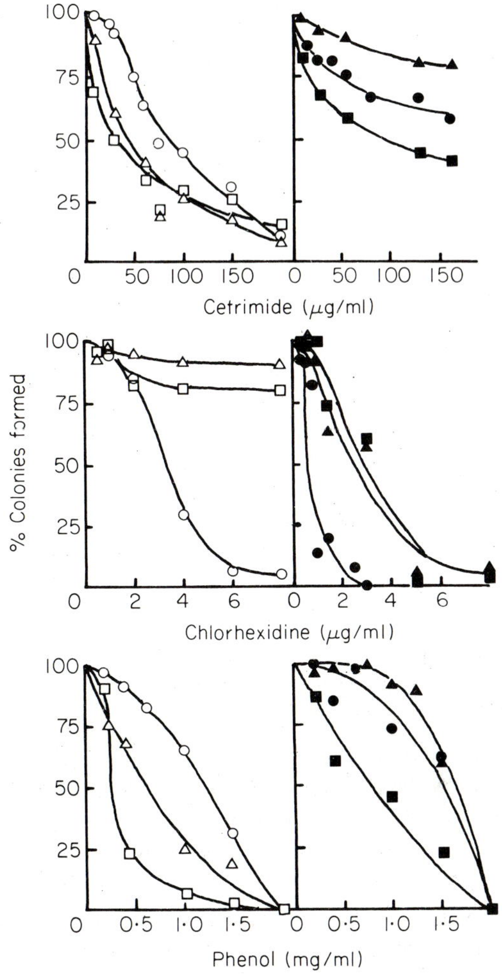

FIG. 5. The formation of colonies by *Escherichia coli* R⁻ (open symbols) and R⁺ (closed symbols) on agar containing cetrimide, chlorhexidine or phenol, following depletion of glucose ●○, Mg^{2+} ▲△, or phosphate ■□. Results are the mean of two experiments (five plates/sample) expressed as a percentage of counts in the absence of disinfectant (from Klemperer *et al.* 1980).

1975). Isoeffective concentrations of benzalkonium chloride against this organism, for example, varied from 2.5×10^{-3} (%w/v) for citrate grown P-dep cells to 5×10^{-2} (%w/v) for glycerol grown C-dep cultures. Variation in preservative sensitivity for the remaining organisms was significant, but with the exception of the thiomersal sensitivity of *Staph. aureus*, never exceeded three times the minimum effective concentration. No universal pattern of nutrient-depletion and preservative sensitivity

TABLE 4. *Effect of various nutrient depletions upon the ability of microorganisms to grow in the presence of preservative*

	Preservative concentration (%w/v × 10^{-4}) required to reduce the number of colony-forming units by 90%								
	Thiomersal			Chlorhexidine			Benzalkonium chloride		
Organism	N-dep	C-dep	P-dep	N-dep	C-dep	P-dep	N-dep	C-dep	P-dep
Escherichia coli	0.53	0.37	0.60	24.0	13.0	20.0	100.0	65.0	82.0
Pseudomonas aeruginosa									
grown on citrate	1.90	1.40	0.25	28.0	24.0	30.0	100.0	450.0	25.0
grown on glycerol	5.00	5.00	3.70	62.0	45.0	41.0	500.0	500.0	180.0
Staphylococcus aureus	0.12	0.54	0.14	8.0	10.5	10.5	0.5	0.6	0.6
Candida albicans	0.009	0.009	0.013	10.5	11.0	10.0	52.0	45.0	35.0
Aspergillus niger	0.025	0.025	0.025	2.3	2.3	2.3	2.5	2.6	2.5

* Al-Hiti & Gilbert (1980).

evolved between species but general patterns emerged for the three preservatives for each organism. Thus, for *Esch. coli* the sensitivity of C-dep, N-dep, P-dep; for *Ps. aeruginosa*, P-dep, N-dep, C-dep and for *Staph. aureus*, N-dep, P-dep, C-dep. The greatest changes in preservative sensitivity observed in this study were between cultures grown upon different carbon sources rather than different nutrient-depletions. Earlier workers have made similar observations by manipulating the carbon-source to cause the accumulation of cell wall and cytoplasmic lipids (Chaplin 1952; Vaczi & Farkas 1961).

An inescapable conclusion to be drawn from the various works on nutrient-availability and drug sensitivity is that the properties of typical laboratory nutrient broth-grown cultures bear little relation to those found in natural habitats.

Correlation of Cell Lipid with Resistance

It is apparent from evidence from various sources that variation in total readily-extractable lipids (REL) may be a contributory factor in bacterial resistance to chemical inactivation (Chaplin 1952; Church *et al.* 1956; Lowick & James 1957; Truby & Bennett 1961; Vaczi & Farkas 1961; Ivanov *et al.* 1964). Chaplin (1951) obtained a 43-fold increase in the resistance of *Esch. coli* to some quaternary ammonium compounds, associated with increased lipophilicity in resistant cells. Enhanced lipophilicity in this instance was demonstrated by staining with Sudan black B. In these and similar studies, gross cellular lipid composition was influenced by growth of the cells in the presence of carbohydrate substrates and/or growth factors known to perturb lipid biosynthesis. In this manner Hugo & Stretton

(1966*a*, *b*) showed that when the lipid content of certain Gram-positive organisms was increased by growth in the presence of glycerol, these 'fattened' organisms became more resistant to the penicillins, quinacillin, benzylpenicillin, methicillin and cloxacillin. In some instances when the normal lipid content of the cells was decreased by treatment of the cells with lipase, then the level of resistance also decreased. This phenomenon was investigated further by assessing sensitivity towards homologous series of 4-alkylphenols (Table 5) in 'fattened' and 'lipid-depleted' cultures of *Staph. aureus* (Hugo & Franklin 1968; Hugo & Davidson 1973). The concept introduced on p. 261 of optimal activity within homologous series of drugs, associated with particular log P_0 values, is applicable to these data. It is unfortunate that the choice of *n*-alkyl phenols in these studies show only one side of a parabolic relationship of activity on *n*-alkyl chain length. One would expect the variously grown cultures to exhibit different log P_0 values (optimally active homologue) in addition to different levels of activity at these optima. The data presented by these authors are compatible with such a hypothesis, in that changes in activity were least for the most hydrophilic compounds and that 'lipid-depletion' increased sensitivity towards the lower members of the series and 'fattening' decreased it. The compound series is not long enough, however, to demonstrate cut-offs in activity, these being generally observed at higher chain lengths (Daoud *et al.* 1983). Bulman & Stretton (1975) perturbed phospholipid biosynthesis in cultures of *Esch. coli*, *Staph. aureus* and *Saccharomyces cerevisiae* with the inhibitor 3-chloro-propane-1,2-diol and assessed sensitivity of these 'lipid-depleted' cultures towards a variety of antimicrobial agents. They observed that the extent of change in sensitivity could be related to drug hydrophilicity.

Interpretation of experiments such as these is difficult because the cells are grown in a wide variety of media. Changes in REL content will only be a reflection of many other changes occurring in the cells and will not detect alteration of the

TABLE 5. *MIC (μg/ml) for* Staphylococcus aureus *(Oxford) grown in nutrient broth (normal), under biotin deficiency ('lipid-depleted') and in the presence of 3% (v/v) glycerol ('fattened')**

Agent	Normal cells	'Fattened' cells	'Lipid-depleted' cells
Phenol	2200	2200	1550
4-methylphenol	1390	1300	1660
4-ethylphenol	640	640	520
4-propylphenol	390	390	220
4-butylphenol	85	90	—
4-pentylphenol	38	44	—
4-hexylphenol	42	52	—

* Data from Hugo & Franklin (1968) and Hugo & Davidson (1973).

protein composition or LPS content of the membranes. These could equally well influence drug permeation. In these respects experiments which utilize various nutrient-depletions in a common chemically defined medium (preceding section) are more likely to lead to interpretable results. Here changes in drug sensitivity have been observed when little, if any, change in total REL has occurred (Tables 2 & 3, Figs. 4 & 5).

Influence of Cellular Growth-rate on Preservative Sensitivity

Cells will grow at their maximal growth-rate when provided with an excess of all necessary nutrients. This can often be as rapid as three divisions per hour. In practice cellular growth rate is regulated by the supply of essential nutrients (p. 262). These can vary from carbon-sources through to oxygen requirements for respiration. Faced with a total absence of a critical nutrient, cells are unable to grow and divide. Such an eventuality cannot arise spontaneously in a natural environment, rather the cells are confronted with diminishing supplies of one material relative to others and growth rate gradually decreases from a high maximal rate cell division every 20 min to that characteristic of early stationary populations of every 20–30 h. During these step-down responses, cells are able to economize in the usage of particular nutrients by reduction in the relative amounts of *m*RNA and ribosomal RNA in the cells and also in the quantities of DNA (Mandelstam & McQuillen 1973). A reduced rate of metabolic growth allows further economies to be made in the number and types of specific transport-proteins associated with the non-limiting nutrients and also in terms of catabolic and anabolic enzymes. Recycling of cytoplasmic constituents can also occur in order to supplement temporarily diminishing nutrient supplies. The gross cellular composition of cells and also their envelope composition when growing at sub-maximal growth rates is, therefore, highly dependent upon that rate of growth and upon the nature of the nutrient restricting growth. These effects will indirectly influence sensitivity towards a wide range of chemical agents.

There are a number of ways in which growth rate can be controlled in the laboratory. These depend either upon changes in the physical environment of the cultures, i.e. pH and temperature, or upon the 'richness' of the medium supplied. In both instances changes can be induced in the cells which are a reflection of the method of controlling growth rate rather than of this attribute *per se*. Thus alteration of growth temperature affects the degree of saturation in the fatty-acid components of the cytoplasmic and OM (Ring 1965) and changes in carbon source can influence lipid composition of the cells (Hugo & Franklin 1968), polyhydroxybutyrate deposition—PHB—(Gilbert & Brown 1978*a*) and protein composition of the cell envelope (Turnowsky *et al.* 1983) unassociated with changes in specific-growth rate. The chemostat (Herbert 1956) offers the facility to control growth rate without any change in the physiochemical environment of the cells. Chemically defined media are used in which one nutrient is limiting whereas all the others are present in excess. The chemostat is a fixed volume fermentor in which the cells are held essentially on the verge of stationary phase, but allowed to grow at a rate proportional to that of the addition of fresh medium. Overflow maintains constant volume in the fermentor which is sparged with air and thoroughly mixed. Provided fresh medium is dispersed rapidly throughout the culture vessel, then the specific growth rate of the culture is directly proportional to its rate of dilution (Herbert 1956). Various workers have used systems such as these to study the effects of growth rate *per se* upon the physiological status of micro-organisms (Tempest & Ellwood 1969; Dean *et al.* 1977). There have been few systematic studies, however, in which changes in cellular physiology associated with growth rate have been examined in terms of sensitivity to antiseptics and preservatives. This is clearly important as in natural environments organisms will generally be starved and thus grow very slowly, if at all. Such slow-growing cells are generally regarded as being particularly resistant to chemical inactivation (Brown 1975; Finch & Brown 1975; Gilbert & Brown 1978*a*, 1980).

In Gram positive organisms the extent of peptidoglycan crosslinking varies markedly with specific growth-rate of the cells (De la Rose *et al.* 1982). Fast growing cultures are particularly sensitive to the actions of lysozyme, certain antibiotics and preservatives (Gilbert & Brown 1980). In a recent study Turnowsky *et al.* (1983) observed that the number of penicillin-binding proteins in cultures of iron-limited *Ps. aeruginosa* varied markedly with growth rate. Finch & Brown (1975) observed that this organism became particularly sensitive towards polymyxin and EDTA with increasing growth rate. This was associated with changes in susceptibility of the cells towards phagocytosis (Finch & Brown 1978) and also in lipopolysaccharide content of the cells (Gilbert & Brown 1978*a*). In the latter paper, the sensitivity of *Ps. aeruginosa* towards some substituted phenols was observed to increase markedly with increasing growth rate and was greater for C-dep than for M-dep cultures. Variation in isoeffective concentrations for these agents in cell suspensions prepared at different growth rates was thought to reflect alteration in the ease of drug penetration through the OM to the cytoplasmic membrane either by the formation of sites possessing high affinity for a drug which reduced availability at the cytoplasmic membrane or by decreased permeability of the OM towards the drug (Fig. 6). Uptake of drugs by cell suspensions has often been correlated with their sensitivity (Judis 1966; Bean & Das 1966; Mathre 1968) and in this instance was found to increase with drug sensitivity both for intact cells and for cells from which REL had been extracted. Total REL and PHB content of the cells remained constant for both limitations over the range of growth rates studied, phospholipid decreased and fatty and neutral lipids increased with increasing growth rate. Neither correlated with the variation in the observed drug sensitivity but would presumably affect sensitivity towards bisbiguanides such as chlorhexidine (Ikeda *et al.* 1984;

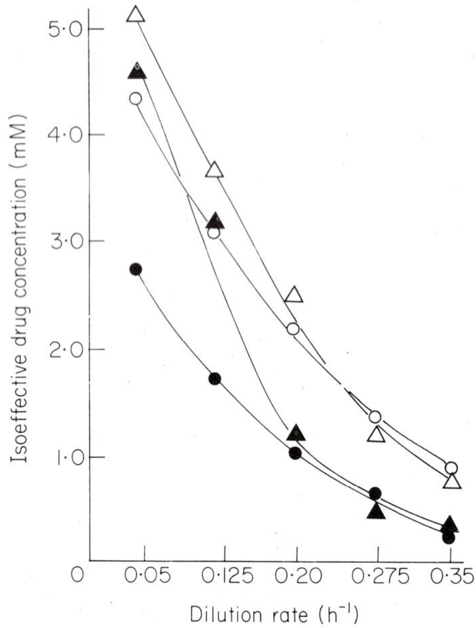

FIG. 6. Effect of nutrient limitations and growth rate on the sensitivity of *Pseudomonas aeruginosa* suspensions to uncoupling concentrations of 3-chlorophenol (▲△) and 4-chlorophenol (●○). Open symbols, magnesium-limited cultures; closed symbols, glucose-limited cultures (from Gilbert & Brown 1978*a*).

Broxton *et al.* 1984). The 3-ketodeoxyoctonic acid (KDO) content, a marker for LPS, varied continuously with the growth rate of the cultures, slower growing cells containing more than fast growing ones. C-dep cultures had lower levels of KDO than M-dep ones at all growth rates. Such variation correlated significantly with the variation in iso-effective drug concentrations (Fig. 7). These results suggest that in this instance the LPS had formed a barrier in the OM against penetration of the drug into the cytoplasm and that increased LPS had resulted in decreased uptake of the drug with an associated increase in iso-effective concentration.

Growth inhibitory and bactericidal activity for the *n*-alkyltrimethylammonium bromides are, for any given cell population, parabolically related to *n*-alkyl chain length and thereby to log P_0 (Al-Taae *et al.* 1983). The chain length at which optimal activity is demonstrated varies between cell types and reflects the lipophilicity and barrier properties of the cell envelope (p. 261 and Hansch & Clayton 1973). Alteration in the envelope through changes in growth rate and nutrient limitation might be expected to produce changes in log P_0 and also in the level of activity demonstrated by the optimally active compound. Such a hypothesis was investigated by Wright & Gilbert (1984) who used chemostat-grown cultures

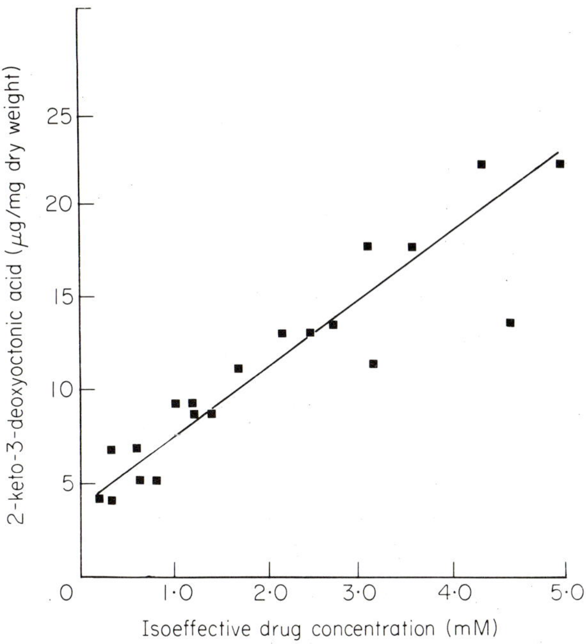

FIG. 7. Interrelationship of the sensitivity of *Pseudomonas aeruginosa* suspensions to chlorinated phenols and their 2-keto-3-deoxyoctonic acid content (from Gilbert & Brown 1978*a*).

of *Esch. coli* under M-dep, C-dep, P-dep or nitrogen-depletion and at growth rates between 0.05 and 0.6/h. In all cases the resistance maximized for growth rates of between 0.1 and 0.23/h and decreased markedly at the faster growth rates. Resistance also decreased, albeit to a lesser extent, for more slowly growing cultures. Data for C-dep cultures are presented in Fig. 8. Generally, P-dep cells were the most sensitive and C-dep and M-dep the least. These effects of nutrient depletion were greatest at the faster growth rates. P-dep cells also showed the most marked change in sensitivity with growth rate. Whilst optimal activity for any given compound was not demonstrated at particular growth rates, the greater dependence of activity upon growth rate for the longer chain-length compounds suggests that changes in log P_0 had occurred. Data such as these underline the difficulties of predicting the activities of preservatives within pharmaceutical products.

Conclusions

The bacterial cell is a remarkably flexible structure and will vary in structure and composition according to the nutritional status of the environment, its physical

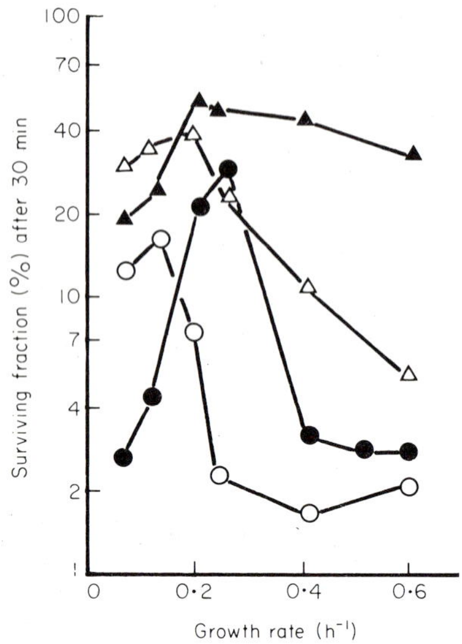

FIG. 8. Effect of growth-rate upon the sensitivity of *Escherichia coli* grown under carbon-limitation towards some *n*-alkyltrimethyl ammonium bromides. *n*-alkyl chain lengths of ▲, C_{12}; △, C_{14}; ●, C_{16} and ○, C_{18} (from Wright & Gilbert 1984).

properties and also upon the growth rate. Typical nutrient broth-grown cultures do not reflect adequately the properties of cells in their natural habitats. Thus organisms are able to resist the action of many antimicrobial agents when challenged *in situ*. It is suggested that whilst little direct evidence is available concerning nutrient availability in particular sites, efforts should be made to use cells closely resembling those occurring naturally for the evaluation and development of antimicrobial agents. In this respect chemostat cultures, depleted of carbon or nitrogen sources, might prove relevant to non-animal habitats and iron-limited cultures to infection states.

References

AL-HITI, M. M. A. & GILBERT, P. 1980 Changes in preservative sensitivity of the USP antimicrobial agents effectiveness test micro-organisms. *Journal of Applied Bacteriology* **49**, 119–126.

AL-TAAE, A. N., DICKINSON, N. A. & GILBERT, P. 1983 Antimicrobial activity of some alkyltrimethylammonium bromides. *Journal of Pharmacy and Pharmacology* Suppl. **35**, 61S.

BASSFORD, P. J., DIEDRICH, D. L., SCHNAITMAN, C. A. & REEVES, P. 1977 Outer-membrane proteins of *Escherichia coli*. VI Protein alterations in bacteriophage resistant mutants. *Journal of Bacteriology* **131**, 608–622.

BEAN, H. S. & DAS, A. 1966 The adsorption of phenols by *Escherichia coli* and their bactericidal activity. *Journal of Pharmacy and Pharmacology* Suppl. **18**, 107–113.

BENZ, R. & HANCOCK, R. E. W. 1981 Properties of large ion permeable pores from protein F of *Pseudomonas aeruginosa* in lipid bilayer membranes. *Biochimica et Biophysica acta* **646**, 298–308.

BOGGIS, W., KENWARD, M. A. & BROWN, M. R. W. 1979 Effects of divalent metal cations in the growth medium upon sensitivity of batch-grown *Pseudomonas aeruginosa* to EDTA or Polymyxin B. *Journal of Applied Bacteriology* **47**, 477–488.

BROUGHTON, A., ANDERSON, I. R. M. & BOWDEN, C. H. 1968 Magnesium-deficiency syndrome in burns. *Lancet* **ii**, 1156–1158.

BROWN, M. R. W. 1975 The role of the envelope in resistance. In *Resistance of Pseudomonas aeruginosa* ed. Brown, M. R. W. pp. 71–107 London: John Wiley & Sons.

BROWN, M. R. W. 1977 Nutrient depletion and antibiotic susceptibility. *Journal of Antimicrobial Chemotherapy* **3**, 198–201.

BROWN, M. R. W. & MELLING, J. 1969*a* Loss of sensitivity to EDTA by *Pseudomonas aeruginosa* grown under conditions of magnesium-limitation. *Journal of General Microbiology* **54**, 439–444.

BROWN, M. R. W. & MELLING, J. 1969*b* Role of divalent cations in the action of polymyxin B and EDTA on *Pseudomonas aeruginosa*. *Journal of General Microbiology* **59**, 263–274.

BROWN, M. R. W. & SCOTT-FOSTER, J. H. 1971 Effect of slime on the sensitivity of *Pseudomonas aeruginosa* to EDTA and polymyxin. *Journal of Pharmacy and Pharmacology* Suppl. **23**, 236S.

BROWN, M. R. W., ANWAR, H. & LAMBERT, P. 1984 Evidence that mucoid *Pseudomonas aeruginosa* in the cystic fibrosis lung grows under iron-restricted conditions. *FEMS Microbiology Letters* **21**, 113–117.

BROXTON, P., WOODCOCK, P. M. & GILBERT, P. 1984. Interaction of some polyhexamethylenebiguanides and membrane phospholipids in *Escherichia coli*. *Journal of Applied Bacteriology* **57**, 115–124.

BULLEN, J. J., ROGERS, H. J. & GRIFFITHS, E. 1974 Bacterial iron metabolism in infection and immunity. In *Microbial iron metabolism: a comprehensive treatise*, 517 ed. Neilands, J. B. pp. 517. London: Academic Press.

BULLMAN, R. A. & STRETTON, R. J. 1975 Influence of 3-chloropropane-1,2-diol on microbial lipid composition and its effect on the activity of some antimicrobial agents. *Microbios* **12**, 7–27.

CHAPLIN, C. E. 1951 Observations on quaternary ammonium disinfectants. *Canadian Journal of Botany* **29**, 373–382.

CHAPLIN, C. E. 1952 Bacterial resistance to quaternary ammonium disinfectants. *Journal of Bacteriology* **63**, 453–458.

CHURCH, B. D., HALVORSAN, H. & RAMSEY, D. S. 1956 Population heterogeneity in the resistance of spores to ethylene oxide. *Journal of Bacteriology* **72**, 242–251.

COSTERTON, J. W. 1977 Cell envelope as a barrier to antibiotics. In *Microbiology* pp. 151–157. Washington, D.C.: American Society of Microbiology.

COSTERTON, J. W., INGRAM, J. M. & CHENG, K.-J. 1974 Structure and function of the cell envelope of Gram negative bacteria. *Bacteriological Reviews* **38**, 87–110.

COSTERTON, J. W., GEESEY, G. G. & CHENG, K.-J. 1978 How bacteria stick. *Scientific American* **238**, 86–95.

DANIELLI, J. F. 1952 In *The permeability of natural membranes* ed. Dawson, H. & Danielli, J. F. pp. 80–84 London: Cambridge University Press.

DAOUD, N. N., DICKINSON, N. A. & GILBERT, P. 1983 Antimicrobial activity and physico-chemical properties of some alkyldimethylbenzylammonium chlorides. *Microbios* **37**, 73–85.

DEAN, A. C. R., ELLWOOD, D. C., MELLING, J. & ROBINSON, A. 1977 The action of antibacterial agents on bacteria grown in continuous culture. In *Continuous culture* 6: *Applications and new fields* ed. Dean, A. C. R., Ellwood, D. C., Evans, C. G. T. & Melling, J. pp. 251–261. Chichester: Ellis Horwood.

DECAD, G. & NIKAIDO, H. 1976 Outer-membrane of Gram negative bacteria. XII Molecular sieving function of cell wall. *Journal of Bacteriology* **128**, 325–336.

DE LA ROSE, E. J., DE PEDRO, M. A. & VAXQUEZ, D. 1982 Modification of penicillin-binding-proteins of *Escherichia coli* associated with changes in the rate of growth of the cells. *FEMS Microbiology Letters* **14**, 91–94.

DE PAMPHILUS, M. C. 1971 Dissociation and reassembly of *Escherichia coli* outer membrane and of lipopolysaccharide and their reassembly onto flagellar basal bodies. *Journal of Bacteriology* **105**, 1184–1199.

ELLWOOD, D. C. & TEMPEST, D. W. 1972 Effects of environment on bacterial wall content and composition. *Advances in Microbial Physiology* **7**, 83–117.

FAVERO, M. S., BOND, W. W., PETERSON, N. J. & COOK, E. H. Jnr. 1983 *Scanning electron microscopic observations of bacteria resistant to iodophor solutions.* Proceedings of the International Symposium on Povidine. pp. 158–166. University of Kentucky.

FINCH, J. E. & BROWN, M. R. W. 1975 The influence of nutrient-limitation in a chemostat on the sensitivity of *Pseudomonas aeruginosa* to polymyxin and EDTA. *Journal of Antimicrobial Chemotherapy* **1**, 379–386.

FINCH, J. E. & BROWN, M. R. W. 1978 Effect of growth environment on *Pseudomonas aeruginosa* killing by rabbit polymorphonuclear leukocytes and cationic proteins. *Infection and Immunity* **20**, 340–346.

GERHARDT, P. & Judge, *J. A. 1964 Porosity of isolated cells walls of* Saccharomyces cerevisiae and Bacillus megaterium. *Journal of Bacteriology* **87**, 945–995.

GILBERT, P. & BROWN, M. R. W. 1978*a* Influence of growth rate and nutrient limitation on the gross cellular composition of *Pseudomonas aeruginosa* and its resistance to some uncoupling phenols. *Journal of Bacteriology* **133**, 1066–1072.

GILBERT, P. & BROWN, M. R. W. 1978*b* Effect of R-plasmid RPI and nutrient-depletion on the gross cellular composition of *Escherichia coli* and its resistance to some uncoupling phenols. *Journal of Bacteriology* **133**, 1062–1065.

GILBERT, P. & BROWN, M. R. W. 1980 Cell-wall mediated changes in sensitivity of *Bacillus megaterium* to chlorhexidine and 2-phenoxyethanol, associated with growth-rate and nutrient-limitation. *Journal of Applied bacteriology* **48**, 223–230.

GILBERT, P., BEVERIDGE, E. G. & CRONE, P. B. 1977*a*. The lethal action of 2-phenoxyethanol and its analogues upon *Escherichia coli* NCTC 5933. *Microbios* **19**, 125–142.

GILBERT, P., BEVERIDGE, E. G. & CRONE, P. B. 1977*b* Effect of 2-phenoxyethanol on permeability of *Escherichia coli* NCTC 5933 to inorganic ions. *Microbios* **19**, 17–26.

GOVAN, J. R. W. 1976 Antibiotic therapy and cystic fibrosis: increased resistance of mucoid *Pseudomonas aeruginosa* to carbenicillin. *Journal of Antimicrobial Chemotherapy* **2**, 215–217.

HAMILTON, W. A. 1970 The mode of action of membrane active antibacterials. *FEBS Symposium* **20**, 71–79.

HAN, M. N. & SILHAVY, T. J. 1981 Genetic analysis of the outer membrane proteins of *Escherichia coli. Annual Reviews of Genetics* **15**, 91–142.

HANCOCK, R. E. W., HANTKE, K. & BRAUN, V. 1976 Iron transport in *Escherichia coli* K12, involvement of the colicin B receptor and of a citrate inducible protein. *Journal of Bacteriology* **127**, 1370–1375.

HANCOCK, R. E. & NIKAIDO, H. 1978 Outer-membranes of Gram negative bacteria XIX. Isolation from *Pseudomonas aeruginosa* PAO1 and use in reconstitution and definition of the permeability barrier. *Journal of Bacteriology* **136**, 381–390.

HANSCH, C. 1973 Quantitative approaches to pharmacological structure-activity relationships. In *Structure activity relationships* Vol. 1 ed. Cavallito, C. J. pp. 1–21. Oxford: Pergamon.

HANSCH, C. & CLAYTON, J. M. 1973 Lipophilic character and biological activity of drugs: the parabolic case. *Journal of Pharmaceutical Science* **62**, 1–32.

HANSCH, C. & FUJITA, T. 1964 A method for the correlation of biological activity and chemical structure. *Journal of American Chemical Society* **86**, 1616–1626.

Hantke, K. 1976 Phage T6—colicin K receptor and nucloside transport in *Escherichia coli. FEBS Letters* **70**, 109–112.

Herbert, D. 1956 The continuous culture of bacteria: a theoretical and experimental study. *Journal of general Microbiology* **14**, 601–622.

Hess, E. L. & Lagg, S. E. 1958 Some physical aspects of the bacterial cell. *Science, N.Y* **128**, 356–358.

Holme, T. 1972 Influence of environment on the content and composition of bacterial envelopes. *Journal of Applied Chemistry & Biotechnology* **22**, 391–399.

Hugo, W. B. & Davidson, J. R. 1973 Effect of cell lipid depletion in *Staphylococcus aureus* upon its resistance to antibacterial drugs. 2. A comparison of the response of normal and lipid-depleted cells of *S. aureus* to antibacterial drugs. *Microbios* 8, 73–84.

Hugo, W. B. & Franklin, I. 1968 Cellular lipid and the antistaphylococcal activity of phenol. *Journal of General Microbiology* **52**, 365–373.

Hugo, W. B. & Longworth, A. R. 1965 The effects of chlorhexidine on the electrophoretic mobility, cytoplasmic contents, dehydrogenase activity and cell walls of *Escherichia coli* and *Staphylococcus aureus. Journal of Pharmacy and Pharmacology* **18**, 569–578.

Hugo, W. B. & Stretton, R. G. 1966*a* The role of cellular lipid in the resistance in Gram positive bacteria to penicillins. *Journal of General Microbiology* **42**, 133–8.

Hugo, W. B. & Stretton, R. G. 1966*b* Effect of cellular lipid on the sensitivity of some Gram positive bacteria to penicillins. *Nature, London* **209**, 940–942.

Ikeda, T., Ledwith, A., Bamford, C. H. & Hann, R. A. 1984 Interaction of a polymeric biguanide with phospholipid membranes. *Biochimica et Biophysica acta* **769**, 57–66.

Ismail, N. T. A. J., Klemperer, R. M. M. & Brown, M. R. W. 1977 Resistance of *Escherichia coli* to disinfectants: influence of R-plasmid RPI and nutrient depletion. *Journal of Applied Bacteriology* **43**, xvi–xvii.

Ivanov, V., Markov, Golovinskii, E. & Kharisanouat, T. 1964 Importance of surface lipids for various biological properties of a *Pseudomonas aeruginosa* strain. *Zeitschrift für Naturforschung* **196**, 604–606.

Judis, J. 1966 Factors influencing binding of phenol derivatives to *Micrococcus lysodeikticus* cells. *Journal of Pharmaceutical Science* **53**, 803–817.

Kenward, M. A., Brown, M. R. W. & Fryer, J. J. 1979 The influence of calcium or manganese on the resistance to EDTA, polymyxin B or cold shock, and the composition of *Pseudomonas aeruginosa* grown in glucose—or magnesium-depleted batch culture. *Journal of Applied Bacteriology* **47**, 489–503.

Klemperer, R. M. W., Al-Dujaili, D. A., Lawson, J. J. & Brown, M. R. W. 1977 The effect of nutrient-depletion on the resistance of *Proteus mirabilis* to antibacterial agents. *Journal of Applied Bacteriology* **43**, xvi.

Klemperer, R. M. M., Gilbert, P., Meier, M. M., Cozens, R. M. & Brown, M. R. W. 1979 Influence of suspending media upon the susceptibility of *Pseudomonas aeruginosa* NCTC 6750 and its sphaeroplasts to polymyxin B. *Antimicrobial Agents and Chemotherapy* **15**, 147–151.

Leive, L. 1974 The barrier function of the Gram negative envelope. *Annals of the N.Y. Academy of Science* **235**, 109–129.

Lien, E. J., Hansch, C. & Anderson, S. M. 1968 Structure-activity correlations for antibacterial agents of Gram positive and Gram negative cells. *Journal of Medical Chemistry* **11**, 430–441.

Lowick, J. H. & James, A. M. 1957 The electrokinetic properties of *Aerobacter aerogenes.* A comparison of normal and crystal violet-trained cells. *Biochemical Journal* **65**, 431–438.

Lugtenberg, B. & van Alphen, L. 1983 Molecular architecture and functioning of the outer-membrane of *Escherichia coli* and other Gram negative bacteria. *Biochimica et Biophysica acta* **737**, 51.

Mandelstam, J. & McQuillen, K. 1973 *Biochemistry of bacterial growth.* 2nd Ed. pp. 143–154. London: Blackwell Scientific.

Marquis, R. E. 1968 Salt induced contraction of bacterial cell walls. *Journal of Bacteriology* **95**, 775–781.

MATHRE, D. E. 1968 Uptake and binding of oxathiin systemic fungicides by resistant and sensitive fungi. *Phytopathology* **58**, 1464–1469.

MEADOW, P. M. 1975 Wall and membrane structures in genus *Pseudomonas*. In *Genetics and biochemistry of* Pseudomonas ed. Clarke, P. H. & Richmond, M. H. pp. 67–98 Chichester: John Wiley & Sons.

MEERS, J. L. & TEMPEST, D. W. 1970 The influence of growth-limiting substrate and medium NaCl concentrations on the synthesis of magnesium-binding sites in the walls of *Bacillus subtilis* var. *niger*. *Journal of general Microbiology* **63**, 325–331.

MINNIKIN, D. E., ABDULRAHIMZADEH, H. & BADDILEY, J. 1971*a* The interrelation of phorsphatidylethanolamine and glycosyldiglycerides in bacterial membranes. *Biochemical Journal* **124**, 447–448.

MINNIKIN, D. E., ABDULRAHIMZADEH, H. & BADDILEY, J. 1971*b* The interrelation of polar lipids in bacterial membranes. *Biochimica et Biophysica acta* **249**, 651–655.

MINNIKIN, D. E., ABDULRAHIMZADEH, H. & BADDILEY, J. 1972 Variation in polar lipid composition of *Bacillus subtilis* (Marburg) with different growth conditions. *FEBS Letters* **27**, 16–18.

MITCHELL, P. 1959 Biochemical cytology of micro-organisms. *Annual Reviews of Microbiology* **13**, 407–440.

NADIR, M. T. & GILBERT, P. 1980 Influence of potassium chloride upon the binding and antibacterial activity of chlorhexidine diacetate and $(ethoxy)_5$ octyl phenol (Triton X45) towards *Bacillus megaterium* KM. *Microbios* **26**, 51–63.

NAKAE, T. 1976*a* Outer membrane of *Salmonella*: isolation of protein complex that produces transmembrane channels. *Journal of Biological Chemistry* **251**, 2176–2178.

NAKAE, T. 1976*b* Identification of the outer membrane protein of *Escherichia coli* that produces transmembrane channels in reconstituted vesicle membranes. *Biochemical Biophysical Research Communications* **71**, 877–884.

NIKAIDO, H. 1976 Outer membrane of *Salmonella typhimurium*: transmembrane diffusion of some hydrophobic substances. *Biochimica et Biophysica acta* **433**, 118–132.

NIKAIDO, H. & NAKAE, T. 1979 The outer membrane of Gram negative bacteria. *Advances in Microbial Physiology* **20**, 163–250.

NIKAIDO, H., LUCKEY, M. & ROSENBERG, E. Y. 1980 Non-specific and specific diffusion channels in the outer-membrane of *Escherichia coli*. *Journal of Supramolecular Structure* **13**, 305–313.

OSBORN, M. J. & WU, H. C. P. 1980 Proteins of the outer membrane of Gram negative bacteria. *Annual Reviews of Microbiology* **34**, 369–422.

PALVA, E. T. & RANDALL, L. L. 1978 Arrangement of protein I in *Escherichia coli* outer membrane: cross-linking study. *Journal of Bacteriology* **133**, 279–286.

PAYNE, J. W. & GILVARG, C. 1968 Size restriction on peptide utilisation in *Escherichia coli*. *Journal of Biological Chemistry* **243**, 6291–6299.

PECHEY, D. T., YAU, A. O. P. & JAMES, A. M. 1974 Total and surface lipid of cells of *Pseudomonas aeruginosa* and their relationship to gentamycin resistance. *Microbios* **11**, 77–86.

RING, K. 1965 The effect of low temperatures on permeability in *Streptomyces hydrogenans*. *Biochemical and Biophysical Research Communications* **19**, 579–581.

ROANTREE, R. J., KUO, T.-T. & MACPHEE, D. G. 1977 The effect of defined lipopolysaccharide core defects upon antibiotic resistances of *Salmonella typhimurium*. *Journal of General Microbiology* **103**, 223–224.

ROSENBUSCH, J. P. 1974 Characterisation of the major envelope protein from *Escherichia coli*: regular arrangement on the peptidoglycan and unusual dodecylsulphate binding. *Journal of Biological Chemistry* **249**, 8109–8029.

SABET, S. F. & SCHNAITMAN, C. A. 1973 Purification and properties of the colicin E receptor of *Escherichia coli*. *Journal of Biological Chemistry* **248**, 1797–1806.

SCHERRER, R. & GERHARDT, P. 1971 Molecular sieving by the *Bacillus megaterium* cell wall and protoplast. *Journal of Bacteriology* **107**, 718–735.

SCHNAITMAN, C. A. 1971*a* Solubilisation of the cytoplasmic membrane of *Escherichia coli* by Triton X45. *Journal of Bacteriology* **108**, 545–552.

SCHNAITMAN, C. A. 1971*b* Effect of ethylenediamine tetra acetic acid, Triton X100 and lysozyme on the morphology and chemical composition of isolated cell walls of *Escherichia coli*. *Journal of Bacteriology* **108**, 533–544.

SOLOMAN, A. K. 1968 Characterisation of biological membranes by equivalent pores. *Journal of General Physiology* **51**, 335–354.

STOCK, J. B., RAUCH, B. & ROSEMAN, S. 1977 Periplasmic space in *Salmonella typhimurium* and *Escherichia coli*. *Journal of Biological Chemistry* **252**, 7850–7861.

TAMAKI, S., SATO, T. & MATSUHASHI, M. 1971 Role of lipopolysaccharides in antibiotic resistance and bacteriophage adsorption of *Escherichia coli* Kz12. *Journal of Bacteriology* **105**, 968–975.

TEMPEST, D. W. & ELLWOOD, D. C. 1969 The influence of growth conditions on the composition of some cell wall components of *Aerobacter aerogenes*. *Biotechnology & Bioengineering* **11**, 775–783.

TEMPEST, D. W., DICKS, J. W. & ELLWOOD, D. C. 1968 Influence of growth conditions on the concentration of potassium in *Bacillus subtilis* var *niger* and its positive relationship to cellular ribonucleic acid, teichoic acid and teichuronic acid. *Biochemical Journal* **106**, 237–243.

TRUBY, C. P. & BENNETT, E. O. 1961 Role of lipid in the protection of *Staphylococcus aureus* against trichlorophenol in mixed culture. *Applied Microbiology* **14**, 769–781.

TURNOWSKY, F., BROWN, M. R. W., ANWAR, H. & LAMBERT, P. A. 1983 Effect of iron limitation of growth rate on the binding of penicillin G to the penicillin binding protein of mucoid and non-mucoid strains of *Pseudomonas aeruginosa*. *FEMS Letters* **17**, 243–245.

UNITED STATES PHARMACOPOEIA 1980 XX edn. Philadelphia: Mack Publishing Co. Easton.

VACZI, F. & FARKAS, L. 1961 Association between lipid metabolism and antibiotic sensitivity. The lipid composition of antibiotic sensitive and resistant *Staphylococcus aureus* strains. *Acta Microbioligica Magyar* **8**, 205–211.

WAGMAN, G. H., BAILEY, J. V. & WEINSTEIN, M. J. 1975 Binding of aminoglycoside antibiotics to filtration materials. *Antimicrobial Agents and Chemotherapy* **7**, 316–319.

WEINBERG, E. D. 1974 Iron and susceptibility to disease. *Science, N.Y.* **184**, 952–956.

WEINBERG, E. D. 1978 Iron and infection. *Microbiological Reviews* **42**, 45–66.

WILKINSON, S. G. 1975 Sensitivity to ethylenediamine tetraacetic acid. In *Resistance of* Pseudomonas aeruginosa ed. Brown, M. R. W. pp. 145–183 London: John Wiley & Sons.

WOOKEY, P. & ROSENBERG, H. 1978 Involvement of inner and outer membrane components in the transport of iron and colicin B action in *Escherichia coli*. *Journal of Bacteriology* **133**, 661–666.

WRIGHT, N. E. & GILBERT, P. 1984 Interrelation of growth-rate and nutrient-limitation with the sensitivity of *Escherichia coli* ATCC 8739 towards some *n*-alkyltrimethylammonium bromides. *Journal of Pharmacy and Pharmacology Supplement* **36**, 55S.

YU, F., ICHIHARA, S. & MIZUSHIMA, S. 1979 A major outer membrane protein (0–8) of *Escherichia coli* K12 exists as a trimer in sodium dodecyl sulphate solution. *FEBS Letters* **100**, 71–74.

ZIMMERMAN, W. & ROSSELET, A. 1977 Function of the outer membrane of *Escherichia coli* as a permeability barrier to *beta*-lactam antibiotics. *Antimicrobial Agents and Chemotherapy* **12**, 368–372.

The Concentration Exponent of Disinfectants and Preservatives (Biocides)*

W. B. HUGO AND S. P. DENYER

Department of Pharmacy, The University Nottingham, UK

Many factors can affect the activities of disinfectants and preservatives (Russell 1982). One of these, which has been extensively studied but very frequently ignored, is the effect of dilution on activity, the so-called concentration exponent or dilution co-efficient denoted by the greek letter η (*eta*). As a result of the failure to appreciate the significance of η many inappropriate use-dilutions have been quoted for biocides, especially the solubilized phenolic type. In clinical practice the dilution of a preserved pharmaceutical product may well result in the inactivation of the preservative. Furthermore, end-point methods of evaluating disinfectants, such as the Rideal-Walker and Chick-Martin methods, by their nature take no account of the effect of dilution on activity.

Historical

A relationship between dilution of a biocide and its activity was elucidated at the end of the last century from the data of Kronig & Paul (1897)—an exemplary paper on the killing of *Bacillus anthracis* by mercury II chloride—which laid the foundations for studies in the field of antimicrobial evaluation and kinetics. A Japanese physical chemist, K. Ikeda, examined their results and, as a consequence of correspondence between the parties, Kronig & Paul showed in section IV of their paper that a relationship existed between disinfectant dilution and activity. This could be described by the use of a constant.

Ikeda had used data based on death times to formulate his notion of a relationship between concentration and activity. Chick (1908) re-examined the problem using

* In the text of this paper, present-day nomenclature for micro-organisms and, where appropriate, chemical substances has been substituted for that used by the authors quoted.

Preservatives in the Food, Pharmaceutical and Environmental Industries

0-632-01727-9

times for 'almost complete disinfection' and derived the following equation:

$$\frac{1}{C_o - C_n} \log \frac{C_n t_n}{C_o t_o} = \text{a constant relating concentration and activity} \quad (1)$$

where t_o was the disinfection time at concentration C_o and t_n was the disinfection time at concentration C_n.

In the same year and journal, Watson (1908), accepting Chick's proposal that the disinfection process was analogous to a chemical reaction, stated the relationship between concentration (C) and disinfection time (t) as:

$$C^{\eta} t = \text{constant} \quad (2)$$

where η would indicate the order of reaction. When activity is considered over a concentration range, this may be written (Tilley 1939) as:

$$\eta = \frac{\log t_2 - \log t_1}{\log C_1 - \log C_2} \quad (3)$$

where t_1 is the time to kill at concentration C_1 and t_2 is the time to kill at concentration C_2.

Watson went on to say that the best way to calculate η was to plot the log of biocide concentration against the log of killing time; η was then determined from the slope of the line so obtained. Equation (3) and Watson's (1908) graphical approach are now accepted methods for calculating η. As Watson's theoretical treatment and Chick's pioneering experimental work are the first documents to introduce a set of values of η to bacteriological science, it is proper to quote them here (Table 1).

The graphical method was employed by Lusher *et al.* (1984) to calculate η for potassium sorbate. These authors used a computer programme to calculate the slope of the best straight line relating log C and log t. This slope gave the value for η. This reference is quoted to enable readers to consult a recent paper and to see the application of Watson's method.

TABLE 1. *η values obtained by Chick (1908) and Watson (1908)*

Organism	Biocide	η
Staphylococcus aureus	Phenol	5.5
Salmonella paratyphi	Phenol	5.5
	Mercury II chloride	3.8
	Silver nitrate	0.86

The Practical Meaning of η

If a biocide has a concentration exponent of 1 its activity will be reduced by the power of 1 on dilution. Thus a three-fold dilution will mean that the disinfectant activity will be reduced by a factor of three or to one-third, i.e., the killing time will increase by a factor of three. If, however, a biocide has a concentration exponent of five a three-fold dilution will reduce the activity by 3^5 or 243 times, i.e., the time for a similar lethal event to that of the undiluted material may increase by a factor of 243.

The effect of the value of the concentration exponent on the activity of a disinfectant, antiseptic or preservative if diluted to one-half or one-third can be seen in Table 2. The figures speak for themselves and, whatever the academic arguments about the determination of η, if a biocide has an η value much above 3 then the rapid loss of activity on dilution should be recognized. Equally if it is wished to dilute out the activity of a biocide for sterility testing or some other reason, then this is easier the higher the value of η.

TABLE 2. *Loss of activity of a biocide on dilution according to its concentration exponent*

	Fraction of activity remaining on dilution to	
η	1/2	1/3
1	1/2	1/3
2	1/4	1/9
3	1/8	1/27
4	1/16	1/81
5	1/32	1/243
6	1/64	1/729
7	1/128	1/2187
8	1/256	1/6560

Experimental Determination of Concentration Exponents

Methodology

Since the three seminal papers leading to the concept and calculation of η—Ikeda in the paper of Kronig & Paul (1897), Chick (1908) and Watson (1908)—numerous values for η have appeared in a scattered literature.

A variety of experimental conditions and the use of both death rates and death times have been applied in the determination of η. Death rates may be determined as decimal reduction times (D values; Hurwitz & McCarthy 1985). Examples of these methods have been summarized by Berry & Michaels (1950). Death rates are acceptable if they remain constant throughout the disinfection process, i.e., there is a linear relationship between the loss of viability and time at each concentration of lethal agent examined. This is not always the case and certainly at concentrations of disinfectant which are not rapidly or very rapidly lethal, departure from the general linear trend is the rule rather than the exception. Accordingly more reliable data can be obtained if death times are used in place of death rates together with a graphical method of calculating η. Where investigators have used death times to determine η, however, another experimental variant has appeared, i.e., the chosen mortality level at which the killing time is taken for a given concentration. Some workers have used the time for 50% kill others have used 84%, 90%, 99%, 99.99%, right up to 99.999999% referred to as a virtual sterilization time (vst), the end point used by Chick.

According to Jordan & Jacobs (1944), death times at 99%, 99.9% and 99.999999% (virtual sterilization) were satisfactory end points to determine η. They have claimed that the use of lower mortality levels (50%) may lead to irregular results, especially at high biocide concentrations, where death times are very short. In their calculation of η, however, Withell (1942) and Berry & Michaels (1950) used probit—probability units; Bliss (1937)—log time regressions to estimate the time taken for 50% kill (LT_{50}). The latter authors took care to use biocide concentrations removed from those giving short LT_{50}s values. From his data, Withell (1942) calculated η for phenol and chlorocresol using both death rates and death times. His values for phenol and chlorocresol against *Escherichia coli* at 20 °C are given in Table 3. Jordan & Jacobs' (1944) values of η for phenol at 35 °C against *Esch. coli* as determined using differing end points are presented in Table 4.

An appraisal of the literature suggests that entirely acceptable results are obtained by calculating death times at 99.9% or greater levels of kill. The methodology therefore depends on the techniques for determining death times in the presence of lethal agents. Classical experimental work in this area with various strains of *Esch. coli* has been carried out by Withell (1942), with phenol and chlorocresol (4-chloro-3-methylphenol), Jordan & Jacobs (1944; 1946), with

TABLE 3. *η values derived from death times (LT_{50}) and death rates of* Escherichia coli

	Phenol	Chlorocresol
LT_{50}	4.3	7.9
Rate	4.5	8.3

TABLE 4. *Effect of differing end points on η and consequent loss of activity of phenol*

End point at % mortality level	η	Loss of activity for 1 in 3 dilution
99.999999	5.8	585
99.9	6.6	1409
99	7.0	2186

phenol, and Berry & Michaels (1950), with a series of glycolmonoalkyl ethers. Other micro-organisms have not received such detailed treatment. As these authors were working with biocides of high η values, inactivation of the antimicrobial agent was achieved by dilution. Successful inactivation is an essential feature of the methodology in order to permit the recovery of damaged but still viable organisms. The need for a specific inactivator for mercury salts was recognized by Kronig & Paul (1897). Inactivators commonly employed are shown in Table 5.

TABLE 5. *Inactivating agents for some antimicrobial agents*

Antimicrobial agent	Possible inactivating methods and agent(s)
Phenolics	Dilution Tweens (polysorbates)
Parabens	Dilution Tweens
Iodine and related compounds Chlorine and hypochlorites	Sodium thiosulphate
Glutaraldehyde	Glycine Dilution
Quaternary ammonium compounds, chlorhexidine	Lubrol + lecithin, Lecithin + Tween (Letheen)
Mercury compounds Organic arsenicals Bronopol	—SH compounds

The effect of biocide concentration

The linear relationship that exists between the logarithms of concentration and death time holds over a wide range of biocide concentration. It is important to recognize, however, that a minimum threshold bactericidal concentration will exist and levels of drug approaching this value will give extremely long death times with a resultant deviation from linearity (Jordan & Jacobs 1944; 1946). In addition, the accuracy with which short death times can be determined experimentally will limit the useable upper level of biocide concentration.

The effect of temperature on η

Tilley (1939) published a table of the effect of temperature on η for a number of phenols and alcohols. In general η decreases with increase in temperature. The examples from Tilley's data (Table 6) illustrate this. Tilley used death times and a modified Rideal-Walker technique to obtain his results. Jordan & Jacobs (1946), who referred to Tilley's work on the effect of temperature on η, declared that η was 'unaffected by temperature within wide limits'. Using phenol against *Esch. coli* and an end point based on 99.999999% kill, Jordan & Jacobs (1946) found the values for η given in Table 7.

Thus over the range 'room temperature' to 'body heat'—again being somewhat pragmatic—η is virtually independent of temperature at least for phenol and *Esch. coli*. Further work on the effect of temperature on η can be found in the work of Berry & Michaels (1950).

TABLE 6. *Effect of temperature on η values of phenols and ethanol**

		η values			
Organism	Biocide	10 °C	20 °C	30 °C	40 °C
	Phenol	7.9	7.5	5.9	5.2
Salmonella typhi	Ethanol	12.7	11.4	8.8	—
	4-cresol	8.9	8.4	6.4	5.3

* Examples taken from Tilley (1939).

TABLE 7. *Effect of temperature on η values of phenol using* Escherichia coli*

T °C	η
20	5.8
25	5.1
30	5.4
35	5.9
39	9.5

* The data of Jordan & Jacobs (1946).

The influence of the organism species

Chick's data gave the value of η for phenol against both *Salmonella paratyphi* and *Staphylococcus aureus* as 5.5. Withell (1942) collected the following data from the literature for η for phenol: *Staphylococcus* spp., 4.0 and 5.5; *Esch. coli*, 4.4; *Salmonella typhi*, 4.0; *Salm. paratyphi*, 5.5 and *Staph. aureus* 5.7. Tilley (1939) made a comparison of η for *Salm. typhi* and *Staph. aureus* for nine biocides (Table 8). Surprisingly the values of η seem remarkably independent of the organism used. Unpublished observations by the present authors using potassium sorbate at 21 °C have shown a similar trend. Their η values were: *Esch. coli*, 3.1; *Pseudomonas aeruginosa*, 3.2, and *Saccharomyces cerevisiae*, 2.6.

TABLE 8. *η values for phenols and alcohols using two test organisms**

Biocide	η value *Salmonella typhi*	η value *Staphylococcus aureus*
phenol	7.3	5.9
2-cresol	7.7	7.5
4-cresol	8	8.8
2-butylphenol	9.9	9.1
4-butylphenol	9.3	9.5
resorcinol	5.0	4.9
n-propylresorcinol	5.3	5.3
ethanol	11.3	11.4
n-butanol	12.0	12.7

*The data of Tilley (1939). A temperature of 20 °C was used and the end point calculated from a modified Rideal-Walker method.

The foregoing summary has attempted to span a period from 1887 to 1984. Many examples of η values have appeared in the literature over this time, some of which have been presented here chiefly to illustrate historical approaches and experimental methods together with the effects of varying experimental conditions on η. A summary of values obtained from the literature is presented in Table 9.

The Physicochemical Significance of η

It will be recalled that Kronig & Paul (1897) had received the scrutiny of the physical chemist Ikeda. He was almost certainly influenced by the emergence of

TABLE 9. *Values for η for a range of biocides**

	Biocides	η values
A.	Hydrogen peroxide	0.5
	Silver nitrate	0.9–1.0
	Mercurials	0.03–3.0
	Iodine	0.9
	Crystal violet	0.9
	Chlorhexidine	2
	Formaldehyde	1
	Quaternary ammonium compounds	0.8–2.5
	Acridines	0.7–1.9
	Formaldehyde donors	0.8–0.96
	Bronopol	0.7–0.92
	Polymeric biguanides	1.5–1.6
B	Parabens	2.5
	Sorbic acid	2.6–3.2
	Potassium laurate	2.3
C	Aliphatic alcohols	6–12.7
	Glycolmonophenyl ethers	5.8–6.4
	Glycolmonoalkyl ethers	6.4–15.9
	Phenolic agents	4–9.9
	Benzyl alcohol	2.6–7.1
	Phenethyl alcohol	9

* Data taken from the literature.
Group A <2; Group B 2–4; Group C >4.

new thinking in theoretical chemistry regarding reaction rates, orders of reaction, solution and surface chemistry and electrochemical interactions by such workers as Arrhenius, Ostwald, Nernst, van't Hoff, Guldberg & Waage and Wilhelmy. Indeed, he was the first to derive an equation for a relationship between dilution and biocidal effect with mercury II chloride. Chick (1908) saw an analogy between a unimolecular reaction and disinfection, and with the help of the chemist J. C. C. Ledingham, derived a logarithmic relationship between activity and dilution. Watson then proposed the formula for η as it is used today.

In only one paper has an attempt been made to elucidate a meaning for the varying values of η found amongst biocides. This was published by Cowles (1939) who contended that high values of η could not be taken as evidence of high orders of chemical reactions. He believed that there was a relationship between the wetting ability of a biocide and its η value. He defined wetting as the displacement of substances at the cell surface by the germicide. The time necessary for such displacement would be a function of the interfacial energy and this in turn would be modified by changes in the concentrations of the germicide. It was the relationship between these factors which determined the value of η.

Cowles (1939) measured wetting by a modified Draves test which consists essentially of measuring the time for a hank of woollen yarn to sink in a solution of the biocide. In addition, the relationship between concentration and death time (the slope of the regression line for the logarithm of these two parameters giving the concentration exponent) and the relationship between concentration and surface tension was compared. Cowles commented on the fact that for phenol and resorcinol the curves obtained by plotting wetting, surface tension and killing against concentration all showed breaks near the same concentration. He concluded that η was related to the effect of change of concentration on surface activity rather than any order of chemical reaction.

Biocide uptake by cells is another physical factor which may shed light on this interpretation of the meaning of η. Several workers have studied the uptake of biocides by micro-organisms [see Hugo (1976) for a summary]. Bean & Das (1966), who measured the uptake of a series of phenols by *Esch. coli*, found a characteristic break in the isotherm as was noted by Cowles (1939) for the parameters he studied. Similar breaks in the probit mortality *versus* log time curves were also reported by Bean & Das (1966). The interesting fact from their paper is that by plotting log death time for 84% mortality against log concentration, a series of almost parallel lines was obtained, the slope of which gave η. When the concentration required to produce a mortality of 84% in 1 min was divided by the concentration at which an inflexion was seen on the absorption isotherm, another constant was obtained. This paper certainly provides further evidence for a physical explanation of η.

Literature values for η for a range of compounds (Table 9), albeit obtained using a variety of methods, may be used to divide biocides into what appears to be three general categories: those with $\eta < 2$, those with η lying between 2 and 4, and finally those with $\eta > 4$. If these categories are examined in the light of Cowles' hypothesis and the known mode of action of the compounds (Hugo 1982), then we can observe a possible relationship between η and the type of interaction that occurs between a biocide and its cellular target.

Thus amongst the compounds with $\eta < 2$ are found: —SH reactors (silver nitrate, mercurials and bronopol); oxidizing agents (hydrogen peroxide and iodine); cationic surface active agents which disrupt the cytoplasmic membrane (quaternary ammonium compounds and chlorhexidine); compounds which react specifically with DNA (acridines), and reactors with amino groups (formaldehyde). Agents in this category, while having widely different modes of action, may be assumed to interact strongly by chemical or ionic bonding with their target. Their η value may therefore represent an order of interaction analogous to the order of a chemical reaction.

At the other extreme ($\eta > 4$) are compounds, believed to be active as membrane disrupters and proton conductors, that are recognized to mediate their antimicrobial effect through a weakly physical interaction with lipophilic components of the envelope cell (p. 260). This is in general agreement with the conclusions of Cowles; chemical reactions of this order are unknown.

In the category with η lying between 2 and 4 are found compounds which act as: promoters of leakage and direct inhibitors of macromolecular synthesis (parabens); an anionic surface active agent causing leakage of cytoplasmic contents (potassium laurate), and a proton conductor and $-SH$ reactor (sorbic acid). Certain of these compounds appear to possess a duality of action, both chemical and weakly physical, thereby justifying their intermediate values.

Conclusions

The relationship between biocide activity and concentration is logarithmic and can be measured by a constant, η—the concentration exponent or the coefficient of dilution. This constant can be determined by measuring the concentrations to achieve lethal effects, plotting one against the other and deducing η from the slope of the regression. The higher the value of η, the greater the loss of activity on dilution. η may indicate the strength and type of interaction which occurs between a biocide and its cellular target.

References*

BEAN, H. S. & DAS, A. 1966 The absorption by *Escherichia coli* of phenols and their bactericidal activity. *Journal of Pharmacy and Pharmacology*, **18**, 107S–113S.

BLISS, C. I. 1937 The calculation of the time mortality curve. *Annals of Applied Biology* **24**, 815–852.

BERRY, H. & MICHAELS, I. 1950 The evaluation of the bactericidal activities of ethylene glycol and some of its monoalkyl ethers against *Bacterium coli*. Part IX. *Journal of Pharmacy and Pharmacology* **2**, 105–114.

CHICK, H. 1908 An investigation of the laws of disinfection. *Journal of Hygiene, Cambridge* **8**, 92–158.

COWLES, P. B. 1939 The disinfectant concentration exponent. *Yale Journal of Biology and Medicine* **12**, 697–704.

HUGO, W. B. 1976 Survival of microbes exposed to chemical stress. In *26th Symposium of the Society for General Microbiology*, ed. Gray, T. G. R. & Postgate, J. R. Cambridge: Cambridge University Press. pp. 383–413.

HUGO, W. B. 1982 Disinfection mechanisms. In *Principles and practice of disinfection, preservation and sterilisation*. ed. Russell, A. D., Hugo, W. B. & Ayliffe, G. A. J., Oxford: Blackwells Scientific Publications. pp. 158–185.

* For authors' use of nomenclature in the text and bibliography, see footnotes on title page (pp 281).

HURWITZ, S. J. & McCARTHY, T. J. 1985 Dynamics of disinfection of selected preservatives against *Escherichia coli*. *Journal of Pharmaceutical Services* **74**, 892–894.

JORDAN, R. C. & JACOBS, S. E. 1944 Studies in the dynamics of disinfection. II. The calculation of the concentration exponent for phenol at 35 °C with *Bacterium coli* as test organism. *Journal of Hygiene, Cambridge* **43**, 363–369.

JORDAN, R. C. & JACOBS, S. E. 1946 Studies in the dynamics of disinfection. VII. The reaction between phenol and *Bacterium coli*: the effect of temperature on the usually accepted concentration exponent and the calculation of a more satisfactory exponent based on theoretical considerations. *Journal of Hygiene, Cambridge* **44**, 421–429.

KRONIG, B. & PAUL, T. 1897 Die chemischen Grundlagen der Lehre von der Giftwerkung und Desinfection. *Zeitschrift fur Hygiene und Infectionkrankheiten* **25**, 1–112.

LUSHER, P., DENYER, S. P. & HUGO, W. B. 1984 A note on the effect of dilution and temperature on the bactericidal activity of potassium sorbate. *Journal of Applied Bacteriology* **57**, 179–181.

RUSSELL, A. D. 1982 Factors influencing the efficacy of antimicrobial agents. In *Principles and practice of disinfection, preservation and sterilisation*. ed. Russell, A. D., Hugo, W. B. & Ayliffe, G. A. J., Oxford: Blackwells Scientific Publications. pp. 107–133.

TILLEY, F. W. 1939 An experimental study of the relation between concentration of disinfectants and time required for disinfection. *Journal of Bacteriology* **38**, 499–510.

WATSON, H. E. 1908 A note on the variation of the rates of disinfection with changes in the concentration of the disinfectant. *Journal of Hygiene, Cambridge* **8**, 536–542.

WITHELL, E. R. 1942 The effect of change of concentration on the relative activity of phenol and *para*chloro*meta*cresol. *Quarterly Journal of Pharmacy and Pharmacology* **15**, 301–313.

Index

THE SOCIETY FOR APPLIED BACTERIOLOGY TECHNICAL SERIES

General Editor: F. A. Skinner

1 Identification Methods for Microbiologists Part A *1966*
Edited by B. M. Gibbs and F. A. Skinner
Out of print

2 Identification Methods for Microbiologists Part B *1968*
Edited by B. M. Gibbs and D. A. Shapton
Out of print

3 Isolation Methods for Microbiologists *1969*
Edited by D. A. Shapton and G. W. Gould

4 Automation, Mechanization and Data Handling in Microbiology *1970*
Edited by Ann Baillie and R. J. Gilbert

5 Isolation of Anaerobes *1971*
Edited by D. A. Shapton and R. G. Board

6 Safety in Microbiology *1972*
Edited by D. A. Shapton and R. G. Board

7 Sampling—Microbiological Monitoring of Environments *1973*
Edited by R. G. Board and D. W. Lovelock

8 Some Methods for Microbiological Assay *1975*
Edited by R. G. Board and D. W. Lovelock

9 Microbial Aspects of the Deterioration of Materials *1975*
Edited by R. J. Gilbert and D. W. Lovelock

10 Microbial Ultrastructure: The Use of the Electron Microscope *1976*
Edited by R. Fuller and D. W. Lovelock

11 Techniques for the Study of Mixed Populations *1978*
Edited by D. W. Lovelock and R. Davies

12 Plant Pathogens *1979*
Edited by D. W. Lovelock

13 Cold Tolerant Microbes in Spoilage and the Environment *1979*
Edited by A. D. Russell and R. Fuller

14 Identification Methods for Microbiologists (2nd Edn) *1979*
Edited by F. A. Skinner and D. W. Lovelock

15 Microbial Growth and Survival in Extremes of Environment *1980*
Edited by G. W. Gould and Janet E. L. Corry

16 Disinfectants: Their Use and Evaluation of Effectiveness *1981*
Edited by C. H. Collins, M. C. Allwood, Sally F. Bloomfield and A. Fox

17 Isolation and Identification Methods for Food Poisoning Organisms *1982*
Edited by Janet E. L. Corry, Diane Roberts and F. A. Skinner

18 Antibiotics: Assessment of Antimicrobial Activity and Resistance *1983*
Edited by A. D. Russell and L. B. Quesnel

19 Microbiological Methods for Environmental Biotechnology *1984*
Edited by J. J. Grainger and J. M. Lynch

20 Chemical Methods in Bacterial Systematics *1985*
Edited by M. Goodfellow and D. E. Minnikin

21 Isolation and Identification of Micro-organisms of Medical and Veterinary Importance *1985*
Edited by C. H. Collins and J. M. Grange

22 Preservatives in the Food, Pharmaceutical and Environmental Industries *1986*
Edited by R. G. Board, M. C. Allwood and J. G. Banks